Applied Probability
Control
Economics
Information and Communication
Modeling and Identification
Numerical Techniques
Optimization

Applications of Mathematics

1

Applications of Mathematics

1 Fleming/Rishel, **Deterministic and Stochastic Optimal Control** (1975)
2 Marchuk, **Methods of Numerical Mathematics,** Second Ed. (1982)
3 Balakrishnan, **Applied Functional Analysis,** Second Ed. (1981)
4 Borovkov, **Stochastic Processes in Queueing Theory** (1976)
5 Lipster/Shiryayev, **Statistics of Random Processes I: General Theory** (1977)
6 Lipster/Shiryayev, **Statistics of Random Processes II: Applications** (1978)
7 Vorob'ev, **Game Theory: Lectures for Economists and Systems Scientists** (1977)
8 Shiryayev, **Optimal Stopping Rules** (1978)
9 Ibragimov/Rozanov, **Gaussian Random Processes** (1978)
10 Wonham, **Linear Multivariable Control: A Geometric Approach** (1979)
11 Hida, **Brownian Motion** (1980)
12 Hestenes, **Conjugate Direction Methods in Optimization** (1980)
13 Kallianpur, **Stochastic Filtering Theory** (1980)
14 Krylov, **Controlled Diffusion Processes** (1980)
15 Prabhu, **Stochastic Storage Processes: Queues, Insurance Risk, and Dams** (1980)
16 Ibragimov/Has'minskii, **Statistical Estimation: Asymptotic Theory** (1981)
17 Cesari, **Optimization: Theory and Applications** (1982)
18 Elliott, **Stochastic Calculus and Applications** (in prep.)

Wendell H. Fleming
Raymond W. Rishel

Deterministic and Stochastic Optimal Control

Springer-Verlag
Berlin Heidelberg New York
1975

Wendell Fleming
Department of Mathematics
Brown University
Providence, Rhode Island 02912

Raymond Rishel
Department of Mathematics
University of Kentucky
Lexington, Kentucky 40506

AMS Subject Classification
49-XX, 93E20

Library of Congress Cataloging in Publication Data

Fleming, Wendell Helms, 1928–
Deterministic and stochastic optimal control.

(Applications of mathematics; 1)
Bibliography: p. 213
Includes index.
1. Control theory. 2. Mathematical optimization. 3. Markov processes. I. Rishel, Raymond W., joint author. II. Title.
QA402.3.F527 629.8'312 75-28391

Printed in the United States of America.

9 8 7 6 5 4 3 (Third printing, 1986)

ISBN 0-387-90155-8 Springer-Verlag New York Heidelberg Berlin

ISBN 3-540-90155-8 Springer-Verlag Berlin Heidelberg New York

Preface

This book may be regarded as consisting of two parts. In Chapters I–IV we present what we regard as essential topics in an introduction to deterministic optimal control theory. This material has been used by the authors for one semester graduate-level courses at Brown University and the University of Kentucky. The simplest problem in calculus of variations is taken as the point of departure, in Chapter I. Chapters II, III, and IV deal with necessary conditions for an optimum, existence and regularity theorems for optimal controls, and the method of dynamic programming. The beginning reader may find it useful first to learn the main results, corollaries, and examples. These tend to be found in the earlier parts of each chapter. We have deliberately postponed some difficult technical proofs to later parts of these chapters.

In the second part of the book we give an introduction to stochastic optimal control for Markov diffusion processes. Our treatment follows the dynamic programming method, and depends on the intimate relationship between second-order partial differential equations of parabolic type and stochastic differential equations. This relationship is reviewed in Chapter V, which may be read independently of Chapters I–IV. Chapter VI is based to a considerable extent on the authors' work in stochastic control since 1961. It also includes two other topics important for applications, namely, the solution to the stochastic linear regulator and the separation principle.

We wish to thank Charles Holland, George Kent, Bruce Klemsrud, Anders Lindquist, and Chun-ping Tsai who read and criticized various chapters. Part of the book was written while one of the authors was at Bell Telephone Laboratories. We wish to thank Bell Telephone Laboratories for supporting this effort.

W. H. Fleming
R. W. Rishel
March 1, 1975

Contents

Chapter I

The Simplest Problem in Calculus of Variations

1. Introduction, 1
2. Minimum Problems on an Abstract Space—Elementary Theory, 2
3. The Euler Equation; Extremals, 5
4. Examples, 9
5. The Jacobi Necessary Condition, 12
6. The Simplest Problem in *n* Dimensions, 15

Chapter II

The Optimal Control Problem

1. Introduction, 20
2. Examples, 21
3. Statement of the Optimal Control Problem, 23
4. Equivalent Problems, 25
5. Statement of Pontryagin's Principle, 26
6. Extremals for the Moon Landing Problem, 28
7. Extremals for the Linear Regulator Problem, 33
8. Extremals for the Simplest Problem in Calculus of Variations, 34
9. General Features of the Moon Landing Problem, 35
10. Summary of Preliminary Results, 37
11. The Free Terminal Point Problem, 39
12. Preliminary Discussion of the Proof of Pontryagin's Principle, 44
13. A Multiplier Rule for an Abstract Nonlinear Programming Problem, 46
14. A Cone of Variations for the Problem of Optimal Control, 48
15. Verification of Pontryagin's Principle, 52

Chapter III

Existence and Continuity Properties of Optimal Controls

1. The Existence Problem, 60
2. An Existence Theorem (Mayer Problem, *U* Compact), 62

3. Proof of Theorem 2.1, 65
4. More Existence Theorems, 68
5. Proof of Theorem 4.1, 69
6. Continuity Properties of Optimal Controls, 74

Chapter IV

Dynamic Programming

1. Introduction, 80
2. The Problem, 81
3. The Value Function, 81
4. The Partial Differential Equation of Dynamic Programming, 83
5. The Linear Regulator Problem, 88
6. Equations of Motion with Discontinuous Feedback Controls, 90
7. Sufficient Conditions for Optimality, 97
8. The Relationship between the Equation of Dynamic Programming and Pontryagin's Principle, 99

Chapter V

Stochastic Differential Equations and Markov Diffusion Processes

1. Introduction, 106
2. Continuous Stochastic Processes; Brownian Motion Processes, 108
3. Ito's Stochastic Integral, 111
4. Stochastic Differential Equations, 117
5. Markov Diffusion Processes, 120
6. Backward Equations, 127
7. Boundary Value Problems, 129
8. Forward Equations, 131
9. Linear System Equations; the Kalman–Bucy Filter, 133
10. Absolutely Continuous Substitution of Probability Measures, 141
11. An Extension of Theorems 5.1, 5.2, 147

Chapter VI

Optimal Control of Markov Diffusion Processes

1. Introduction, 151
2. The Dynamic Programming Equation for Controlled Markov Processes, 152
3. Controlled Diffusion Processes, 155
4. The Dynamic Programming Equation for Controlled Diffusions; a Verification Theorem, 159
5. The Linear Regulator Problem (Complete Observations of System States), 165
6. Existence Theorems, 166
7. Dependence of Optimal Performance on y and σ, 172

8. Generalized Solutions of the Dynamic Programming Equation, 177
9. Stochastic Approximation to the Deterministic Control Problem, 181
10. Problems with Partial Observations, 187
11. The Separation Principle, 188

Appendices

A. Gronwall–Bellman Inequality, 198
B. Selecting a Measurable Function, 199
C. Convex Sets and Convex Functions, 200
D. Review of Basic Probability, 202
E. Results about Parabolic Equations, 205
F. A General Position Lemma, 211

Bibliography, 213

Index, 221

Chapter I. The Simplest Problem in Calculus of Variations

§ 1. Introduction

In calculus one studies the problem of minimizing $f(x)$, where x is real or more generally $x=(x_1, \ldots, x_n)$ denotes an n-tuple of real numbers. However, in many problems of interest the domain of the function to be miminized is not a portion of some finite-dimensional space. Rather, the function is defined on some portion of an infinite-dimensional space $\mathscr{V}$. We shall begin by outlining the elementary theory of minima of a function J on an abstract space $\mathscr{V}$ (§ 2).

The oldest examples of minimum problems on abstract spaces occur in calculus of variations. The main purpose of this chapter is to give a brief introduction to that subject. In calculus of variations one seeks a curve making a certain function J of the curve minimum (or maximum). We consider a particular kind of problem of this type, called the simplest problem in calculus of variations (§ 3). We admit curves in the plane given by an equation $x=x(t)$, where $x(\cdot)$ is a sufficiently well-behaved function on an interval $[t_0, t_1]$. The endpoints of the curves admitted are prescribed. The function J is given by an integral over $[t_0, t_1]$ involving $x(t)$ and its derivative $\dot{x}(t)$. An example is

$$J(x)=\int_{t_0}^{t_1}\left[a(x(t))^2+b(\dot{x}(t))^2\right] dt.$$

In this instance the integrand is quadratic in x and $\dot{x}$. This and other examples are treated in §4. An example with a trivial solution is the length of the curve

$$J(x)=\int_{t_0}^{t_1}[1+\dot{x}(t)^2]^{1/2}\, dt.$$

The minimum is attained by the straight line segment joining the given endpoints.

Calculus of variations has a long history, starting with the brachistochrone problem solved by the Bernoullis nearly 300 years ago (see Bliss [1] for comments about the early history). The simplest problem was generalized in various ways. Most of these are particular cases of the general problem of Bolza, treated for instance in Bliss [2].

The original motivations for calculus of variations came from classical physics (mechanics, optics) and geometry. Over the years the inspiration derived from these applications became greatly diffused. By around 1950 calculus of variations seemed to have lost its vitality, and seemed destined no longer to be a living part of mathematics. However, since 1950 many new applications have been found. (This is perhaps not surprising. Calculus of variations is, after all, an extension of calculus. Surely no one could in the beginning have predicted the multitude of applications of calculus.)

To deal with the new applications the theory had to be extended. This has revitalized the whole subject. Some of these new applications were to problems in aerospace sciences, industrial process control, and mathematical economics. Typically, these latter problems are not quite of the type considered in calculus of variations, since some of the variables which appear are subject to inequality constraints. They are, however, of a type now called problems of optimal control. These are treated in Chap. II and succeeding chapters.

Note. The historical comments above apply, strictly speaking, to single integral problems. Multiple integral problems in calculus of variations are not discussed in this book. That branch of the subject has been active throughout the 20th century, and has recently experienced remarkable progress. See Morrey [1].

§ 2. Minimum Problems on an Abstract Space—Elementary Theory

In calculus one considers the problem of minimizing a function f defined on some interval I of the real numbers R. Let us suppose, for definiteness, that $I=[a,b]$ is a closed finite interval. The following elementary results are well known.

(i) *Necessary conditions for a minimum at* x^*. Suppose that $f(x^*)\leqq f(x)$ for $a\leqq x\leqq b$. Then

(2.1) $\quad f'(x^*)=0,\quad f''(x^*)\geqq 0,\quad$ if $a<x^*<b$ (interior minimum)

(2.2) $\quad f'(x^*)\geqq 0,\quad$ if $x^*=a$,

(2.3) $\quad f'(x^*)\leqq 0,\quad$ if $x^*=b$,

provided the indicated derivatives exist. The derivatives at a and b are to be interpreted as one-sided derivatives.

(ii) *Sufficient conditions for a local minimum at* x^*. Suppose that the inequalities on f', f'' in (i) are strict. Then there exists a neighborhood N of x^* such that $f(x^*)<f(x)$ for all $x\in N\cap[a,b]$, $x\neq x^*$.

(iii) *Existence of a minimum.* If f is continuous on $[a,b]$, then f has a minimum there. (Actually, it suffices that f be lower semicontinuous on $[a,b]$ rather than continuous. See Graves [1, p. 65].)

(iv) *Uniqueness of a minimum.* If $f(x)$ is a strictly convex[1] function on $[a,b]$ it has a minimum at a unique point x^* in $[a,b]$. A sufficient condition for $f(x)$ to be strictly convex is that $f''(x)>0$ on $[a,b]$.

In this section we show that (i), (iv) generalize in a straightforward way to minimum problems for a function defined on an abstract space. The problems

[1] In Appendix C we review ideas about convex sets and convex functions.

(ii), (iii) of sufficient conditions for a local minimum and existence will be treated later, in the setting of calculus of variations and optimal control theory (Chap. III, IV).

Consider the general optimization problem: given a set $\mathscr{K}$ and real-valued function J defined on $\mathscr{K}$, find an element $u^* \in \mathscr{K}$ such that $J(u^*) \leqq J(u)$ for all $u \in \mathscr{K}$.

Necessary conditions for a minimum are obtained by considering functions $\zeta = \zeta(\varepsilon)$ from some interval $a \leqq \varepsilon \leqq b$, such that the composite function $f(\varepsilon) = J[\zeta(\varepsilon)]$ is differentiable. In fact, we obtain immediately from (2.1)–(2.3):

(i′) *Necessary conditions for a minimum.* Suppose $u^* \in \mathscr{K}$ and $J(u^*) \leqq J(u)$ for any $u \in \mathscr{K}$. If $\zeta = \zeta(\varepsilon)$ is a function from an interval $[a, b]$ into $\mathscr{K}$, such that $\zeta(\varepsilon^*) = u^*$, then

$$\text{(2.4)} \qquad \frac{d}{d\varepsilon} J(\zeta(\varepsilon))\big|_{\varepsilon=\varepsilon^*} = 0, \qquad \frac{d^2}{d\varepsilon^2} J(\zeta(\varepsilon))\big|_{\varepsilon=\varepsilon^*} \geqq 0 \qquad \text{if } a < \varepsilon^* < b$$

$$\text{(2.5)} \qquad \frac{d}{d\varepsilon} J(\zeta(\varepsilon))\big|_{\varepsilon=\varepsilon^*} \geqq 0 \qquad \text{if } \varepsilon^* = a,$$

$$\text{(2.6)} \qquad \frac{d}{d\varepsilon} J(\zeta(\varepsilon))\big|_{\varepsilon=\varepsilon^*} \leqq 0 \qquad \text{if } \varepsilon^* = b,$$

provided that the indicated derivatives exist. (In what follows we shall usually arrange that $\varepsilon^* = 0$.)

In many cases $\mathscr{K}$ is a subset of a vector space $\mathscr{V}$. When this is so, the following definitions can be made.

Definition. Let $u \in \mathscr{K}$ and $v \in \mathscr{V}$. The point u is an *internal point* of $\mathscr{K}$ in the direction v if there exists $\varepsilon(v) > 0$ such that $u + \varepsilon v \in \mathscr{K}$ for $|\varepsilon| < \varepsilon(v)$.

Definition. Let $u \in \mathscr{K}$ and $v \in \mathscr{V}$. The point u is a *radial point* of $\mathscr{K}$ in the direction v if there exists $\varepsilon(v) > 0$ such that $u + \varepsilon v \in \mathscr{K}$ for $0 \leqq \varepsilon < \varepsilon(v)$.

If u is respectively an internal point of $\mathscr{K}$ or a radial point of $\mathscr{K}$ in the direction v, then

$$\zeta(\varepsilon) = u + \varepsilon v$$

maps into $\mathscr{K}$ for respectively $|\varepsilon| < \varepsilon(v)$ or $0 \leqq \varepsilon < \varepsilon(v)$. The notation

$$\delta J(u; v) = \frac{d}{d\varepsilon} J(u + \varepsilon v)\big|_{\varepsilon=0}$$

will be used, and will be called the derivative of J at u in the direction v (two-sided or one-sided derivative according as u is an internal or radial point in the direction v).

Definition. J is *Gateau-differentiable* at u if u is an internal point in the direction v and $\delta J(u; v)$ exists, for every $v \in \mathscr{V}$.

The notation $\delta^2 J(u; v)$ is used to denote

$$\delta^2 J(u; v) = \frac{d^2}{d\varepsilon^2} J(u + \varepsilon v)\big|_{\varepsilon=0}$$

provided the indicated second derivative exists. (In calculus of variations $\delta J(u; v)$, $\delta^2 J(u; v)$ are called the first and second variations. We shall calculate them in Sects. 3 and 5.)

The simplest problem in calculus of variations will be discussed in this chapter using the notion of Gateau-differentiability. In Chap. II.11 the family of functions $\zeta(\varepsilon)=u+\varepsilon v$ and another family of functions will be used to obtain necessary conditions for optimality in a special case of the optimal control problem. Later in Chap. II (Sects. 5, 12–15) we prove a general necessary condition for the optimal control problem, by more sophisticated methods.

We shall suppose from now on that $\mathscr{K}$ is a subset of a vector space $\mathscr{V}$. The following theorem is immediate from (2.4).

Theorem 2.1. *If J has a minimum on $\mathscr{K}$ at an internal point u^* of $\mathscr{K}$ in the direction v and $\delta J(u^*; v)$, $\delta^2 J(u^*; v)$ exist, then*

$$\delta J(u^*; v)=0, \qquad \delta^2 J(u^*; v)\geqq 0.$$

Similarly, we obtain from (2.5):

Theorem 2.2. *If J has a minimum on $\mathscr{K}$ at a radial point u^* of $\mathscr{K}$ in the direction v and $\delta J(u^*; v)$ exists, then $\delta J(u^*; v)\geqq 0$.*

In calculus of variations and control problems, J is not usually a convex function. However, when J happens to be convex the following two theorems can be applied. The first states that the necessary condition in Theorem 2.2 together with convexity of J are sufficient for a (global) minimum on $\mathscr{K}$. The second states that, if the convexity is strict, then the minimizing u^* is unique.

Theorem 2.3. *Let $\mathscr{K}$ be convex, let J be convex on $\mathscr{K}$ and $u^*\in\mathscr{K}$. If $\delta J(u^*; v)\geqq 0$ for all v such that $u^*+v\in\mathscr{K}$, then J has a minimum on $\mathscr{K}$ at u^*.*

Proof. If $u\in\mathscr{K}$ and $0<\varepsilon<1$, it follows from the definition of convex functions that

$$J[(1-\varepsilon)u^*+\varepsilon u]\leqq(1-\varepsilon)J(u^*)+\varepsilon J(u);$$

this can be rewritten

$$\varepsilon^{-1}\{J[u^*+\varepsilon(u-u^*)]-J(u^*)\}\leqq J(u)-J(u^*).$$

Let ε tend to 0. Then

$$\delta J(u^*; u-u^*)\leqq J(u)-J(u^*).$$

From our hypothesis, with $v=u-u^*$, we have $J(u^*)\leqq J(u)$ for any $u\in\mathscr{K}$. □

Theorem 2.4. *Let $\mathscr{K}$ be convex, and let J be strictly convex on $\mathscr{K}$. Then there exists at most one $u^*\in\mathscr{K}$ such that J has a minimum at u^*.*

Proof. Suppose that J had a minimum at both u^* and u, where $u^*\neq u$. Then $\tilde{u}=\frac{1}{2}(u^*+u)$ is in $\mathscr{K}$, and

$$J(\tilde{u})<\tfrac{1}{2}J(u^*)+\tfrac{1}{2}J(u).$$

Thus $J(\tilde{u})$ is less than the minimum value of J on $\mathscr{K}$, a contradiction. □

§ 3. The Euler Equation; Extremals

The remainder of Chap. I is concerned with a particular kind of minimum problem, often called the simplest problem in calculus of variations. Roughly speaking, it is to find among curves $x=x(t)$ with given endpoints one for which an integral of the type

$$J(x)=\int_{t_0}^{t_1} L[t, x(t), \dot{x}(t)]\,dt \tag{3.1}$$

is minimum. Here $\dot{x}(t)$ is the derivative dx/dt.

To formulate the simplest problem in calculus of variations precisely, let us assume that $L(t, x, \dot{x})$ is a real-valued function of three real variables. Such a function L is called a *variational integrand.* Partial derivatives of L are denoted by L_t, L_x, $L_{\dot{x}}$, $L_{x\dot{x}}$, etc. We say that L is of class C^r if all partial derivatives of L of orders $\leqq r$ are continuous. We suppose throughout this chapter that L is at least of class C^2.

Example. Let $L(x, \dot{x})=g(x)(1+\dot{x}^2)^{1/2}$. Then

$$L_x=g'(x)(1+\dot{x}^2)^{1/2}, \qquad L_{\dot{x}}=g(x)\dot{x}(1+\dot{x}^2)^{-1/2}.$$

Two classical examples of this type are the brachistochrone for which $g(x)=(x-\alpha)^{-1/2}$ and the minimal surface of revolution for which $g(x)=x$. See Bliss [1]. The latter is also discussed in §4.

Remarks on Notation. The above notation is standard in the calculus of variations. However, it can be confusing. At first the symbols x, $\dot{x}$ merely denote the second and third variables of the function L. However, in expressions such as (3.1) the composite function $L(t, x(t), \dot{x}(t))$ will be formed by substituting the values $x(t)$ of a function and of its derivative $\dot{x}(t)$.

Strictly speaking, one must distinguish notationally between a function and its values. We sometimes denote a function on $[t_0, t_1]$ by $x(\cdot)$; its value at t is $x(t)$. However, we ignore this distinction when no confusion can arise. For brevity, we often also write simply x for a function. Thus in (3.1) we have written $J(x)$ instead of $J(x(\cdot))$. Whether x stands for a function or a real number should be clear from the context.

A function u is called *piecewise continuous* on $[t_0, t_1]$ if $u(t)$ is defined and continuous except at a finite number of points $t'_1, \ldots, t'_m$ which are interior to $[t_0, t_1]$ and $u(t)$ has right- and left-hand limits (finite) at each t'_i. A function x is called *piecewise* C^1 on $[t_0, t_1]$ if $x(t)$ is continuous and the derivative $\dot{x}(t)$ is piecewise continuous on $[t_0, t_1]$. At the points t' of discontinuity of $\dot{x}(t)$, the right- and left-hand derivatives $\dot{x}(t'_+)$, $\dot{x}(t'_-)$ exist.

A function x is called *of class* C^r on $[t_0, t_1]$ if all its derivatives of orders $\leqq r$ are continuous there. One may think of x as the curve $\{(t, x(t))\colon t_0 \leqq t \leqq t_1\}$ in R^2. When x is piecewise C^1, the curve has a continuously turning tangent except for corners at the points $(t', x(t'))$ where $\dot{x}(t)$ is discontinuous.

We now formulate the simplest problem of calculus of variations. Let $\mathscr{X}$ denote the vector space of all real-valued piecewise C^1 functions on a given interval $[t_0, t_1]$. For x_0, x_1 given real numbers, let $\mathscr{X}_e$ denote the subset consisting of those $x \in \mathscr{X}$ which satisfy the endpoint conditions

$$(3.2) \qquad x(t_0) = x_0, \qquad x(t_1) = x_1.$$

The simplest problem of calculus of variations is to find an x^* *at which* J *has a minimum on* $\mathscr{X}_e$.

Remarks. (1) By admitting only piecewise C^1 functions x the integral (3.1) may be taken in the elementary (Riemann) sense. By using Lebesgue integrals in (3.1), less restrictive assumptions on x could equally well be imposed; see Problem 11, end of chapter. However, if the variational integrand L is regular in the sense defined below, any minimizing x^* is necessarily a C^2 function. In Chap. III we use Lebesgue integration theory in proving existence theorems.

(2) The simplest problem of calculus of variations could be rephrased as the problem of choosing the piecewise continuous function $u = \dot{x}$. The initial value x_0 and the function u determine

$$x(t) = x_0 + \int_{t_0}^{t} u(t)\, dt.$$

Only those u for which $x(t_1) = x_1$ are admitted. When rephrased in this way, the simplest problem becomes a special case of the optimal control problem, to be formulated in Chap. II. In the notation there $J(x)$ is written as $J(u)$.

Besides $\mathscr{X}$ let us consider the space $\mathscr{Y}$ of all piecewise C^1 real valued functions y on $[t_0, t_1]$ such that

$$(3.3) \qquad y(t_0) = y(t_1) = 0.$$

Such functions y are called *admissible variations.* $\mathscr{Y}$ is a vector space. Moreover, if $x \in \mathscr{X}_e$ and $y \in \mathscr{Y}$, then $x + \varepsilon y \in \mathscr{X}_e$ for any real ε. This implies x is an internal point of $\mathscr{X}_e$ in the direction y, for any such y.

To obtain a first necessary condition for a minimum we need to calculate the first variation of J:

$$\delta J(x; y) = \frac{d}{d\varepsilon} J(x + \varepsilon y)\big|_{\varepsilon = 0}.$$

Lemma 3.1. (Differentiation under an integral sign). *Let* $F(t, \varepsilon)$ *and the partial derivative* $F_\varepsilon(t, \varepsilon)$ *be continuous for* $t' \leqq t \leqq t''$, $-\varepsilon_0 < \varepsilon < \varepsilon_0$, *where* $\varepsilon_0 > 0$. *Then on* $(-\varepsilon_0, \varepsilon_0)$

$$\frac{d}{d\varepsilon} \int_{t'}^{t''} F(t, \varepsilon)\, dt = \int_{t'}^{t''} F_\varepsilon(t, \varepsilon)\, dt.$$

This can be proved by showing that both sides have the same integral over any closed subinterval of $(-\varepsilon_0, \varepsilon_0)$. See Fleming [1, p. 199].

Lemma 3.2. *Let* $P(t) = \int_{t_0}^{t} L_x\, d\tau$. *Then, for all* $y \in \mathscr{Y}$, $\delta J(x; y)$ *exists and*

$$(3.4) \qquad \delta J(x; y) = \int_{t_0}^{t_1} (-P + L_{\dot{x}})\, \dot{y}\, dt.$$

In this lemma, L_x is short for $L_x(\tau, x(\tau), \dot{x}(\tau))$ and $L_{\dot{x}}$ short for $L_{\dot{x}}(t, x(t), \dot{x}(t))$.

Proof of Lemma 3.2. Let

$$F(t,\varepsilon)=L\big(t,x(t)+\varepsilon y(t),\dot{x}(t)+\varepsilon \dot{y}(t)\big),$$

and $t_1',\ldots,t_m'$ be those points where either $\dot{x}(t)$ or $\dot{y}(t)$ is discontinuous, $t_0'=t_0$, $t_{m+1}'=t_1$. We apply Lemma 3.1 on each of the intervals $[t_1',t_{i+1}']$, $i=0,1,\ldots,m$ and sum from 0 to m. This gives

$$\delta J(x;y)=\int_{t_0}^{t_1}(L_x y+L_{\dot{x}}\dot{y})\,dt.$$

However, $\dot{P}=L_x$. An integration by parts, taking account of (3.3), gives (3.4). □

Lemma 3.3. *Let ϕ be piecewise continuous on $[t_0,t_1]$ and $\int_{t_0}^{t_1}\phi(t)z(t)\,dt=0$ for all piecewise continuous z on $[t_0,t_1]$ such that $\int_{t_0}^{t_1}z(t)\,dt=0$. Then $\phi(t)$ is constant on $[t_0,t_1]$.*

Proof. Consider the average of ϕ:

$$\bar{\phi}=\frac{1}{t_1-t_0}\int_{t_0}^{t_1}\phi(t)\,dt.$$

Note that

$$\int_{t_0}^{t_1}[\phi(t)-\bar{\phi}]\,z(t)\,dt=0$$

for all such z. In particular, take $z=\phi-\bar{\phi}$. Then $\int_{t_0}^{t_1}z^2\,dt=0$, which implies $z(t)=0$ on $[t_0,t_1]$. □

In the next theorem we again use the notation of Lemma 3.2, with now $x=x^*$.

Theorem 3.1. *Let J have a minimum on $\mathscr{X}_e$ at x^*. Then*

(3.5) $$-\int_{t_0}^{t}L_x\,d\tau+L_{\dot{x}}=\text{constant on } [t_0,t_1].$$

Proof. By Lemma 3.2, $\delta J(x^*;y)$ satisfies (3.4) with $x=x^*$. In Theorem 2.1 we take $\mathscr{V}=\mathscr{X}$, $\mathscr{K}=\mathscr{X}_e$ to conclude that $\delta J(x^*,y)=0$ for all $y\in\mathscr{Y}$. In Lemma 3.3, let

$$\phi(t)=-P(t)+L_{\dot{x}}\big(t,x^*(t),\dot{x}^*(t)\big).$$

For any piecewise continuous z satisfying $\int_{t_0}^{t_1}z(t)\,dt=0$, the function

$$y(t)=\int_{t_0}^{t}z(\tau)\,d\tau,\qquad t_0\leqq t\leqq t_1$$

is in $\mathscr{Y}$. By Lemma 3.3, $\phi(t)$ is constant on $[t_0,t_1]$. □

Theorem 3.1 is the first necessary condition for a minimum. It has a number of important consequences.

Definition. Any piecewise C^1 function x^* satisfying (3.5) is called an *extremal.*

Every minimizing x^* is an extremal. In many problems extremals which do not minimize J also occur.

By taking derivatives in (3.5) we obtain

Corollary 3.1. *Every extremal x^* satisfies the differential equation*

$$L_x = \frac{d}{dt} L_{\dot{x}}. \tag{3.6}$$

Eq. (3.6) is called *Euler's equation.* Both sides are evaluated at $(t, x^*(t), \dot{x}^*(t))$. If $\dot{x}^*(t)$ has a discontinuity at $t = t'$, then (3.6) is satisfied by right- and left-hand derivatives.

Since the term $\int_{t_0}^{t} L_x \, d\tau$ in (3.5) is continuous everywhere on $[t_0, t_1]$,

$$L_{\dot{x}}(t, x^*(t), \dot{x}^*(t))$$

is also continuous. In particular, this is true whenever a corner $(t', x^*(t'))$ occurs.

Corollary 3.2. (Weierstrass-Erdmann corner condition). *Let $\dot{x}^*(t'_-)$, $\dot{x}^*(t'_+)$ denote right- and left-hand derivatives at a point t' of discontinuity of $\dot{x}^*$. Then, if x^* is an extremal,*

$$L_{\dot{x}}(t', x^*(t'), \dot{x}^*(t'_-)) = L_{\dot{x}}(t', x^*(t'), \dot{x}^*(t'_+)). \tag{3.7}$$

Definition. If $L_{\dot{x}\dot{x}} > 0$ for all $(t, x, \dot{x})$, then the variational integrand L is called *regular.*

Regularity implies that, for fixed (t, x), $L(t, x, \cdot)$ is a strictly convex function of its third variable, which is equivalent to saying that $L_{\dot{x}}(t, x, \cdot)$ is a strictly increasing function. When L is regular, extremals can have no corners. The following stronger statement can, in fact, be made.

Corollary 3.3. *Let L be a regular variational integrand of class C^r, $r \geqq 2$. Then any extremal x^* is of class C^r on $[t_0, t_1]$.*

Proof. Since $L_{\dot{x}}(t, x^*(t), \cdot)$ is a strictly increasing function, $\dot{x}^*$ is continuous by Corollary 3.2. Thus x^* is C^1 on $[t_0, t_1]$. Proceeding by induction, let us suppose that x^* is C^j on $[t_0, t_1]$, $1 \leqq j \leqq r-1$. Then

$$P(t) = \int_{t_0}^{t} L_x(\tau, x^*(\tau), \dot{x}^*(\tau)) \, d\tau$$

is C^j. By (3.5), $-P(t) + L_{\dot{x}}(t, x^*(t), \dot{x}^*(t)) = c$ on $[t_0, t_1]$ for some constant c. Consider the function of two variables

$$\Phi(t, \dot{x}) = -P(t) + L_{\dot{x}}(t, x^*(t), \dot{x}).$$

Since P and x^* are C^j, $L_{\dot{x}}$ is C^{r-1}, and $j \leqq r-1$, Φ is also C^j. By regularity $\Phi_{\dot{x}} > 0$. The function $\dot{x}^*$ satisfies $\Phi(t, \dot{x}^*(t)) = c$. By the implicit function theorem, $\dot{x}^*$ is C^j. Hence x^* is C^{j+1} for $j \leqq r-1$. □

Remark. In some problems L is not regular, but there exists a two-dimensional open set D such that $L_{\dot{x}\dot{x}} > 0$ for all $(t, x) \in D$ and all $\dot{x}$. In that case we call L *regular in D.* If L is C^r and regular in D, then x^* is C^r on the set of those t such that $(t, x^*(t)) \in D$.

Let L be regular. After performing the differentiation on the right side of (3.6), we obtain another form of the Euler equation:

(3.6′)
$$\ddot{x}^* = \frac{L_x - L_{\dot{x}t} - L_{\dot{x}x}\dot{x}^*}{L_{\dot{x}\dot{x}}}.$$

The problem of finding the extremals x^* passing through the given endpoints (t_0, x_0), (t_1, x_1) is an example of a two-point boundary problem. There may be no, one or many extremals through the given endpoints. Examples are given in § 4. The problem of numerical solution of the two-point boundary problem for equation (3.6′) is not easy. We shall not treat it in this book.

An easier problem is the initial-value problem for (3.6′), in which $x^*(t_0)$, $\dot{x}^*(t_0)$ are given but no restrictions are imposed at t_1. A standard theorem about ordinary differential equations states that the initial-value problem has a unique solution x^* provided L is regular and C^3 and $t_1 - t_0$ is sufficiently small. This local result is often quoted in stating variational principles in classical mechanics and in differential geometry. It also plays a role in the method of characteristics in Chap. IV.

§ 4. Examples

Calculus of variations is a subject with a long history and with a wide variety of applications. Hence, interesting examples abound. We limit ourselves here to a few simple ones to indicate some of the possibilities.

Example 4.1. Let $L = a x^2 + b \dot{x}^2$, where $b > 0$. Since $L_{\dot{x}\dot{x}} = 2b$, L is regular. The Euler equation $-2ax + 2b\ddot{x} = 0$ is linear of order 2. If $a > 0$, then there is a unique extremal x^* satisfying the end conditions $x(t_0) = x_0$, $x(t_1) = x_1$. (Uniqueness of x^* in this case can also be explained by the fact that J is a strictly convex function; see Remark 5.1.)

If $a < 0$, then there need not be a unique extremal x^* through given end points. For instance, suppose that $(t_0, x_0) = (0, 0)$. The extremals through $(0, 0)$ are

$$x^*(t) = A \sin k t, \qquad k = (|a| b^{-1})^{1/2}.$$

If $t_1 = \pi k^{-1} m$, $m = 1, 2, \ldots$, then all of these pass through $(t_1, 0)$; thus none pass through (t_1, x_1) if $x_1 \neq 0$. This example corresponds to the principle of least action for the oscillating spring; see § 6.

If $a = 0$, the extremals are straight lines.

Example 4.2. Suppose that L is linear in $\dot{x}$, namely $L = M(t, x) + N(t, x)\dot{x}$. Since $L_{\dot{x}\dot{x}} = 0$, L is not regular. With L we can associate the linear differential form $M\,dt + N\,dx$. In fact, from (3.1),

$$J(x) = \int_{t_0}^{t_1} (M\,dt + N\,dx).$$

The values of J on two curves with the same endpoints can than be compared using Green's theorem; see Problem 3. This technique is of interest in certain aerospace applications; see Miele [1].

The Euler equation (3.6) becomes $M_x(t, x^*(t)) = N_t(t, x^*(t))$. This is an implicit equation for x^*, not a differential equation. It generally has few solutions; thus we may not generally expect that there is a solution joining given points (t_0, x_0) and (t_1, x_1).

The following lemma allows us to find extremals from a first-order differential equation rather than the second-order equation (3.6), provided the variable t does not appear in L.

Lemma 4.1. *Let $L = L(x, \dot{x})$ be regular. Then every extremal x^* satisfies*

$$L - \dot{x}^* L_{\dot{x}} = \text{constant}. \tag{4.1}$$

Conversely, any solution of (4.1) *with $\dot{x}^* = 0$ only at isolated points is an extremal.*

Proof. By Corollary 3.3, x^* is C^2. Then

$$\frac{d}{dt}(L - \dot{x}^* L_{\dot{x}}) = \dot{x}^* \left(L_x - \frac{d}{dt} L_{\dot{x}}\right) = 0. \quad \square$$

Example 4.3. Minimal surfaces of revolution. A problem which has been studied from many points of view is to find surfaces of minimum area with given boundary. This is called *Plateau's problem.* See Federer [2, Chap. V], Rado [1]. Area is expressed as a double integral. However, if the boundary is invariant under rotations about a line l, then it suffices to admit surfaces of revolution about l. The area is then given by a single integral.

Let us suppose that line l is the t-axis, and that the given boundary consists of a pair of circles, in planes $t = t_0$, $t = t_1$ perpendicular to l, passing through (t_0, x_0), (t_1, x_1) with $x_0 > 0$, $x_1 > 0$. If the surface is obtained by rotating a curve $x = x(t)$, with $x(t) > 0$, about the t-axis, then the area is

$$A = 2\pi \int_{t_0}^{t_1} x(t)\, ds$$

where s is arc length. A curve x^* minimizing $A/2\pi$ also minimizes A. Hence we may take $L(x, \dot{x}) = x(1 + \dot{x}^2)^{1/2}$. The regularity condition

$$L_{\dot{x}\dot{x}} = \frac{x}{(1 + \dot{x}^2)^{3/2}} > 0$$

holds in the upper half $(x > 0)$ of the (t, x) plane.

In Example 4.3 we may, without loss of generality, take $(t_0, x_0) = (0, 1)$. Let us find the extremals x^* passing through $(0, 1)$, with $x^*(t) > 0$. The catenaries

$$x^*(t) = b \cosh\left(\frac{t - a}{b}\right) \tag{4.2}$$

by inspection satisfy, for any positive a, b,

$$L - \dot{x}^* L_{\dot{x}} = \frac{x^*}{\sqrt{1 + (\dot{x}^*)^2}} = \text{constant},$$

and pass through $(0, 1)$ when

$$1/b = \cosh(a/b). \tag{4.3}$$

It is convenient to introduce the parameter $\alpha = -ab^{-1}$. Then (4.2), (4.3) define a family of extremals through (0, 1) which we denote by $x^*(\cdot, \alpha)$. Since $\dot{x}^*(0, \alpha) = \sinh \alpha$ assumes all real values, by uniqueness of the extremal with given initial data $x^*(0)$, $\dot{x}^*(0)$ every extremal through (0, 1) belongs to this family. The family $x^*(\cdot, \alpha)$ has an envelope e as shown in Fig. I.1. The envelope is found by solving the equation $\partial x^*/\partial \alpha = 0$. This is shown in Bliss [1], where the present example (as well as the brachistochrone) is discussed in complete detail.

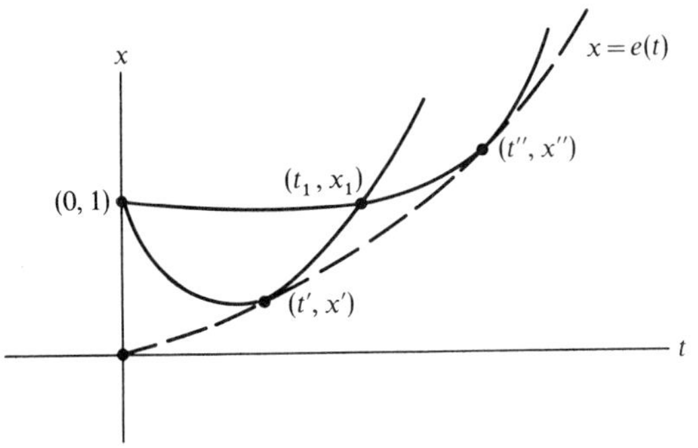

Fig. I.1

Suppose that (t_1, x_1) is above the envelope. Then two extremals pass through (0, 1), and (t_1, x_1). One of these, which we call $\tilde{x}^*$, touches the envelope at a point (t', x') between (0, 1) and (t_1, x_1). In § 5 we shall see from the Jacobi condition that the extremal $\tilde{x}^*$ does not minimize $J(x)$.

Let D denote the region $x > e(t)$ above the envelope. It can be shown using Theorem IV.7.1, Chap. IV, that the extremal x^* which touches the envelope at (t'', x''), with $t'' > t_1$, minimizes $J(x)$ among all curves joining (0, 1) and (t_1, x_1) and lying in D.

There remains the question whether the catenoid S_1 generated by rotating x^* about l has minimum area among all surfaces bounded by the two given circles. When this is so, we say that S_1 gives *absolute minimum* area. Another candidate is the surface S_2 consisting of two circular discs bounded by the given circles, one disc in the plane $t = 0$ and the other in the plane $t = t_1$. S_2 is called the Goldschmidt solution. The following can be shown [Bliss [1], p. 118]: there exists another function e_1, with $e_1(t) > e(t)$ for $t > 0$, $e_1(0) = e(0) = 0$, such that

a) If $x_1 \geqq e_1(t_1)$, then the catenoid S_1 gives absolute minimum area;

b) If $e(t_1) < x_1 \leqq e_1(t_1)$, then S_2 gives an absolute minimum; however (as noted above), S_1 minimizes among surfaces of revolution generated by curves in D;

c) If $x_1 < e(t_1)$, then S_2 gives the absolute minimum; S_1 does not exist.

When $x_1 = e_1(t_1)$, then S_1 and S_2 have the same area. When $x_1 = e(t_1)$, then S_1 minimizes among surfaces of revolution generated by curves x lying in D except for the point (t_1, x_1) Bliss [1, p. 112].

§ 5. The Jacobi Necessary Condition

In §3 we studied the first necessary condition $\delta J(x^*; y)=0$ for a minimum. A second necessary condition for a minimum is $\delta^2 J(x^*; y) \geqq 0$ for all y in the space $\mathscr{Y}$ of admissible variations. This leads to the condition of Jacobi (Theorem 5.2).

In this section we suppose that L is C^4. Let x^* be an extremal of class C^4, such that $L_{\dot{x}\dot{x}}(t, x^*(t), \dot{x}^*(t))>0$ for $t_0 \leqq t \leqq t_1$. In particular, every extremal x^* has these properties if L is regular. Let us set

$$2\,\Omega(t, y, \dot{y}) = L_{xx}\, y^2 + 2\,L_{x\dot{x}}\, y\,\dot{y} + L_{\dot{x}\dot{x}}\,\dot{y}^2,$$

where L_{xx}, $L_{x\dot{x}}$, $L_{\dot{x}\dot{x}}$ are evaluated at $(t, x^*(t), \dot{x}^*(t))$, and

$$Q(x^*; y) = \int_{t_0}^{t_1} 2\,\Omega(t, y(t), \dot{y}(t))\, dt.$$

Using the notation of the proof of Lemma 3.2,

$$\frac{d^2}{d\varepsilon^2} J(x^* + \varepsilon\, y) = \int_{t_0}^{t_1} F_{\varepsilon\varepsilon}\, dt.$$

Since $F_{\varepsilon\varepsilon} = 2\,\Omega$ when $\varepsilon = 0$, we get

$$\delta^2 J(x^*; y) = Q(x^*; y). \tag{5.1}$$

Lemma 5.1. *If J has a minimum on $\mathscr{X}_e$ at x^*, then $Q(x^*; y) \geqq 0$ for all $y \in \mathscr{Y}$.*

Proof. Use formula (5.1). □

The problem of minimizing $Q(x^*; y)$ among $y \in \mathscr{Y}$ is called the *secondary minimum problem.* The variational integrand for the secondary minimum problem is $2\,\Omega$. Since

$$\Omega_{\dot{y}\dot{y}} = L_{\dot{x}\dot{x}}(t, x^*(t), \dot{x}^*(t)) > 0,$$

$2\,\Omega$ is regular.

We will begin by developing a connection between the secondary minimum problem and the behavior of a family of extremals through a given point illustrated in Example 4.3. For notational simplicity we denote such a family of extremals by $x(\cdot, \alpha)$ rather than $x^*(\cdot, \alpha)$.

Theorem 5.1. *Let L be regular, and for $\alpha_1 < \alpha < \alpha_2$ let $x(\cdot, \alpha)$ be an extremal with $x(t_0, \alpha) = x_0$, $\dfrac{\partial \dot{x}}{\partial \alpha}(t_0, \alpha) \neq 0$. If $t' = t'(\alpha)$ is such that $\dfrac{\partial x}{\partial \alpha}(t', \alpha) = 0$, then $y = \partial x / \partial \alpha$ is a solution of the secondary Euler equations*

$$\Omega_y = \frac{d}{dt}\,\Omega_{\dot{y}} \tag{5.2}$$

with $y(t_0) = y(t') = 0$ but y is not the zero function $(y(t) \not\equiv 0)$.

Conversely, if there is a nonzero solution y *of* (5.2) *with* $y(t_0)=y(t')=0$, *then* $\frac{\partial x}{\partial \alpha}(t',\alpha)=0$.

Proof. Let us take $\partial/\partial\alpha$ in the Euler equation (3.6) and interchange orders of differentiation in t and α:

$$L_{xx}\frac{\partial x}{\partial \alpha}+L_{x\dot{x}}\frac{\partial}{\partial t}\left(\frac{\partial x}{\partial \alpha}\right)=\frac{\partial}{\partial t}\left[L_{\dot{x}x}\frac{\partial x}{\partial \alpha}+L_{\dot{x}\dot{x}}\frac{\partial}{\partial t}\left(\frac{\partial x}{\partial \alpha}\right)\right].$$

This is just (5.2) for $y=\partial x/\partial\alpha$. Since $x(t_0,\alpha)$ is the constant x_0, $y(t_0)=0$; and $y(t')=0$ by assumption. Since $\dot{y}(t_0)=\frac{\partial \dot{x}}{\partial \alpha}(t_0,\alpha)\neq 0$, $y(t)\not\equiv 0$.

To prove the converse, since Ω is C^2 and regular, the solutions of the second-order linear differential Eq. (5.2) with the initial condition $y(t_0)=0$ form a one-dimensional space. As shown above, this space contains $\partial x/\partial\alpha$; in fact, the nontrivial solutions are all scalar multiples of $\partial x/\partial\alpha$. From this the converse follows. □

Motivated by Theorem 5.1, the following definition is made:

Definition. Let x^* be an extremal. A point $(t',x^*(t'))$ is called *conjugate* to (t_0,x_0) if there is a nonzero secondary extremal y^* with $y^*(t_0)=y^*(t')=0$, $t'>t_0$. (By secondary extremal we mean a piecewise C^1 solution of (5.2).)

Theorem 5.2. *(Jacobi necessary condition). If* J *has a minimum on* $\mathscr{X}_e$ *at* x^*, *then there are no conjugate points to* (t_0,x_0) *with* $t_0<t'<t_1$.

A preliminary lemma will be given before proving Theorem 5.2.

Lemma 5.2. *Let* y^* *be an extremal for the secondary minimum problem with* $y^*\in\mathscr{Y}$. *Then* $Q(x^*;y^*)=0$.

Proof. Since Ω is quadratic in $(y,\dot{y})$

$$2\Omega=y^*\Omega_y+\dot{y}^*\Omega_{\dot{y}},$$

$$2\Omega=y^*\frac{d}{dt}\Omega_{\dot{y}}+\dot{y}^*\Omega_{\dot{y}}=\frac{d}{dt}(y^*\Omega_{\dot{y}}),$$

$$Q(x^*;y^*)=\int_{t_0}^{t_1}\frac{d}{dt}(y^*\Omega_{\dot{y}})\,dt=0.$$

In the last equality we use $y^*(t_0)=y^*(t_1)=0$. □

Proof of Theorem 5.2. Obviously, the trivial variation $y(t)\equiv 0$ is in $\mathscr{Y}$, and $Q(x^*;0)=0$. Therefore, by Lemma 5.1, the minimum on $\mathscr{Y}$ of $Q(x^*;y)$ is 0. Suppose there were a conjugate point with $t_0<t'<t_1$, with y^* as in the definition above. Then $\dot{y}^*(t')\neq 0$; otherwise we would have $y^*(t)\equiv 0$ since $y^*(t')=0$, by the uniqueness theorem for solutions of (5.2). Let

$$y(t)=y^*(t)\quad \text{if } t_0\leqq t\leqq t'$$
$$y(t)=0\quad \text{if } t'\leqq t\leqq t_1.$$

Then $y \in \mathscr{Y}$; and $\dot{y}$ has a discontinuity when $t = t'$. By Lemma 5.2, applied on the interval $[t_0, t']$,

$$Q(x^*; y) = \int_{t_0}^{t'} 2\,\Omega\,dt = \int_{t_0}^{t_1} 2\,\Omega\,dt = 0.$$

Thus, y is minimizing in the secondary problem. Since Ω is regular, $\dot{y}$ cannot have any discontinuities by Corollary 3.2 (applied to the secondary problem). This is a contradiction. □

Theorem 5.2 implies, in Example 4.3, that the points conjugate to (0, 1) are precisely those lying on the envelope e. By Theorem 5.2, the catenary $\tilde{x}^*$ which touches the envelope between (0, 1) at (t_1, x_1) cannot minimize area.

In Example 4.1, the points $(\pi k^{-1} m, 0)$ are conjugate to $(0, 0)$ if $m = 1, 2, \ldots$.

Remark 5.1. The regularity condition $L_{\dot{x}\dot{x}} > 0$ implies strict convexity of $L(t, x, \cdot)$; regularity does not exclude conjugate points, as is shown in Example 4.3. However, suppose that the stronger condition

$$L_{xx}\, y^2 + 2\, L_{x\dot{x}}\, y\, \dot{y} + L_{\dot{x}\dot{x}}\, \dot{y}^2 > 0 \tag{5.3}$$

holds for all $(t, x, \dot{x})$ and for all $(y, \dot{y}) \neq (0, 0)$. Then L is strictly convex in $(x, \dot{x})$. This implies J is strictly convex on $\mathscr{X}$ (Problem 6). By Theorems 2.3 and 2.4, there is at most one extremal x^* with the given end conditions (3.2); if this extremal x^* exists, then x^* minimizes J on $\mathscr{X}_e$. When (5.3) holds, $Q(x^*; y) > 0$ unless $y(t) \equiv 0$. There are no conjugate points.

In Example 4.1, (5.3) holds if $a > 0$.

Remark 5.2. (Local vs. absolute minima). Theorems 3.1 and 5.2 give two necessary conditions for a minimum on $\mathscr{X}_e$, in other words an absolute minimum, at x^*. However, these necessary conditions also hold for a local minimum, and no changes are needed in the proofs. Let us consider the following norm $\|\ \|_1$ on $\mathscr{X}$:

$$\|x\|_1 = \|x\| + \|\dot{x}\|,$$

where

$$\|u\| = \sup_{t_0 \leqq t \leqq t_1} |u(t)|.$$

A set $\{x \in \mathscr{X}: \|x - x^*\|_1 < a\}$, $a > 0$, is called a *weak* neighborhood of x^*. Suppose that $J(x^*) \leqq J(x)$ for all $x \in \mathscr{X}_e \cap \mathscr{N}_a$, where $\mathscr{N}_a$ is a weak neighborhood of x^*. Given $y \in \mathscr{Y}$, $x^* + \varepsilon\, y \in \mathscr{X}_e \cap \mathscr{N}_a$ for $0 \leqq |\varepsilon| < \varepsilon(y)$. This implies $\delta J(x^*; y) = 0$, $\delta^2 J(x^*; y) \geqq 0$. The proofs of Theorems 3.1, 5.1 used only these two properties of first and second variations. Hence, these necessary conditions hold also for local minima.

A local minimum in this sense is called *weak*. The Legendre condition (Problem 10) is another necessary condition for a weak local minimum. A set $\{x \in \mathscr{X}: \|x - x^*\| < a\}$, $a > 0$, is called a *strong* neighborhood of x^*. Note that nearness of x and x^* in the $\|\ \|$ norm does not require nearness of $\dot{x}(t)$ and $\dot{x}^*(t)$. If $J(x^*) \leqq J(x)$ for all $x \in \mathscr{X}_e \cap \mathscr{N}_a'$, where $\mathscr{N}_a'$ is a strong neighborhood of x^*, then J has a *strong* local minimum at x^*. There is another necessary condition, due to Weierstrass, for a strong local minimum. It will be obtained in Chap. II.8 as a corollary of Pontryagin's principle.

§ 6. The Simplest Problem in n Dimensions

Let E^n denote the space of n tuples $x=(x_1, \ldots, x_n)$ of real numbers. Consider the problem of finding n unknown functions $x_1(\cdot), \ldots, x_n(\cdot)$ such that

$$J=\int_{t_0}^{t_1} L[t, x_1(t), \ldots, x_n(t), \dot{x}_1(t), \ldots, \dot{x}_n(t)]\, dt$$

is minimum, subject to the end conditions

$$x_i(t_0)=x_{0_i}, \qquad x_i(t_1)=x_{1_i}, \qquad i=1, \ldots, n.$$

The discussion of this problem is almost the same as for $n=1$ above. We merely indicate the changes needed. Let us write

$$x(t)=(x_1(t), \ldots, x_n(t)), \qquad \dot{x}(t)=(\dot{x}_1(t), \ldots, \dot{x}_n(t)).$$

Thus $x(t)$ has values in E^n, and $\dot{x}(t)$ is the derivative of the vector-valued function x. Then $J=J(x)$ again has the form (3.1). If we write $x_j=(x_{j1}, \ldots, x_{jn}), j=0, 1$, then the end conditions are again (3.2). By x piecewise C^1 we mean that each x_i is piecewise C^1. Let

$$L_x=(L_{x_1}, \ldots, L_{x_n}), \qquad L_{\dot{x}}=(L_{\dot{x}_1}, \ldots, L_{\dot{x}_n})$$

denote the gradients of $L(t, x, \dot{x})$ in the variables $x_i, \dot{x}_i, i=1, \ldots, n$. In § 3 we need merely reinterpret the various formulas. Products of vectors mean scalar products, for instance, $L_x y=\sum_i L_{x_i} y_i$. Eq. (3.5) involves a vector integral, and is equivalent to

$$(6.1) \qquad -\int_{t_0}^{t} L_{x_i}\, d\tau + L_{\dot{x}_i}=c_i, \qquad i=1, \ldots, n,$$

for suitable constants c_i. Eq. (3.6) stands for the system of n Euler equations

$$(6.2) \qquad L_{x_i}=\frac{d}{dt} L_{\dot{x}_i}, \qquad i=1, \ldots, n,$$

for the functions $x_1^*, \ldots, x_n^*$.

The regularity condition $L_{\dot{x}\dot{x}}>0$ now means that the matrix of partial derivatives $(L_{\dot{x}_i\dot{x}_j})$, $i, j=1, \ldots, n$, is positive definite for all $(t, x, \dot{x})$. In § 5 we now take

$$(6.3) \qquad 2\Omega=\sum_{i,j=1}^{n} [L_{x_i x_j} y_i y_j + 2 L_{x_i \dot{x}_i} y_i \dot{y}_j + L_{\dot{x}_i \dot{x}_j} \dot{y}_i \dot{y}_j].$$

The secondary Euler equations

$$(6.4) \qquad \Omega_{y_i}=\frac{d}{dt}\Omega_{\dot{y}_i}, \qquad i=1, \ldots, n,$$

form a linear system of second order. The set $\mathscr{Y}_0$ of solutions y of this system with $y(t_0)=0$ is a vector space of dimension n. The existence of a conjugate point for $t=t'$ is equivalent to the fact that the space $\{y(t'): y\in\mathscr{Y}_0\}$ is a proper subspace of E^n (in other words, has dimension $<n$).

In terms of Theorem 5.1, suppose that $x(\cdot,\alpha)$ is an extremal for each $\alpha=(\alpha_1,\dots,\alpha_n)$ in some open set $\mathscr{A}$ with $x(t_0,\alpha)=x_0$, and $\partial\dot{x}/\partial\alpha_1,\dots,\partial\dot{x}/\partial\alpha_n$ linearly independent vectors for $t=t_0$. Then a point conjugate to (t_0,x_0) will occur on the extremal $x(\cdot,\alpha)$ when $t=t(\alpha)$ if and only if $t(\alpha)$ is a solution of the equation $\det\partial x/\partial\alpha=0$. Here $\det\partial x/\partial\alpha$ is the determinant of the $n\times n$ matrix $(\partial x_i/\partial\alpha_j)$. This is discussed in more detail, for instance, in Bliss [2].

Remark. The *multiplicity* of a conjugate point $(t', x^*(t'))$ is the dimension of the vector space of solutions of the secondary Euler equation with the end conditions $y(t_0)=y(t')=0$. The number of conjugate points with $t_0<t'<t_1$ is finite. The sum of their multiplicities is called the *index*. It can be shown that the index equals the largest dimension of subspaces $\tilde{\mathscr{Y}}$ of the space $\mathscr{Y}$ of admissible variations (satisfying $y(t_0)=y(t_1)=0$) on which Q is negative definite; negative definite on $\tilde{\mathscr{Y}}$ means $Q(x^*;y)<0$ for all $y\in\tilde{\mathscr{Y}}$, $y(t)\not\equiv 0$.

This fact is important for applying calculus of variations to the global study of geometric and topological questions on manifolds, using Morse theory. See Milnor [1], Morse [1].

Example 6.1. The principle of least action. Suppose that the state of a mechanical system at time t is described, in some coordinate system, as a vector $x(t)=(x_1(t),\dots,x_n(t))$. Suppose, moreover, that the system has at time t

$$\text{Kinetic energy} = \sum_{i,j=1}^{n} a_{ij}[x(t)]\,\dot{x}_i(t)\,\dot{x}_j(t)$$

$$\text{Potential energy} = U[x(t)].$$

The matrix $(a_{ij}(x))$ is assumed to be symmetric and positive definite for each $x\in E^n$.

Consider the *action integral*

$$J(x)=\int_{t_0}^{t_1} L[x(t),\dot{x}(t)]\,dt, \tag{6.5}$$

where

$$L(x,\dot{x})=\sum_{i,j=1}^{n} a_{ij}(x)\,\dot{x}_i\,\dot{x}_j - U(x).$$

The principle of least action states that the path x^* followed by the motion of the system is extremal; since the action integrand L is regular, this means that x^* is a solution of the system of Euler equations (6.2). The extremal x^* is not generally minimizing. Hence, the principle should perhaps be called one of extremal (or stationary) action rather than least action. However, it can be shown that x^* does minimize action if the time interval $[t_0,t_1]$ is sufficiently short.

Example 4.1, with $a<0$, corresponds to the principle of least action for an oscillating spring obeying Hooke's law. Notice that extremals in Example 4.1 do not minimize the action integral if $(t_0,x_0)=(0,0)$, $t_1>t'=\pi k^{-1}$ for in this case there is a conjugate point $(t',0)$ to $(0,0)$ with $t_0<t'<t_1$ and the Jacobi necessary condition is violated.

Example 6.2. Geodesics on Riemannian manifolds. Suppose that M is a space such that a notion of length is defined for curves lying on M. A curve which makes the length an extremum is called a geodesic. Let us suppose that M is a Riemannian manifold, of dimension n. Then M is covered by a collection $\{\theta\}$ of coordinate patches. With each coordinate patch θ is associated a set of local coordinates $x=(x_1, \ldots, x_n)$ for θ. The length of a curve contained in θ is given by

$$(6.6) \qquad J(x)=\int_{t_0}^{t_1}\left(\sum_{i,j=1}^{n} g_{ij}[x(t)]\,\dot{x}_i(t)\,\dot{x}_j(t)\right)^{1/2} dt,$$

where the matrices $(g_{ij}(x))$ are positive definite and symmetric; (6.6) is equivalent to the formula $(ds)^2=\sum_{i,j} g_{ij}\,dx_i\,dx_j$, where s is arc length.

In this case the variational integrand is

$$L(x, \dot{x})=\left[\sum_{i,j=1}^{n} g_{ij}(x)\,\dot{x}_i\,\dot{x}_j\right]^{1/2}.$$

Suppose that $x=x^*(t)$ describes, in local coordinates, a geodesic lying in θ. This means that the Euler equations (6.2) hold. (The example of great circles on a sphere shows that geodesics need not minimize length; however, sufficiently short pieces of a geodesic have minimum length.)

Length is by its nature invariant under changes of parameter t. We choose parameters so that for the geodesic in question the parameter is arc length s. Then

$$L[x^*(s), \dot{x}^*(s)]=1,$$

where now $\dot{x}=d/ds$. The Euler equations (6.2) then become

$$\frac{1}{2}\sum_{i,j=1}^{n}\frac{\partial g_{ij}}{\partial x_k}\,\dot{x}_i^*\,\dot{x}_j^*=\frac{d}{ds}\sum_{i=1}^{n} g_{ik}\,\dot{x}_i^*, \qquad k=1, \ldots, n.$$

These equations can be rewritten

$$(6.7) \qquad \ddot{x}_k^*+\sum_{i,j=1}^{n}\Gamma_{ij}^k(x)\,\dot{x}_i^*\,\dot{x}_j^*, \qquad k=1, \ldots, n,$$

for suitable coefficients $\Gamma_{ij}^k(x)$ called the Christoffel symbols. See for instance Auslander [1].

If instead of L we use the simpler integrand L^2, we also arrive at (6.7). This observation is often used for studying geodesics, in particular for applying Morse theory to estimate the number of closed geodesics on a compact Riemannian manifold M.

Problems—Chapter I

(1) In each case find the extremals. Which pairs of endpoints (t_0, x_0), (t_1, x_1) have an extremal joining them?

(a) $L=x^2+2kx\dot{x}+\dot{x}^2$.

(b) $L=-x^2+2kx\dot{x}+\dot{x}^2$.

(c) $L=t\dot{x}^2$.

(d) $L=\dot{x}^2(1-\dot{x}^2)$.

In (a), (b) k is a constant. Why are the extremals in (a), (b) the same as in Example 4.1 with $a=\pm 1$, $b=1$?

(2) Find the geodesics on the cone $x^2=k^2(y^2+z^2)$ in E^3. Show that if a nappe of the cone is cut along a generator and the cone is flattened, then geodesics become straight lines. *Hint.* Coordinatize the cone by polar coordinates in the plane where the flattened cone lies.

(3) Suppose that $x(t)\leqq x^*(t)$ for $t_0\leqq t\leqq t_1$, with equality when $t=t_0, t_1$. Also suppose that in Example 4.2 $N_t-M_x\geqq 0$ when $x(t)\leqq x\leqq x^*(t)$, $t_0\leqq t\leqq t_1$. Show that $J(x)\geqq J(x^*)$.

(4) Let $L=x^4+x\dot{x}+\frac{1}{2}$.

(a) Verify that the only extremal is $x^*(t)\equiv 0$.

(b) For $(t_0, x_0)=(0,1)$, $(t_1, x_1)=(1,0)$, show that the infimum on $\mathscr{X}_e$ of $J(x)$ is 0 but that $J(x)$ has no minimum on $\mathscr{X}_e$. *Hint.* For $r=1,2,\ldots$ consider $x(t)=1-rt$ if $0\leqq t\leqq r^{-1}$ and $x(t)=0$ for $r^{-1}\leqq t\leqq 1$.

(5) Let $J(x)=\int_{t_0}^{t_1}\int_{t_0}^{t_1} K(s,t)\,x(s)\,x(t)\,ds\,dt+\int_{t_0}^{t_1}[x(t)]^2\,dt-\int_{t_0}^{t_1} x(t) f(t)\,dt$, where f and K are given and $K(t,s)=K(s,t)$.

(a) Find $\delta J(x;y)$. Show that a necessary condition for $J(x)$ to be minimum at x^*, among all continuous functions on $[t_0, t_1]$, is that x^* be a solution of the integral equation

$$2\int_{t_0}^{t_1} K(s,t)\,x(s)\,ds+2x(t)=f(t).$$

(b) Compute $\delta^2 J(x;y)$. Give a condition on K under which Theorem 2.4 applies.

(6) Prove that (5.3) implies J strictly convex on $\mathscr{X}$.

(7) Let $L=g(x)+\frac{1}{2}\dot{x}^2$, where $g''(x)>0$ for $-\infty<x<\infty$. Use Problem 6 and Theorems 2.3, 2.4 to show that the differential equation $\ddot{x}=g'(x)$ has at most one solution through given points (t_0, x_0), (t_1, x_1).

(8) In Example 6.1 show that the sum of kinetic and potential energy is constant in t, if x^* is an extremal. *Hint.* Use the n-dimensional version of Lemma 4.1.

(9) Suppose that, in Example 6.1, $a_{ii}=\frac{1}{2}m$ and $a_{ij}=0$ for $i\neq j$. Show that the Euler equations become $m\ddot{x}^*=-U_x$.

(10) Show that if $J(x)$ has a weak local minimum at x^*, then

$$L_{\dot{x}\dot{x}}(t, x^*(t), \dot{x}^*(t))\geqq 0 \qquad \text{for } t_0\leqq t\leqq t_1.$$

This is the *Legendre* condition. *Hint.* If this condition fails at some t', consider $y(t)=1-r|t-t'|$ if $|t-t'|\leqq r^{-1}$ and $y(t)=0$ otherwise. Show that $Q(x^*, y)<0$ if r is large enough.

(11) (a) Let x be an absolutely continuous function on $[t_0, t_1]$ such that $|\dot{x}(t)|\leqq M$ almost everywhere, for some constant M. Show that there exists a

sequence $x^1, x^2, \ldots$ of C^1 functions with $x^j(t_i)=x(t_i)$ for $i=0, 1$, $j=1, 2, \ldots$, such that x^j converges to x uniformly on $[t_0, t_1]$ and $J(x^j) \to J(x)$ as $j \to \infty$. *Hint.* First find such a sequence x^j which may not satisfy the right hand end condition $x^j(t_1)=x(t_1)$; then add a small linear function.

(b) Prove the same result without the restriction $|\dot{x}(t)| \leqq M$, provided the following holds: there exist $a>0$ and an increasing function $g(r)$ such that

$$\int_{t_0}^{t_1} g(|\dot{x}(t)|+a)\,dt<\infty, \qquad |L(t, \tilde{x}(t), \dot{\tilde{x}}(t))| \leqq g(|\dot{\tilde{x}}(t)|)$$

for $t_0 \leqq t \leqq t_1$, whenever $\|\tilde{x}-x\|<a$.

Chapter II. The Optimal Control Problem

§ 1. Introduction

In this chapter we shall discuss an optimization problem that we will call "the optimal control problem." In the 1950's, motivated especially by aerospace problems, engineers became interested in the problem of controlling a system governed by a set of differential equations. In many of the problems it was natural to want to control the system so that a given performance index would be minimized. In some aerospace problems large savings in cost could be obtained with a small improvement in performance so that optimal operation became very important. As techniques were developed which were practical for computation and implementation of optimal controls the use of this theory became common in a large number of fields. References which illustrate work typical in applying optimal control to economic problems are Burmeister-Döbell [1], Pindyck [1], Shell [1].

In classical calculus of variations there are three equivalent optimization problems called the Bolza problem, the Lagrange problem, and the Mayer problem, which deal with minimizing a performance index of a system governed by a set of differential equations. See Bliss [2] for a discussion of these problems. There are two differences between these problems and the optimal control problem. The differential equations involved are of a slightly less general type in the optimal control problem and certain variables called control variables are required to lie in a closed set for the optimal control problem while for the problems of Bolza, Lagrange, and Mayer the variables of the problem are assumed to lie in an open set. In fact, the optimal control problem which we shall formulate in §3 is a Mayer problem with the added condition that the control variables are restricted to lie in a closed set. Conversely, except that the form of the differential equation involved in the optimal control problem is slightly less general, the optimal control problem includes the Bolza problem, the Lagrange problem, and the Mayer problem as special cases.

Necessary conditions for optimality for the optimal control problem were derived by Pontryagin, Boltyanskii and Gamkrelidze. It has become common terminology to call these necessary conditions "Pontryagin's Principle." It has been shown by Berkovitz [3], using constructions of Valentine [2], that Pontryagin's principle can be derived from the necessary conditions for optimality of the Bolza problem. Since there is considerable conceptional simplicity in a direct proof of Pontryagin's Principle and the optimal control problem nearly includes the Bolza, Lagrange, and Mayer problems as special cases, we will only discuss the optimal control problem.

The chapter begins by giving several examples of optimal control problems. Then in §3 the optimal control problem is formally stated. The optimality conditions of Pontryagin's principle are stated in §5. They are applied to compute extremal control laws for three of the examples in §6, §7, and §8. A number of the topics and concepts which arise in optimal control problems are discussed in §9. To motivate the proof of Pontryagin's principle and to illustrate the role of the methods of Chap. I in the proof of Pontryagin's principle, two derivations of optimality conditions are given in a special case of the optimal control problem. In this case the derivation is especially simple and conceptually clear. The proof of Pontryagin's principle in its general form is given in §13, §14, and §15.

We shall mention some notation which will be used in the remainder of this book. Vectors will be identified with column matrices. Components of a vector will be indicated by subscripts. Sequences of vectors will be indicated by superscripts. The symbol A' will denote the transpose of a matrix A. Thus $x(t)'$ is the row vector whose elements are the corresponding elements of the column vector $x(t)$. Whenever matrices are written adjacent to each other it will be understood that this indicates the product of the matrices involved. Thus $x(t)' L(t)\, x(t)$ is the product of the three matrices involved which agrees with the quadratic form in the components of $x(t)$

$$\sum_{i,j=1}^{n} x_i(t)\, L_{ij}(t)\, x_j(t).$$

Matrices of partial derivatives of vector functions will be indicated using a subscript which denotes the vector of variables involved in the partial differentiation. Thus if $f(x, y, z)$ is a vector function of the vector variables x, y, z, $f_y(x, y, z)$ will denote the matrix whose ij-th element is the partial derivative of the i-th component of f with respect to the variable y_j. To coincide with this notation we will make the following exception to our identification of vectors with column matrices. If, say, $f(x, y, z)$ is a scalar function of vector variables, we shall identify a vector of partial derivatives such as $f_y(x, y, z)$ with the corresponding row matrix rather than the corresponding column matrix.

§ 2. Examples

We shall begin our discussion of the optimal control problem by giving several examples of this type of problem. The following is a simple example from the aerospace field.

Example 2.1. Consider the problem of a spacecraft attempting to make a soft landing on the moon using a minimum amount of fuel. To define a simplified version of this problem, let m denote the mass, h and v denote the height and vertical velocity of the spacecraft above the moon, and u denote the thrust of the spacecraft's engine. Let M denote the mass of the spacecraft without fuel, h_0 and

v_0 the initial height and vertical velocity of the spacecraft, F the initial amount of fuel, α the maximum thrust attainable by the spacecraft's engine, k a constant, and g the gravitational acceleration of the moon.

The gravitational acceleration g may be considered constant near the moon. The equations of motion of the spacecraft are

$$\begin{aligned} \dot{h} &= v \\ \dot{v} &= -g + m^{-1} u \\ \dot{m} &= -k u. \end{aligned} \tag{2.1}$$

The thrust $u(t)$ of the spacecraft's engine is the control for the problem. Suppose the class $\mathscr{U}$ of control functions is all piecewise continuous functions $u(t)$ defined on an interval $[t_0, t_1]$ such that

$$0 \leqq u(t) \leqq \alpha. \tag{2.2}$$

It is natural to take initial time, $t_0 = 0$, and terminal time t_1 equal to the first time the spacecraft reaches the moon. End conditions which must be satisfied at the initial time and terminal time are

$$\begin{aligned} &h(0) - h_0 = 0, \quad v(0) - v_0 = 0, \quad m(0) - M - F = 0 \\ &h(t_1) = 0, \quad v(t_1) = 0. \end{aligned} \tag{2.3}$$

The problem is to land using a minimum amount of fuel or equivalently to minimize $-m(t_1)$ over the class $\mathscr{U}$.

Example 2.2. In the economics the following situation is called a Ramsey model for a one sector economy. The output rate $y(t)$ of the economy and the capital $K(t)$ are related through a production function $y(t) = F(K(t))$. The rate of consumption $c(t)$ is a proportion $u(t)$ of a function $G(K(t))$ of the capital. That is

$$c(t) = u(t)\, G(K(t)) \tag{2.4}$$

and

$$0 \leqq u(t) \leqq 1. \tag{2.5}$$

Then the rate of change of capital is given by

$$\dot{K}(t) = F(K(t)) - u(t)\, G(K(t)). \tag{2.6}$$

Let $H(c)$ be a utility function which represents the utility to the system of consuming at rate c. Consider the performance of the economy to be given by the integral of the utility function

$$\int_{t_0}^{t_1} H(u(t)\, G(K(t)))\, dt. \tag{2.7}$$

Formulate the problem of choosing the proportion to be consumed, $u(t)$, so the capital satisfying (2.6) changes from an initial value $K(t_0) = K_0$ to a desired terminal value $K(t_1) = K_1$ and the performance (2.7) is maximized.

Example 2.3. The following is an optimization problem called the linear regulator problem which is applied to a large number of design problems in engineering. Let $A(t)$, $M(t)$ and D be $n\times n$ matrices and $B(t)$, $n\times m$, and $N(t)$, $m\times m$, matrices of continuous functions. Let $u(t)$ be an m-dimensional piecewise continuous vector function defined on a fixed interval $[t_0, t_1]$. Let $x(t)$ be the n-dimensional vector function which is the corresponding solution of

$$\dot{x}(t)=A(t)\,x(t)+B(t)\,u(t) \tag{2.8}$$

with initial condition $x(t_0)=x_0$. Suppose $M(t)$, $N(t)$, and D are symmetric with $M(t)$, D non negative definite and $N(t)$ positive definite. Consider the problem of choosing $u(t)$ so that

$$x(t_1)'\,D\,x(t_1)+\int_{t_0}^{t_1}[x(t)'\,M(t)\,x(t)+u(t)'\,N(t)\,u(t)]\,dt \tag{2.9}$$

is minimized.

The optimal control for the linear regulator problem is a linear function of $x(t)$. (See Theorem IV.5.1.) This is particularly convenient for implementation. Because of this, controls have been designed for many nonlinear problems as well as linear problems, using the solution of the linear regulator problem. Nonlinear problems may be linearized (see Theorem 10.2, particularly formula (10.7)) and the performance criterion approximated by a quadratic criterion for the linearized equations. See Athans, [1], [2] for examples of this type of application.

Example 2.4. The simplest problem in calculus of variations formulated in I.3 can be rewritten as an optimal control problem. The optimal control problem with equations of motion,

$$\dot{x}_1=u(t)$$
$$\dot{x}_2=L(t, x_1(t), u(t))$$

control set $U=E^1$, performance index $x_2(b)$, and end conditions

$$x_1(a)=c, \quad x_2(a)=0, \quad x_1(b)=d$$

for fixed a, b, c, d is equivalent to the simple problem in calculus of variations of minimizing

$$J(x)=\int_a^b L(t, x(t), \dot{x}(t))\,dt$$

over the class of curves $x(t)$ with piecewise continuous derivatives which satisfy

$$x(a)=c, x(b)=d.$$

§ 3. Statement of the Optimal Control Problem

To state the optimal control problem we will make the following notational conventions. Let U be a closed subset of E^m; t, x, u variables respectively in

$E^1, E^n, E^m; f(t,x,u)$ a vector function

$$f: E^1 \times E^n \times E^m \to E^n$$

which is continuous and has continuous first partial derivatives with respect to the coordinates of x. Let $\phi(t_0, t_1, x_0, x_1)$ be a vector function

$$\phi: E^1 \times E^1 \times E^n \times E^n \to E^k$$

which is of class C^1.

Let $\mathscr{U}$ be a set of piecewise continuous functions $u(t)$ with values in U, each function $u(t)$ being defined on some interval $[t_0, t_1]$, which may differ for different elements of $\mathscr{U}$. A function $u(t)$ in $\mathscr{U}$ will be called a *control*. The set $\mathscr{U}$ of controls will be assumed to have the following property. If $u(t)$ defined on $t_0 \leqq t \leqq t_1$ is in $\mathscr{U}$ and for $i=1, \ldots, p$, $v_i \in U$ and $\tau_i - h_i < t \leqq \tau_i$ are non-overlapping intervals intersecting $[t_0, t_1]$ then

$$\tilde{u}(t) = \begin{cases} v_i & \text{if } \tau_i - h_i < t \leqq \tau_i \\ u(t) & \text{if } t \in [t_0, t_1] \text{ and } \notin \text{ one of the intervals } \tau_i - h_i < t \leqq \tau_i \end{cases}$$

is in $\mathscr{U}$.

For a control $u(t)$ defined on $[t_0, t_1]$ the solution $x(t)$ of the differential equation

$$\dot{x} = f\big(t, x(t), u(t)\big) \tag{3.1}$$

on the interval $[t_0, t_1]$ with initial condition $x(t_0) = x_0$ will be called the *trajectory* corresponding to the control $u(t)$ and initial condition x_0. The value of $x(t)$ at time t is called the *state* of the system at time t. Eq. (3.1) is called the *equation of motion* of the system. Frequently our discussions will involve controls and their corresponding trajectories. If $x(t)$ appears without mention in a formula it is always understood that a control $u(t)$ and initial condition x_0 have been specified and $x(t)$ is the trajectory corresponding to $u(t)$ and x_0 through Eq. (3.1).

The first component of ϕ evaluated at $\big(t_0, t_1, x(t_0), x(t_1)\big)$, where $x(t)$ is a solution of (3.1),

$$\phi_1\big(t_0, t_1, x(t_0), x(t_1)\big) \tag{3.2}$$

is the *performance index* or *performance criterion* of the system. To indicate the performance's dependence on the initial state $x_0 = x(t_0)$ and control $u(t)$ we shall denote the performance (3.2) by

$$J(x_0, u) = \phi_1\big(t_0, t_1, x(t_0), x(t_1)\big).$$

The next k-1 components of ϕ define through the equations

$$\phi_j\big(t_0, t_1, x(t_0), x(t_1)\big) = 0 \qquad j = 2, \ldots, k \tag{3.3}$$

end conditions for the trajectories of the system. A pair (x_0, u), of an initial condition x_0 and a control $u = u(t)$, will be called *feasible*[1] if there is a solution $x(t)$ of (3.1) on $[t_0, t_1]$ with initial condition $x(t_0) = x_0$ and the end conditions (3.3) are satisfied by $x(t)$. Let $\mathscr{F}$ denote the class of feasible pairs (x_0, u).

[1] Notice that a feasible pair also depends on t_0, t_1 the end points of the interval $[t_0, t_1]$ on which the control $u(t)$ is defined.

The optimal control problem is to find in the class $\mathscr{F}$ an element (x_0, u) such that the corresponding performance index (3.2) is minimized. A pair (x_0, u) of $\mathscr{F}$ which achieves this minimum will be called an *optimal initial condition* and an *optimal control.*

Notice that the trajectory $x(t)$ and the performance index (3.2) are unchanged if the corresponding control $u(t)$ is altered so as to be continuous from the left. Hence in the future we shall assume without loss of generality that $\mathscr{U}$ is a class of left continuous, piecewise continuous, control functions. For brevity the notation e will be used to denote a $(2n+2)$-tuple of end points.

$$e=\big(t_0, t_1, x(t_0), x(t_1)\big).$$

§ 4. Equivalent Problems

If U was taken to be an open subset of E^m rather than a closed subset the optimal control problem would be an optimization problem discussed in classical calculus of variations, Bliss [2], called a Mayer problem. Extending this terminology from calculus of variations we shall also call the optimal control problem defined in § 3 a *Mayer problem.*

Let $L(t, x, u)$ denote a continuous function

$$L\colon\ E^1 \times E^n \times E^m \to E^1$$

of class C^1 in (x, u). If instead of a performance index

$$J(x_0, u)=\phi_1(e) \tag{4.1}$$

the performance index is

$$J(x_0, u)=\int_{t_0}^{t_1} L\big(t, x(t), u(t)\big)\,dt \tag{4.2}$$

the optimization problem is called a *Lagrange problem.* If the performance index is

$$J(x_0, u)=\phi_1(e)+\int_{t_0}^{t_1} L\big(t, x(t), u(t)\big)\,dt \tag{4.3}$$

the problem is called a *Bolza problem.* These are again names used in classical calculus of variations, Bliss [2].

The three optimization problems as formulated with performances (4.1), (4.2) and (4.3) are equivalent in that each can be formulated as one of the other forms. To see this notice that a Lagrange problem can be formulated as a Mayer problem by adding another component differential equation

$$\dot{x}_{n+1}(t)=L\big(t, x(t), u(t)\big)$$

with initial condition $x_{n+1}(t_0)=0$ to those of (3.1). Then the performance (4.2) is given by

$$J(x_0, u)=\phi_1(e)=x_{n+1}(t_1)=\int_{t_0}^{t_1} L\big(t, x(t), u(t)\big)\,dt$$

as a performance of Mayer form. A Bolza problem can be converted to a Mayer problem in a similar fashion with (4.3) rewritten as $\phi_1(e)+x_{n+1}(t_1)$. A Mayer problem can be converted to a Lagrange problem by adding an extra coordinate and differential equation

$$\dot{x}_{n+1}=0$$

to (3.1) and adding an extra component $\phi_{k+1}(e)$ to the $\phi_2(e), \ldots, \phi_k(e)$ with $\phi_{k+1}(e)$ defined by

$$\phi_{k+1}(e)=x_{n+1}(t_1)-\frac{\phi_1(e)}{t_1-t_0}.$$

Then $\phi_{k+1}(e)=0$ implies

$$\int_{t_0}^{t_1} x_{n+1}(t)\,dt=\phi_1(e).$$

Thus (4.1) has been expressed in the form (4.2) with $L(t,x,u)=x_{n+1}$.

While the three optimization problems are equivalent, it is natural to suspect that a more general theorem could be established for the problem expressed in one form than another. For necessary conditions for optimality this is not the case. In Chap. III existence theorems for optimal controls are obtained for problems in both Mayer form and Bolza form. The existence theorem obtained directly for the problem in Bolza form is obtained under weaker hypotheses than the theorem which could have been obtained by converting the Bolza form to Mayer form and applying the existence theorem obtained for the problem in Mayer form.

Considerable notational simplicity is achieved in Chap. II and IV by using the optimal control problem in Mayer form. If examples arise naturally in Lagrange or Bolza form we shall assume without comment that they satisfy theorems established for problems in Mayer form by making use of the correspondence between the problems mentioned above.

§ 5. Statement of Pontryagin's Principle

The derivation of necessary conditions for optimality for the optimal control problem, that is the proof of "Pontryagin's principle", is quite long. In § 11 we shall try to motivate the proof by giving a discussion of the derivation of these conditions and alternative approaches in a simple special case. The derivation of Pontryagin's principle is given in § 12–15. However before doing this we shall state "Pontryagin's principle" and illustrate its application by computing controls for Examples 2.1, 2.3 and 2.4.

The conditions of Pontryagin's principle reduce the computation of an optimal control to the solution of a two point boundary problem for a set of differential equations together with a minimization side condition. In many important applications optimal controls have been computed through use of Pontryagin's principle. However in complicated examples solution of the two point boundary value problem can be difficult. Computing methods which make a direct numerical attack on the optimization problem have been devised and widely used, Kelly [1], Bryson Denham [1], McGill [1]. The theory and use of these direct numerical methods is an important part of the subject. However, we shall not discuss them in this book. See Falb DeJong [1], Dyer McReynolds [1], Polak [1], [2] for a discussion of these techniques.

Theorem 5.1. (Pontryagin's Principle). *Necessary conditions that* $(x_0^*, u^*(t))$ *be an optimal initial condition and optimal control for the optimal control problem are the existence of a nonzero k-dimensional vector* λ *with* $\lambda_1 \leqq 0$ *and an n-dimensional vector function* $P(t)$ *such that for* $t \in [t_0, t_1]$:

$$\dot{P}(t)' = -P(t)' f_x(t, x^*(t), u^*(t)); \tag{5.1}$$

for $t \in (t_0, t_1)$ *and* $u \in U$

$$P(t)'[f(t, x^*(t), u) - f(t, x^*(t), u^*(t))] \leqq 0; \tag{5.2}$$

$$P(t_1)' = \lambda' \phi_{x_1}(e); \tag{5.3}$$

$$P(t_0)' = -\lambda' \phi_{x_0}(e); \tag{5.4}$$

$$P(t_1)' f(t_1, x^*(t_1), u^*(t_1)) = -\lambda' \phi_{t_1}(e); \tag{5.5}$$

$$P(t_0)' f(t_0, x^*(t_0), u^*(t_0)) = \lambda' \phi_{t_0}(e). \tag{5.6}$$

If $f(t, x, u)$ *has a continuous partial derivative* $f_t(t, x, u)$, *then the condition*

$$P(t)' f(t, x^*(t), u^*(t)) = \lambda' \phi_{t_0}(t_0, t_1, x^*(t_0), x^*(t_1)) + \int_{t_0}^{t} P(s)' f_t(s, x^*(s), u^*(s))\, ds \tag{5.7}$$

holds for each $t \in [t_0, t_1]$.

The quantity

$$H(t, x, u) = P(t)' f(t, x, u) \tag{5.8}$$

is generally called the *Hamiltonian* in analogy with a corresponding quantity occuring in classical mechanics. See for instance, Goldstein [1] for a discussion of this concept in mechanics. Condition (5.2) can be expressed as

$$\max_{u \in U} \{H(t, x^*(t), u)\} = H(t, x^*(t), u^*(t)) \tag{5.9}$$

and is called Pontryagin's maximum principle. Conditions (5.3)–(5.6) are generalizations of conditions found in calculus of variations and optimal control problems Bliss [2], p. 202, Pontryagin *et al.* [1], p. 49 called *transversality conditions.* Eq. (5.1) are called the *adjoint equations.*

A control will be called an *extremal* if (5.1)–(5.7) are satisfied and the corresponding trajectory satisfies the end conditions (3.3). Since the conditions of Pontryagin's principle are necessary conditions for optimality each optimal control must be an extremal; however, since the conditions need not be sufficient for optimality there may be extremal controls which are not optimal.

It will be seen in the proof of Pontryagin's principle that condition (5.7) is implied by condition (5.2) and (5.6). Thus in checking to see if a trajectory is extremal it is not necessary to verify (5.7). Formula (5.7) implies $H(t, x^*(t), u^*(t))$ is continuous even though the control $u^*(t)$ may be discontinuous.

Remark 5.1. If the statement of Theorem 5.1 holds for $P(t)$, and $P(t)$ is replaced by $\tilde{P}(t)=\gamma P(t)$ and λ replaced by $\tilde{\lambda}=\gamma\lambda$ where γ is a positive real number, then Eq. (5.1)–(5.6) still hold. If it can be determined from other considerations that $\lambda_1<0$, then a constant γ may be selected so that $\lambda_1=-1$. Hence in the case in which $\lambda_1<0$ it is no loss of generality to assume $\lambda_1=-1$.

There are problems in which λ_1 must equal zero. These problems are called *abnormal*. In § 12 abnormal problems are discussed again and an example of one is given. Suppose a control $u(t)$ is an extremal control with $\lambda_1=0$. Notice, from Eq. (5.1)–(5.7), if the original control problem is replaced by a control problem with the same equations of motion, control set, end conditons, and any other performance index $\phi_1(e)$, that the same control $u(t)$ is extremal for the new problem. Thus for abnormal problems the necessary conditions of Pontryagin's principle do not involve the performance index, but are already specified by the equations of motion, control set, and end conditions.

§ 6. Extremals for the Moon Landing Problem

Let us apply Pontryagin's Principle to compute the extremal control for the moon landing problem of § 2. This problem was first solved by Miele [1], [2]. Our discussion is close to that of Meditch [1]. The computation is somewhat long and complicated, however, it illustrates well many of the features involved in the computation of extremal controls.

Referring to the statement of the moonlanding problem given in § 2 let $(x_1, x_2, x_3)'=(h, v, m)'$. Fix $t_0=0$ and $x_0=(x_{01}, x_{02}, x_{03})'=(h_0, v_0, M+F)'$. Terminal values of h, v are 0, that is $x_{11}=0$, $x_{12}=0$. The vector function f of the equation of motion is

$$f(t,x,u)=\begin{pmatrix} x_2 \\ -g+x_3^{-1}u \\ -k\,u \end{pmatrix}. \tag{6.1}$$

Hence Eqs. (5.1) are

$$(P_1^{\cdot}(t), P_2^{\cdot}(t), P_3^{\cdot}(t))=-(P_1(t), P_2(t), P_3(t))\begin{pmatrix} 0 & 1 & 0 \\ 0 & 0 & -x_3(t)^{-2}u(t) \\ 0 & 0 & 0 \end{pmatrix} \tag{6.2}$$

or

$$\dot{P}_1(t)=0, \quad \dot{P}_2(t)=-P_1(t), \quad \dot{P}_3(t)=P_2(t)\,x_3(t)^{-2}\,u(t). \tag{6.2}$$

Inequality (5.2) in the form (5.9) is

$$\begin{aligned}\max_{u\in U}\{P_1(t)\,x_2(t)+P_2(t)(x_3(t)^{-1}u-g)-P_3(t)\,k\,u\}\\ =P_1(t)\,x_2(t)+P_2(t)(x_3(t)^{-1}u(t)-g)-P_3(t)\,k\,u(t).\end{aligned} \tag{6.3}$$

Defining the vector function $\phi(t_0, t_1, x_{01}, x_{02}, x_{03}, x_{11}, x_{12}, x_{13})$ corresponding to the performance function (3.2) and end conditions (3.3), we have

(6.4)
$$\phi(t_0, t_1, x_{01}, x_{02}, x_{03}, x_{11}, x_{12}, x_{13}) = \begin{pmatrix} -x_{13} \\ t_0 \\ x_{01} - h_0 \\ x_{02} - v_0 \\ x_{03} - M - F \\ x_{11} \\ x_{12} \end{pmatrix}.$$

Hence (5.3)–(5.7) are

(6.5) $$(P_1(t_1), P_2(t_1), P_3(t_1)) = (\lambda_6, \lambda_7, -\lambda_1)$$

(6.6) $$(P_1(0), P_2(0), P_3(0)) = -(\lambda_3, \lambda_4, \lambda_5)$$

(6.7) $$P_1(t_1)\, x_2(t_1) + P_2(t_1)[x_3(t_1)^{-1} u(t_1) - g] - P_3(t_1)\, k\, u(t_1) = 0$$

(6.8) $$P_1(0)\, x_2(0) + P_2(0)[x_3(0)^{-1} u(0) - g] - P_3(0)\, k\, u(0) = \lambda_2$$

(6.9) $$P_1(t)\, x_2(t) + P_2(t)[x_3(t)^{-1} u(t) - g] - P_3(t)\, k\, u(t) = \text{constant}$$

for all $t \in [0, t_1]$. From (6.7) and (6.8) we see that the constant in (6.9) is λ_2, and that $\lambda_2 = 0$.

Conditions (6.3) and (2.2) imply an extremal control $u(t)$ must satisfy

(6.10) $$u(t) = \begin{cases} 0 & \text{if } P_2(t)\, x_3(t)^{-1} - k\, P_3(t) < 0 \\ \alpha & \text{if } P_2(t)\, x_3(t)^{-1} - k\, P_3(t) > 0. \end{cases}$$

First we shall show that a control for which $u(t)$ is zero over an initial interval and then switches to α over an appropriately chosen terminal interval is extremal. Later it will be shown that this is the only extremal control.

On an interval $[\tau, \tilde{\tau}]$ on which $u(t) = 0$ the solutions of (2.1) are

(6.11)
$$\begin{aligned} h(t) &= -\frac{g(t-\tau)^2}{2} + v(\tau)(t-\tau) + h(\tau) \\ v(t) &= -g(t-\tau) + v(\tau) \\ m(t) &= m(\tau). \end{aligned}$$

On an interval $[\tau, \tilde{\tau}]$ on which $u(t) = \alpha$ solutions of (2.1) are

(6.12)
$$\begin{aligned} h(t) &= -\tfrac{1}{2} g(t-\tau)^2 + \frac{m(\tau) - k\,\alpha(t-\tau)}{k^2\,\alpha} \ln \frac{m(\tau) - k\,\alpha(t-\tau)}{m(\tau)} + \frac{t-\tau}{k} + v(\tau)(t-\tau) + h(\tau) \\ v(t) &= -g(t-\tau) - \frac{1}{k} \ln \frac{m(\tau) - k\,\alpha(t-\tau)}{m(\tau)} + v(\tau) \\ m(t) &= -k\,\alpha(t-\tau) + m(\tau). \end{aligned}$$

Let us determine the locus of points, (h, v) which can be initial points of a segment on which $u(t) = \alpha$ and $h = 0$, $v = 0$ is satisfied at the terminal time. Setting

$h(\tilde{\tau})=0$, $v(\tilde{\tau})=0$, $h(\tau)=h$, $v(\tau)=v$, $m(\tau)=m=M+F$, $\tilde{\tau}-\tau=s$ in Eq. (6.12) gives

$$h=\tfrac{1}{2}g s^2-\frac{m-k\alpha s}{k^2\alpha}\ln\frac{m-k\alpha s}{m}-\frac{s}{k}-v s \tag{6.13}$$

$$v=g s+\frac{1}{k}\ln\frac{m-k\alpha s}{m}. \tag{6.14}$$

Substituting (6.14) in (6.13) gives

$$h=-\tfrac{1}{2}g s^2-\frac{m}{k^2\alpha}\ln\frac{m-k\alpha s}{m}-\frac{s}{k} \tag{6.15}$$

$$v=g s+\frac{1}{k}\ln\frac{m-k\alpha s}{m}. \tag{6.14}$$

This describes the initial height and velocity from which a soft landing $h=0$, $v=0$ is possible by using full thrust α for time s.

For the problem to be realistic the relation

$$\frac{\alpha}{M+F}>g \tag{6.16}$$

should hold. This implies that if thrusting is begun at maximum rate the thrust to initial mass ratio is greater than the gravitational acceleration so that braking can take place.

If the spacecraft burns fuel at rate $k\alpha$ the total amount of fuel will be burned in time $\frac{F}{k\alpha}$. Plotting (6.14), (6.15) under the condition (6.16) for $0\leqq s\leqq\frac{F}{k\alpha}$ gives the curve in the second quadrant of the (v,h)-plane indicated in Fig. II.1. This curve will be called the switching curve. From the construction of this curve, its interpretation is that if the spacecraft is at a (v,h)-point on the curve corresponding to the parameter s and if it thrusts at maximum rate α then it will arrive at $v=0$, $h=0$ in time s.

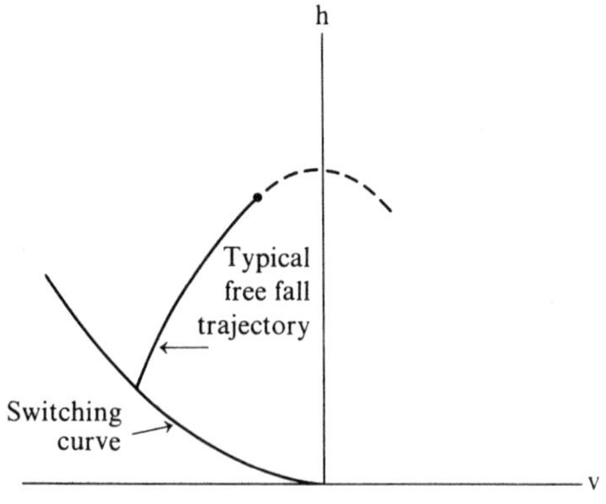

Fig. II.1. Switching curve and free fall trajectory

If the spacecraft free falls it will follow (6.11). If at $t=0$ it is at v_0, h_0 and begins a free fall trajectory, the first two equations of (6.11) imply that it follows the parabola

$$h=h_0-\frac{1}{2g}[v^2-v_0^2] \tag{6.17}$$

in the direction of decreasing v from (v_0, h_0).

Let (v_0, h_0) be a point in the (v, h) plane so that (6.17) intersects the switching curve at a (v, h) point parameterized by a value of s for which $0 \leqq s \leqq \frac{F}{k\alpha}$. Let τ denote the time the spacecraft would spend on this free fall trajectory. We shall show that the control

$$u(t)=\begin{cases} 0 & \text{if } 0 \leqq t \leqq \tau \\ \alpha & \text{if } \tau < t \leqq \tau+s \end{cases} \tag{6.18}$$

is an extremal control. This will follow if it is shown that a choice of λ can be made so that (6.2), (6.3), (6.5)–(6.9) and the end conditions are satisfied.

Refering to Fig. II.1 we see that the velocity at time τ of a spacecraft using control (6.18) is negative and is given by $v_0 - g\tau$. Conditions (6.2), (6.5) and (6.6) are satisfied by:

$$\begin{aligned} P_1(t)&=-\lambda_3=\lambda_6 \\ P_2(t)&=\lambda_3 t-\lambda_4=\lambda_6(t_1-t)+\lambda_7 \\ P_3(t)&=\begin{cases} -\lambda_5, & \text{if } t_0 \leqq t \leqq \tau \\ -\lambda_5+\int_\tau^t \frac{(\lambda_3 \tilde{t}-\lambda_4)\alpha}{m(\tilde{t})^2}\, d\tilde{t} & \text{if } \tau \leqq t \leqq t_1. \end{cases} \end{aligned} \tag{6.19}$$

Define $r(t)$ by

$$r(t)=P_2(t)\, x_3(t)^{-1}-k P_3(t). \tag{6.20}$$

From (2.1) (6.2) and (6.6)

$$\dot{r}(t)=\lambda_3\, x_3(t)^{-1}. \tag{6.21}$$

From (6.19) and (6.10) we must have

$$r(\tau)=(\lambda_3 \tau-\lambda_4)(M+F)^{-1}+\lambda_5 k=0. \tag{6.22}$$

From (6.8) (6.9) (6.7) and (6.18)

$$-\lambda_3 v_0+\lambda_4 g=0. \tag{6.23}$$

Eqs. (6.22) (6.23) have the solution

$$\lambda_3=\frac{\lambda_5 k g(M+F)}{v_0-g\tau}, \qquad \lambda_4=\frac{\lambda_5 k v_0(M+F)}{v_0-g\tau}. \tag{6.24}$$

When (6.24) are substituted in (6.19)

$$P_3(t)=\lambda_5\left[-1+\frac{k(M+F)}{v_0-g\tau}\int_\tau^t \frac{(g\tilde{t}-v_0)}{(m(\tilde{t}))^2}\,\alpha\, d\tilde{t}\right], \quad \text{if } \tau \leqq t \leqq t_1. \tag{6.25}$$

If the choices, $\lambda_5=-1$, $\lambda_2=0$ are made; λ_3 and λ_4 given by (6.24); $P_1(t)$, $P_2(t)$, $P_3(t)$, λ_6 and λ_7 given by (6.19), we see that (6.2), (6.5), (6.6) and (6.8) are satisfied. Notice that the integrand in (6.25) is positive and the coefficient of the integral is negative because $v_0-g\tau<0$. Thus if $P_3(t_1)=-\lambda_1$ then $\lambda_1<0$. The choice $\lambda_5=-1$ and (6.24) imply $\lambda_3>0$, hence by (6.21), $r(t)$ is strictly increasing. By (6.22), $r(\tau)=0$, thus (6.10) and (6.3) are satisfied. As mentioned in §5, (6.9) and consequently (6.7) are implied by (6.2) and (6.8). Thus we have verified that (6.18) is an extremal control.

Next it will be shown that (6.18) is the only extremal control law. To begin this it will be shown that an extremal control law can switch at most once from zero to α or from α to zero. If $\lambda_3\neq 0$ this will follow from (6.10) and (6.21) since in this case (6.20) implies $r(t)$ is strictly monotonic. If $\lambda_3=0$, (6.21) implies $r(t)$ is constant. The assertion will follow again from (6.10) if $r(t)\neq 0$, since in this case (6.10) implies $u(t)$ is constant. Thus we need only show that $\lambda_3=0$ and $r(t)=0$ cannot both occur.

If $\lambda_3=0$, by (6.2) and (6.6) $P_1(t)=0$. If $P_1(t)=0$ and $r(t)=0$, from (6.9) $P_2(t)=0$. If $P_2(t)=0$ and $r(t)=0$, (6.19) implies $P_3(t)=0$. Thus $(P_1(t), P_2(t), P_3(t))$ is the zero vector. This implies through (6.2) that this vector is identically zero and through (6.5)–(6.8) that $\lambda=0$, contradicting $\lambda\neq 0$.

Next we will show that if v_0 and h_0 are above the switching curve a control for which $u(t)=\alpha$ and then $u(t)=0$ cannot satisfy the end conditions. If there were such a control the segment of the trajectory on which $u(t)=0$ is of the form (6.11). This trajectory has $v(t)=0$ only at the time

$$t=\frac{v(\tau)}{g}+\tau.$$

If $h(t)=0$ at this time the relation

$$h(\tau)=-\frac{v(\tau)^2}{2g}$$

must be satisfied. That is the initial height of this segment, and in fact the entire segment, is below the surface of the moon. It is not difficult to show, that for initial height and velocity above the switching curve, the trajectory of the form (6.12), on which $u(t)=\alpha$, lies above the surface of the moon. Thus these two segments do not meet to form a single trajectory satisfying the end conditions. Thus if the initial height and velocity are above the switching curve the control of the form (6.18) is the only extremal for which the end conditions are satisfied.

Notice that if initially the spacecraft was at a height and velocity (v_0, h_0) below the switching curve a soft landing is not possible, because even by using full thrust for the entire trajectory the spacecraft will impact the moon with a nonzero velocity.

If the initial height and velocity are above the switching curve (6.14), (6.15), but such that the parabola (6.17) intersects this curve at an (h, s) corresponding to a value of s greater than $F/k\alpha$, a soft landing cannot be achieved. This can be inferred from the following argument. The existence theorems given in Chap. III will apply to this problem. They assert that if it is possible to have a soft landing at all, then there is a control which makes a soft landing using minimum fuel.

Such a control must be an extremal and hence of the form (6.18). For initial conditions for which the parabola (6.17) intersects the switching curve at a point corresponding to a value of s greater than $F/k\alpha$ the spacecraft using a control of the type (6.18) runs out of fuel before reaching the moon. Thus there is a final interval of zero thrust contradicting that the control has the form (6.18).

§ 7. Extremals for the Linear Regulator Problem

The linear regulator problem is formulated in Bolza form. As in §4 reduce it to Mayer form by introducing an extra coordinate x_{n+1} and differential equation

$$\dot{x}_{n+1}(t)=x(t)'M(t)x(t)+u(t)'N(t)u(t).$$

Then the performance $\phi_1(e)$ becomes

$$\phi_1(e)=x(t_1)'Dx(t_1)+x_{n+1}(t_1).$$

To write out the conditions of Pontryagin's principle let

$$P(t)'=(\tilde{P}(t)',P_{n+1}(t))$$

where $P_{n+1}(t)$ is a scalar function and $\tilde{P}(t)$ is an n-dimensional vector function. Using formulas for differentiating quadratic forms, see Problem [16], the adjoint Eqs. (5.1) are

$$\begin{aligned}\dot{\tilde{P}}(t)'&=-\tilde{P}(t)'A(t)-2P_{n+1}(t)x(t)'M(t)\\ \dot{P}_{n+1}(t)&=0.\end{aligned}\tag{7.1}$$

The transversality conditions (5.3)–(5.6) are

$$\begin{aligned}\tilde{P}(t_1)'&=2\lambda_1 x(t_1)'D\\ P_{n+1}(t_1)&=\lambda_1\end{aligned}\tag{7.2}$$

$$\begin{aligned}\tilde{P}(t_0)'&=-(\lambda_4,\ldots,\lambda_{n+3})\\ P_{n+1}(t_0)&=-\lambda_{n+4}\end{aligned}\tag{7.3}$$

$$\begin{aligned}H(t_0,x(t_0),u(t_0))&=\lambda_2\\ H(t_1,x(t_1),u(t_1))&=-\lambda_3.\end{aligned}\tag{7.4}$$

Notice that λ_1 cannot be zero, for if it was, by (7.2) $P(t_1)=0$ and hence by (7.1) $P(t)\equiv 0$. This would imply by (7.2)–(7.4) that $\lambda=0$ which is a contradiction. Since the equations of Pontryagin's principle are homogeneous in λ and $\lambda_1\neq 0$ we may divide these equations by an appropriate positive number to obtain $\lambda_1=-\frac{1}{2}$. With this normalization

$$P_{n+1}(t)=-\tfrac{1}{2}.\tag{7.5}$$

The Hamiltonian is

$$H(t, x, u) = \tilde{P}(t)'[A(t)x + B(t)u] - \tfrac{1}{2}[x'M(t)x + u'N(t)u]. \tag{7.6}$$

The unique u which maximizes $H(t, x(t), u)$ can be found by standard methods to be

$$u(t) = N(t)^{-1} B(t)' \tilde{P}(t). \tag{7.7}$$

Substituting (7.5) into (7.1) and (7.7) into (2.8) replaces the problem of finding extremal controls for the linear regulator problem to finding solutions $x(t)$, $\tilde{P}(t)$ of the system

$$\begin{aligned} \dot{x}(t) &= A(t)x(t) + B(t)N(t)^{-1} B(t)' \tilde{P}(t) \\ \dot{\tilde{P}}(t) &= -A(t)' \tilde{P}(t) + M(t)x(t) \end{aligned} \tag{7.8}$$

with initial condition for $x(t)$ and terminal condition for $\tilde{P}(t)$ given by

$$\begin{aligned} x(t_0) &= x_0 \\ \tilde{P}(t_1) &= -D x(t_1). \end{aligned} \tag{7.9}$$

Given a solution of (7.8) with boundary conditions (7.9) the extremal control is given by (7.7).

Later in Chap. IV a different form is obtained for the optimal control law. See Problems IV.4 and IV.5 for the connection between these two representations of the optimal control.

§ 8. Extremals for the Simplest Problem in Calculus of Variations

Referring to Example 2.4, for this problem Eq. (5.1) is

$$(\dot{P}_1(t), \dot{P}_2(t)) = -(P_1(t), P_2(t)) \begin{pmatrix} 0 & 0 \\ L_x(t, x_1(t), u(t)) & 0 \end{pmatrix} \tag{8.1}$$

or

$$\dot{P}_1(t) = -P_2(t)\, L_x(t, x_1(t), u(t)), \qquad \dot{P}_2(t) = 0. \tag{8.2}$$

Inequality (5.2) is

$$P_1(t)u + P_2(t)L(t, x_1(t), u) - [P_1(t)u(t) + P_2(t)L(t, x_1(t), u(t))] \leqq 0. \tag{8.3}$$

The vector function ϕ giving the performance and end conditions is

$$\phi(t_0, t_1, x_{01}, x_{02}, x_{11}, x_{12}) = \begin{pmatrix} x_{12} \\ t_0 - a \\ t_1 - b \\ x_{01} - c \\ x_{02} \\ x_{11} - d \end{pmatrix}. \tag{8.4}$$

Hence conditions (5.3)–(5.6) are

$$(8.5)\qquad \begin{aligned} (P_1(t_1), P_2(t_1)) &= (\lambda_6, \lambda_1) \\ (P_1(t_0), P_2(t_0)) &= -(\lambda_4, \lambda_5) \\ H(t_1, x(t_1), u(t_1)) &= -\lambda_3 \\ H(t_0, x(t_0), u(t_0)) &= \lambda_2 . \end{aligned}$$

Condition (5.7) is

$$(8.6)\qquad P_1(t)u(t) + P_2(t)L(t, x_1(t), u(t)) = \lambda_2 + \int_{t_0}^{t} P_2(s)L_t(s, x_1(s), u(s))\,ds.$$

Notice since (8.3) implies $H(t, x(t), u)$ has a maximum at $u = u(t)$, the control set is all of E^1, and $L(t, x, u)$ is differentiable in u we must have that

$$(8.7)\qquad 0 = \frac{\partial}{\partial u} H(t, x(t), u(t)) = P_1(t) + P_2(t)L_u(t, x_1(t), u(t)).$$

Eq. (8.2) implies $P_2(t)$ is a constant P_2. From (8.7) if $P_2 = 0$, $P_1(t) \equiv 0$. This would imply by (8.5) that $\lambda = 0$ contradicting $\lambda \neq 0$. Hence $P_2 \neq 0$. By Remark 5.1 we may assume $P_2 = -1$. Hence from (8.7)

$$(8.8)\qquad P_1(t) = L_u(t, x_1(t), u(t)).$$

Putting $P_2(t) = -1$ in (8.2) and integrating using the end condition (8.5) we have

$$(8.9)\qquad P_1(t) = -\lambda_4 + \int_{t_0}^{t} L_x(s, x(s), u(s))\,ds.$$

Combining (8.8) and (8.9) and identifying $x(t) = x_1(t)$, $\dot{x}(t) = u(t)$ we obtain the Euler equation I (3.5)

$$(8.10)\qquad L_{\dot{x}}(t, x(t), \dot{x}(t)) - \int_{t_0}^{t} L_x(s, x(s), \dot{x}(s))\,ds = -\lambda_4 .$$

Formula (8.6) can be similarly rewritten as

$$L_{\dot{x}}\dot{x} - L = \lambda_2 - \int_{t_0}^{t} L_t(s, x(s), \dot{x}(s))\,ds.$$

This is a generalization of Lemma I.4.1. Similarly formula (8.3) may be rewritten as the inequality

$$L(t, x(t), u) - L(t, x(t), \dot{x}(t)) - L_{\dot{x}}(t, x(t), \dot{x}(t))(u - \dot{x}(t)) \geqq 0$$

for each u in E^1. This is a necessary condition for optimality in classical calculus of variations called the *Weierstrass condition.*

§ 9. General Features of the Moon Landing Problem

We shall discuss a number of topics and concepts important for optimal control problems in general by illustrating how they arise in Example 2.1.

Notice that the requirement that the spacecraft make a soft landing, that is have velocity $v=0$ at the first time $h=0$, restricts trajectories to those which satisfy $h \geqq 0$. This is a constraint on the state of the system. Generally we shall not consider problems with state space constraints in this book. In this problem the state space constraint was taken into account by allowing the controls to violate the constraint and determining that in this wider class the optimum satisfied the constraint $h>0$ except at the terminal time. In some control problems with state space constraints part of the optimal trajectory will satisfy the constraint as an equality. In this case some controls of the type $\tilde{u}(t)$ defined in §3 may violate the state space constraints. Since the class of controls satisfying the state space constraints may not be closed under adding controls of the form $\tilde{u}(t)$ other conditions arise which are necessary for a control to be optimal. See Pontryagin *et. al.* [1] for a discussion of these conditions.

Notice that the terminal condition $v=0$ at the first time that $h=0$ could only be satisfied for certain initial conditions. Referring to Fig. II.2 these initial conditions are the portion of the region $h>0$ below and on the parabola (6.17) which intersects the switching curve at the point (v, h) parameterized by $s \leqq F/k\alpha$. This set is called the set of initial conditions from which the terminal conditions are reachable or more briefly the *reachable set*.

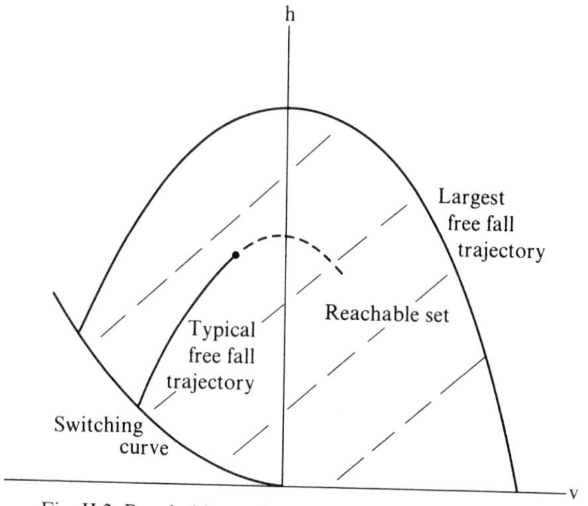

Fig. II.2. Reachable set. For moon landing problem

Actually in determining the extremal controls (6.18) we did not determine extremal controls for a single optimization problem but for a whole family of optimization problems. This follows because for each initial (v, h) in the reachable set, the controls (6.18) are extremal for the optimization problem with this (v, h) as initial conditions. The control (6.18) could be expressed as a function of (v, h) rather than as a function of t as in (6.18). The control is zero if (v, h) is a point of the reachable set above the switching curve and α if (v, h) lies on the switching curve. A control expressed in this form is said to be synthesized in *feedback form* or more briefly a *feedback control*. It is called feedback form because the control is expressed as a function of the state of the system and the values of the state are thought of as being observed and fed back into the control. The control in feedback form is extremal for any of the optimization problems with initial conditions in the reachable set. Controls in feedback form will play an important role in the sufficiency conditions for optimality given in Chap. IV. They are closely related to a concept called a "field of extremals" in classical calculus of variations.

In the computation of the extremal control law (6.18) an argument was given to show that the terms of the Hamiltonian which multiplied the control $u(t)$ were not zero. In the case in which the Hamiltonian can be zero on an interval for every value of u in the control set U the optimization problem is called *singular*. In this case the maximization of the Hamiltonian gives no information about the value of the control. The possibility of the control problem being singular is quite common in optimal control problems. Even when the problem is not singular, often a lengthy argument showing that singularity does not occur may be required. In some problems portions of optimal trajectories are singular, see Problems (9), (10). In the problem of the motion of a rocket in an inverse square field, a singular extremal called Lawden's spiral (after its discoverer Derick Lawden [1]) occurs. A great deal of work was done to determine if Lawden's spiral could be a portion of an optimal trajectory, Kelley [2], Gurley [1], Robbins [1], Kopp and Moyer [1], Goh [2]. Further optimality conditions than those of §5 which are necessary for a singular arc to be a portion of an optimal trajectory are given in Johnson-Gibson [1], Kelley *et al.* [1], Goh [1], [2], Gabasov [1], Gabsov-Kirillova [1], Jacobson [1].

§ 10. Summary of Preliminary Results

Before beginning a discussion of optimality conditions for the optimal control problem, we will state some results from differential equations which are needed in the discussion of these conditions.

Theorem 10.1. *Let $A(t)$ be an $n\times n$ matrix, $G(t)$ an n-dimensional vector of piecewise continuous functions defined on an interval $[t_0, t_1]$, and y_0 an n-dimensional vector. Then if $\tau\in[t_0, t_1]$ there is a unique piecewise continuously differentiable solution of the vector differential equation*

$$\dot{y}=A(t)\,y+G(t) \tag{10.1}$$

on the interval $[t_0, t_1]$ which satisfies the condition

$$y(\tau)=y_0. \tag{10.2}$$

The system of vector differential equations

$$\dot{P}=-A(t)'P \tag{10.3}$$

is called the system of differential equations adjoint to Eqs. (10.1) or more briefly the *adjoint equations*. By Theorem 10.1 it has a unique solution for a given boundary condition. If $y(t)$ is a solution of (10.1) and $P(t)$ a solution of (10.3),

(10.4) $$\frac{d}{dt} P(t)' y(t) = -P(t)' A(t) y(t) + P(t)' A(t) y(t) + P(t)' G(t).$$

For any two times τ_1 and τ_2 an integration from τ_1 to τ_2 of (10.4) establishes the important formula

(10.5) $$P(\tau_2)' y(\tau_2) - P(\tau_1)' y(\tau_1) = \int_{\tau_1}^{\tau_2} P(t)' G(t)\, dt.$$

The next theorems express the dependence of corresponding trajectories on the parameters of various mappings.

Theorem 10.2. *Let $f_u(t, x, u)$ be continuous. For $0 \leqq \varepsilon \leqq \eta$, let $x^\varepsilon(t)$ be the solutions of (3.1) corresponding to controls $u(t) + \varepsilon v(t)$ with the same initial condition $x^\varepsilon(t_0) = x_0$. Then*

(10.6) $$x^\varepsilon(t) = x(t) + \varepsilon\, \delta x(t) + o(t, \varepsilon)$$

where $\delta x(t)$ is the solution of

(10.7) $$\delta \dot{x}(t) = f_x(t, x(t), u(t))\, \delta x(t) + f_u(t, x(t), u(t))\, v(t)$$

with initial condition

(10.8) $$\delta x(t_0) = 0.$$

Theorem 10.3. *For a real variable ε, let $x^\varepsilon(t)$ be the solution of (3.1) on $[t_0, t_1]$ corresponding to the control $u(t)$ with initial condition*

(10.9) $$x^\varepsilon(t_0) = x_0 + \varepsilon y_0 + o(\varepsilon).$$

Then:

(10.10) $$x^\varepsilon(t) = x(t) + \varepsilon\, \delta x(t) + o(t, \varepsilon)$$

where $\delta x(t)$ is the solution of

(10.11) $$\delta \dot{x}(t) = f_x(t, x(t), u(t))\, \delta x(t)$$

on $[t_0, t_1]$ with initial condition

(10.12) $$\delta x(t_0) = y_0.$$

An n-dimensional version of Theorem 10.3 is given by Theorem 10.4.

Theorem 10.4. *Let ρ denote a vector variable in E^n, M an $n \times n$ matrix, and $x^\rho(t)$ the solution of (3.1) on $[t_0, t_1]$ corresponding to the control $u(t)$ with initial condition*

(10.13) $$x^\rho(t_0) = x_0 + M \rho + o(\rho).$$

Then

(10.14) $$x^\rho(t) = x(t) + \delta x(t)\, \rho + o(t, \rho)$$

where $\delta x(t)$ is the $n \times n$ matrix which is the solution of

(10.15) $$\delta \dot{x}(t) = f_x(t, x(t), u(t))\, \delta x(t)$$

on $[t_0, t_1]$ with initial condition

(10.16) $$\delta x(t_0) = M.$$

§ 11. The Free Terminal Point Problem

To motivate the proof of Pontryagin's Principle, we will carry out two derivations of necessary conditions for optimality in the special case of the problem of optimal control in which the initial time and state and the final time are fixed and there are no conditions on the final state. This problem is called the *free terminal point problem.* The derivation of Pontryagin's principle in this case is especially simple and conceptually clear. These derivations illustrate the role the adjoint equations play in evaluating directional derivatives of the mappings involved. They also compare results obtained using the two different families of mappings.

Let t_0 and t_1 denote the fixed initial and terminal times. In these derivations the space $\mathscr{V}$ described in Chap. I will be taken to be the space of all left continuous piecewise continuous functions defined on $[t_0, t_1]$ with values in E^m. Call this the space of control functions. The subset $\mathscr{K}$ will be the set of control functions u such that $u(t) \in U$ for each $t \in [t_0, t_1]$ and (x_0, u) is feasible pair for the fixed initial state x_0. Being a feasible pair in this case merely requires that (3.1) with control u and initial condition x_0 have a solution on the entire interval $[t_0, t_1]$.

There are many ways of constructing mappings ζ into $\mathscr{K}$ as discussed in Chap. I, § 2. We will consider two such mappings. First let us consider a family of mappings similar to those used in Chap. I. If U is convex and $u(t)$ and $u(t)+v(t)$ are in U for each $t \in [t_0, t_1]$ then $u(t) + \varepsilon\, v(t)$ is in U for each $t \in [t_0, t_1]$ and $0 \leqq \varepsilon \leqq 1$. Theorems on continuous dependence of solutions of differential equations on parameters imply $(x_0, u + \varepsilon\, v)$ is a feasible pair for $0 \leqq \varepsilon \leqq \eta$ for some $\eta > 0$. Therefore, we may construct the mapping

(11.1) $$\zeta\colon\ 0 \leqq \varepsilon \leqq \eta \to \mathscr{K}$$

into the controls u such that (x_0, u) is feasible, by defining

(11.2) $$\zeta(\varepsilon) = u + \varepsilon\, v.$$

Let x^ε denote the solution of (3.1) with initial condition $x^\varepsilon(t_0) = x_0$ corresponding to $\zeta(\varepsilon)$ and x the similar solution corresponding to u. Then x^ε for small ε, remains in a weak neighborhood of x as defined in Remark I.5.2. Consequently (11.2) is called a *weak variation* of the control.

Another family of mappings is the following. Suppose u is an element of $\mathscr{K}$, v a fixed vector in U and $\tau \in (t_0, t_1]$. For this fixed v, small enough η, and $0 < \varepsilon < \eta$ existence theorems for differential equations imply there will be a solution x^ε of

(11.3) $$\dot{x} = f(t, x, v)$$

on the interval $\tau-\varepsilon \leqq t \leqq \tau$ with the condition

(11.4) $$x^{\varepsilon}(\tau-\varepsilon)=x(\tau-\varepsilon).$$

The theorem for continuous dependence of solutions of differential equations on initial conditions implies that for small enough η, (3.1) has a solution on $[\tau, t_1]$ with initial condition $x^{\varepsilon}(\tau)$. Define the control u^{ε} by

(11.5) $$u^{\varepsilon}(t)=\begin{cases} v & \text{if } \tau-\varepsilon<t \leqq \tau \\ u(t) & \text{for other values of } [t_0, t_1]. \end{cases}$$

The previous remarks imply that for small enough η we may define a mapping

$$\zeta\colon 0 \leqq \varepsilon \leqq \eta \rightarrow \mathscr{K}$$

into the controls which are feasible for x_0 by defining

(11.6) $$\zeta(\varepsilon)=u^{\varepsilon}.$$

For the mapping (11.6), x^{ε} for small ε need not be contained in a weak neighborhood of x. However x^{ε}, for small ε, is contained in a strong neighborhood of x. (See Theorem 11.3.) Thus (11.5) is called a *strong variation* of the control.

We will carry out two derivations of necessary conditions for optimality, one using weak variations, the other strong variations of the control. The derivation using strong variations will obtain Pontryagin's principle for the free endpoint problem. The one using weak variations will achieve a slightly weaker result.

Notice, since the quantities $(t_0, t_1, x(t_0))$ are fixed, for the free terminal point problem we may as well assume the performance index

$$\phi_1(t_0, t_1, x(t_0), x(t_1))$$

is of the form $\phi_1(x(t_1))$. Since the other components of the vector ϕ defined in § 3 will not play a role in our discussion of the free terminal point problem we will also drop the subscript one and write the performance index as

$$\phi(x(t_1)).$$

In the notation of Chap. I.2 the performance index determines the real valued function of the control

$$J(u)=\phi(x(t_1)).$$

The following theorem evaluates the directional derivative of the mapping $J(\zeta(\varepsilon))$ when $\zeta(\varepsilon)$ is the control corresponding to a weak variation of the type (11.2).

Theorem 11.1. *Let U be convex and $f_u(t, x, u)$ be continuous. Let $P(t)$ be the solution of the adjoint equations of Eq.* (10.7), *that is,*

(11.7) $$\dot{P}(t)'=-P(t)' f_x(t, x(t), u(t)),$$

with boundary condition

(11.8) $$P(t_1)'=-\phi_x(x(t_1)).$$

Then

$$\delta J(u,v) = -\int_{t_0}^{t_1} P(t)' f_u(t, x(t), u(t))\, v(t)\, dt. \tag{11.9}$$

Proof. If $\delta x(t)$ is the solution of (10.7) with initial condition (10.8), formula (10.5) implies

$$\phi_x(x(t_1))\,\delta x(t_1) = -\int_{t_0}^{t_1} P(t)' f_u(t, x(t), u(t))\, v(t)\, dt. \tag{11.10}$$

Theorem 10.2 and the chain rule for differentiation imply that

$$\delta J(u,v) = \phi_x(x(t_1))\,\delta x(t_1). \tag{11.11}$$

Thus (11.11) and (11.10) imply (11.9). □

Theorem 11.2. *Under the hypotheses of Theorem* 11.1, *a necessary condition that a control u be optimal for the free terminal point problem is that for each fixed* $t \in (t_0, t_1]$

$$P(t)' f_u(t, x(t), u(t))\, v \leqq 0 \tag{11.12}$$

for each $v \in U$ *such that* $u(t)+v \in U$.

Proof. Since u may be assumed left continuous there is some interval $t_0 < t-\delta < s \leqq t$ on which $u(s)$ is continuous. Define a function w on $[t_0, t_1]$ by

$$w(s) = \begin{cases} u(t)+v-u(s) & \text{if } t-\delta < s \leqq t \\ 0 & \text{otherwise} \end{cases} \tag{11.13}$$

Considerations similar to those of (11.3) through (11.6) imply for δ small enough $u(s)+w(s) \in \mathscr{K}$. Hence for δ small enough by Theorem 11.1 and Theorem I.2.2

$$\delta J(u,w) = -\int_{t-\delta}^{t} P(s)' f_u(s, x(s), u(s))\,[u(t)+v-u(s)]\, ds \geqq 0. \tag{11.14}$$

Since the integrand is left continuous dividing (11.14) by δ and taking the limit as δ approaches zero implies (11.12). □

Next we will consider the analogs of Theorems 11.1 and 11.2 when strong variations of the control are considered.

Theorem 11.3. *If* x^ε *are solutions of* (3.1) *corresponding to controls* u^ε *defined by* (11.5) *with the same initial condition* $x^\varepsilon(t_0) = x_0$ *then*

$$x^\varepsilon(t) = x(t) + \varepsilon\,\delta x(t) + o(t,\varepsilon) \tag{11.15}$$

where $\delta x(t)$ *is the solution of*

$$\delta x(t) = \begin{cases} 0 & \text{if } t_0 \leqq t < \tau \\ [f(\tau, x(\tau), v) - f(\tau, x(\tau), u(\tau))] + \int_{\tau}^{t} f_x(s, x(s), u(s))\,\delta x(s)\, ds & \text{if } \tau \leqq t \leqq t_1 \end{cases} \tag{11.16}$$

Proof. From (11.5) $u^\varepsilon(t)=u(t)$ if $t_0 \leqq t \leqq \tau-\varepsilon$. Hence $x^\varepsilon(t)=x(t)$ if $t_0 \leqq t \leqq \tau-\varepsilon$. Thus (11.16) is true with $\delta x(t)=0$ if $t_0 \leqq t \leqq \tau-\varepsilon$.

We have if $\tau-\varepsilon \leqq t \leqq \tau$ that

(11.17) $$x^\varepsilon(t)=x(t)-\int_{\tau-\varepsilon}^{t} f(s, x(s), u(s))\,ds+\int_{\tau-\varepsilon}^{t} f(s, x^\varepsilon(s), v)\,ds.$$

Let for $\eta>0$

(11.18) $$A=\{(\varepsilon, t)\colon 0 \leqq \varepsilon \leqq \eta,\ \tau-\varepsilon \leqq t \leqq \tau\}.$$

The theorem of continuous dependence of solutions of differential equations on initial conditions implies for some $\eta>0$ that the mapping

$$(\varepsilon, t) \rightarrow x^\varepsilon(t)$$

of A into E^n is continuous. Hence

$$|f(s, x(s), u(s))-f(s, x^\varepsilon(s), v)|$$

is bounded on A. Thus (11.17) implies

(11.19) $$x^\varepsilon(t)=x(t)+O(t, \varepsilon) \quad \text{if } \tau-\varepsilon \leqq t \leqq \tau.$$

Applying (11.19) in (11.17) and using the left continuity of $f(t, x(t), u(t))$ gives

(11.20) $$x^\varepsilon(\tau)=x(\tau)+[f(\tau, x(\tau), v)-f(\tau, x(\tau), u(\tau))]\varepsilon+o(\tau, \varepsilon).$$

Now an application of Theorem 10.3 and (11.20) completes the derivation of (11.16). □

Theorem 11.4. *For* $0 \leqq \varepsilon \leqq \eta$, *let*

(11.21) $$J(u^\varepsilon)=\phi(x^\varepsilon(t_1))$$

where $u^\varepsilon(t)$ *is the control defined by* (11.5). *Let* $P(t)$ *be the solution of* (11.7) *with boundary condition* (11.8). *Then*

(11.22) $$\frac{d}{d\varepsilon} J(u^\varepsilon)\Big|_{\varepsilon=0}=-P(\tau)'[f(\tau, x(\tau), v)-f(\tau, x(\tau), u(\tau))].$$

Proof. Formulas (10.5), (11.7), (11.8) and (11.16) imply

(11.23) $$-P(\tau)'[f(\tau, x(\tau), v)-f(\tau, x(\tau), u(\tau))]=\phi_x(x(t_1))\,\delta x(t_1)$$

where $\delta x(t_1)$ is the solution of (11.16). Formulas (11.21), (11.15), and the chain rule for differentiation imply (11.22). □

Formula I (2.5) and Theorem 11.4 imply Theorem 11.5.

Theorem 11.5. (Pontryagin's Principle for the Free Terminal Point Problem). *A necessary condition for optimality of a control* u *for the Free Terminal Point*

Problem is that

$$P(t)'\left[f(t,x(t),v)-f(t,x(t),u(t))\right]\leqq 0 \tag{11.24}$$

for each $v\in U$ *and* $t\in(t_0,t_1]$, *where* $P(t)'$ *is the solution of*

$$\dot{P}(t)'=-P(t)'f_x(t,x(t),u(t)) \tag{11.25}$$

with boundary condition

$$P(t_1)'=-\phi_x(x(t_1)). \tag{11.26}$$

It is natural to ask if the conditions of Theorem 11.2 or 11.5 are sufficient conditions for optimality as well as necessary ones. They are not sufficient in general. However Theorem I.2.3 implies that the conditions of Theorem 11.2 are sufficient when $\mathscr{K}$ is convex and the mapping

$$J(u)=\phi(x(t_1)) \tag{11.27}$$

is a convex function on $\mathscr{K}$. Theorem 11.6 discusses a situation in which this happens.

Let U be convex. Let the equations of motion of the system be given by the linear differential equation

$$\dot{x}=A(t)\,x(t)+B(t)\,u(t) \tag{11.28}$$

where $A(t)$ and $B(t)$ are appropriate dimensional matrices of continuous functions. Let the performance index be given by

$$J(u)=\int_{t_0}^{t_1} L(t,x(t),u(t))\,dt+\psi(x(t_1)), \tag{11.29}$$

where L is a continuous real valued function continuously differentiable and convex in (x,u) and ψ is a continuously differentiable convex function of x. Under these assumptions the conditions of Theorem 11.2 can be rewritten.

Theorem 11.6. *A necessary and sufficient condition for optimality of a control* $u(t)$ *for the free terminal point problem with system Eq.* (11.28) *and performance index* (11.29) *is that for* $t\in(t_0,t_1]$

$$-L_u(t,x(t),u(t))\,v+\tilde{P}(t)'\,B(t)\,v\leqq 0 \tag{11.30}$$

for each $v\in U$ *such that* $u(t)+v\in U$, *where* $\tilde{P}(t)$ *is the solution of*

$$\dot{\tilde{P}}(t)'=-\tilde{P}(t)'\,A(t)+L_x(t,x(t),u(t)) \tag{11.31}$$

$$\tilde{P}(t_1)'=-\psi_x(x(t_1)). \tag{11.32}$$

Moreover if in addition $L(t,x,u)$ *is strictly convex in* (x,u) *for each fixed* t, *the optimal control* $u(t)$ *is unique.*

Proof. As mentioned the necessity part is just a restatement of Theorem 11.2 using the technique of § 4 to rewrite the problem as one with performance index of the Mayer type (4.1). Since (11.28) is a linear system of differential equations, the set $\mathscr{K}$ of controls such that $u(t)\in U$ and $(x_0, u(t))$ is a feasible pair is all piecewise continuous functions such that $u(t)\in U$. This is a convex set. Let $u^0(t)$ and $u^1(t)$ be controls in $\mathscr{K}$ and $x^0(t)$ and $x^1(t)$ corresponding solutions of (11.28) with initial condition $x(t_0)=x_0$. If $0<\alpha<1$, the convexity of $L(t,x,u)$ and $\psi(x)$ implies

$$\begin{aligned}\alpha J(u^0)+(1-\alpha)J(u^1)&=\alpha\left[\int_{t_0}^{t_1} L(t,x^0(t),u^0(t))\,dt+\psi(x^0(t_1))\right]\\&\quad+(1-\alpha)\left[\int_{t_0}^{t_1} L(t,x^1(t),u^1(t))\,dt+\psi(x^1(t_1))\right]\\&\geqq\int_{t_0}^{t_1} L(t,\alpha x^0(t)+(1-\alpha)x^1(t),\alpha u^0(t)+(1-\alpha)u^1(t))\,dt\\&\quad+\psi(\alpha x^0(t_1)+(1-\alpha)x^1(t_1))=J(\alpha u^0+(1-\alpha)u^1).\end{aligned}\tag{11.33}$$

The last equality of (11.33) follows because $\alpha x^0(t)+(1-\alpha)x^1(t)$ is the solution (11.28) corresponding to $\alpha u^0(t)+(1-\alpha)u^1(t)$. Thus $J(u)$ is a convex function defined on a convex set $\mathscr{K}$. Theorem 11.1 and condition (11.30) imply for each $v(t)$ which satisfies $u(t)+v(t)\in U$ for each $t\in[t_0,t_1]$ that

$$\delta J(u,v)=\int_{t_0}^{t_1}\left[L_u(t,x(t),u(t))\,v(t)-\tilde{P}(t)'B(t)\,v(t)\right]dt\geqq 0.$$

Hence by Theorem I.2.3 $J(u)$ has a minimum at $u=u(t)$.

To show uniqueness of the minimum, note that if $L(t,x,u)$ is strictly convex the inequality in (11.33) is strict. Thus $J(u)$ is a strictly convex function on $\mathscr{K}$ and by Theorem I.2.4 the minimum is unique. □

§12. Preliminary Discussion of the Proof of Pontryagin's Principle

In trying to establish necessary conditions for the general problem of optimal control, it might seem natural to consider the performance index as a function

$$J(x_0,u(t))=\phi_1(t_0,t_1,x(t_0),x(t_1))$$

on the space $\mathscr{F}$ defined in §3, and to consider the optimization problem as one of minimizing this function on this space. To proceed as in Chap. I, mappings into the space $\mathscr{F}$

$$\zeta: 0\leqq\varepsilon\leqq\eta\to\mathscr{F}$$

such that the composition of these mappings with the mapping J taking $\mathscr{F}$ into the performance index would be constructed. A problem occurs in that there are examples, one is given below, in which the space $\mathscr{F}$ is a single element. While this is no difficulty in determining the optimum which must be the single element, it hinders the development of the theory in that it precludes the construction of mappings of the type given by (11.2) or (11.5) into the space $\mathscr{F}$.

The following is an example in which $\mathscr{F}$ has only one element. Consider the optimal control problem with scalar differential equation

$$\dot{x} = u^2,$$

terminal conditions,

$$x(0)=0, \qquad x(1)=0,$$

control set

$$U = \{u: \ -1 \leqq u \leqq 1\},$$

and performance criterion

$$\int_0^1 L(x(t), u(t))\, dt.$$

The only control whose corresponding trajectory satisfies the terminal conditions is

$$u(t) \equiv 0 \quad \text{if } 0 \leqq t \leqq 1.$$

Hence this is the only element of $\mathscr{F}$.

In deriving an optimality theory for the Bolza problem in classical calculus of variations, a concept called normality was introduced and assumed. This condition assured that there were enough nontrival mappings into the space $\mathscr{F}$ so that conditions for optimality could be developed. Problems such as the example just given are abnormal. The need for making normality assumptions to derive optimality conditions for the Bolza problem was removed in a paper by McShane [6]. Similar constructions were carried out by Pontryagin, Boltyanskii and Gamkrelidze [1] in their proof of Pontryagin's principle. We shall give a proof of Pontryagin's Principle which combines features of proofs by McShane [6]; Pontryagin, Boltyanskii, and Gamkrelidze [1]; Halkin [2]; and Hestenes [1].

The basic idea in overcoming the difficulty mentioned above will be to consider the set $\mathscr{G}$ of pairs (x_0, u) such that $u \in \mathscr{U}$ and there is a solution of (3.1) on the interval $[t_0, t_1]$ corresponding to the control u with initial condition $x(t_0) = x_0$. Trajectories corresponding to pairs of $\mathscr{G}$ do not necessarily satisfy the end conditions $\phi_2(e)=0, \ldots, \phi_k(e)=0$. On the set $\mathscr{G}$ consider the mapping

$$\mathscr{J}: \mathscr{G} \to E^k$$

defined by

$$\mathscr{J}(x_0, u) = \phi(e).$$

Let the components of $\mathscr{J}$ be denoted by $\mathscr{J} = (J_1, \ldots, J_k)$. In terms of the mapping $\mathscr{J}$ the problem of optimal control becomes the abstract nonlinear problem of finding the element of $\mathscr{G}$ such that the first component $J_1(x_0, u)$ of $\mathscr{J}$ is minimized subject to the satisfaction of the conditions $J_i(x_0, u) = 0$, $i = 2, \ldots, k$. Mappings of the type (11.2) or (11.5) into the space $\mathscr{G}$ can be constructed. In § 14 a mapping more general than, but analogous to, those of (11.5) will be constructed.

The proof of Pontryagin's Principle will be broken up into three parts. First optimality conditions will be deduced for an abstract nonlinear programming problem (§ 13). Then it will be shown that the hypotheses of this problem are satisfied for the problem of optimal control (§ 14). The final part interprets the conditions obtained in the programming problem to obtain Pontryagin's Principle (§ 15).

§13. A Multiplier Rule for an Abstract Nonlinear Programming Problem

Let $\mathscr{S}$ be a set and $\mathscr{J}(s)$ a mapping

$$\mathscr{J}: \mathscr{S} \to E^k. \tag{13.1}$$

Consider the problem of finding an element s^* of $\mathscr{S}$ that minimizes $J_1(s)$ on the set $\mathscr{S} \cap \{s: J_i(s)=0, i=2, \ldots, k\}$.

To develop conditions for optimality for this problem we will need a notion of differentiation slightly stronger than could be obtained by considering derivatives of $\mathscr{J}$ along curves mapping into the set $\mathscr{S}$. Roughly what we will need will be a continuous directional derivative in a k-dimensional cone of directions.

Let e^i, $i=1, \ldots, k$ denote the standard basis vectors of E^k; i.e., $e^1=(1, 0, \ldots, 0)$, etc. Let C denote the convex hull of $\{0, e^1, \ldots, e^k\}$. We shall call C the standard k-dimensional simplex of E^k. Let ηC denote the convex hull of $\{0, \eta e^1, \ldots, \eta e^k\}$. Instead of curves mapping into the space $\mathscr{S}$, we shall consider mappings $\zeta(\rho)$ defined on ηC for some $\eta>0$, that is,

$$\zeta: \eta C \to \mathscr{S}. \tag{13.2}$$

We shall say that a $k \times k$ matrix M is the *conical differential of* $\mathscr{J}$ *at* s *along* ζ if $\zeta(0)=s$ and for ρ in ηC

$$\mathscr{J}(\zeta(\rho))=\mathscr{J}(s)+M\rho+o(\rho). \tag{13.3}$$

Let D be a convex cone with vertex zero contained in E^k. The set D will be called a *cone of variations of* $\mathscr{J}$ *at* s if for each linearly independent set $\{d^1, \ldots, d^k\}$ of vectors in D there is for some $\eta>0$ a mapping $\zeta: \eta C \to \mathscr{S}$ such that the composition mapping $\mathscr{J}(\zeta(\rho))$ is continuous on ηC and the matrix M whose columns are $d^1, \ldots, d^k$ is the conical differential of $\mathscr{J}$ at s along ζ.

Theorem 13.1. (Abstract Multiplier Rule). *If s^* is a solution to the abstract nonlinear programming problem and if D is a cone of variations of $\mathscr{J}$ at s^*, then there is a nonzero vector λ in E^k such that $\lambda_1 \leqq 0$ and $\lambda' d \leqq 0$ for all d in D.*

Proof. Let

$$L=\{y \in E^k: y_1<0, y_i=0, i=2, \ldots, k\}. \tag{13.4}$$

The conditions of the theorem are equivalent to the statement: There exists a hyperplane $\lambda' y=0$ such that $\lambda' y \geqq 0$, $y \in L$; $\lambda' y \leqq 0$, $y \in D$. Thus we must show the existence of such a hyperplane separating L and D.

We shall suppose L and D are not separated and show this contradicts the optimality of s^*. If L and D are not separated the vector $(-1, 0, \ldots, 0)$ which generates L must be interior to D. If it were not the separation theorem for convex sets would imply that there was a hyperplane separating L and D. Thus there must be a sphere about $(-1, 0, \ldots, 0)$ contained in D. Let y_1 denote the first coordinate of a vector $(y_1, \ldots, y_k)$ in E^k. Consider the hyperplane $y_1=-1$. In this hyperplane place a suitable scaled and translated standard $k-1$ dimensional simplex so that it is contained in the intersection of the sphere and the hyperplane and $(-1, 0, \ldots, 0)$ is in the interior of the simplex in the relative topology of the hyperplane. Draw k vectors $d_1, \ldots, d_k$ from the origin of E_k to the vertices of this simplex. Then these k vectors are linearly independent, contained in the interior of D, and all have first coordinate -1.

Let M denote the matrix whose columns are $\{d_1, \ldots, d_k\}$. Let Δ denote the convex hull of $\{0, d_1, \ldots, d_k\}$. Since D is a cone of variations of $\mathscr{J}$ at s^* there is for some $\gamma > 0$ a mapping $\zeta\colon \gamma C \to \mathscr{S}$ such that

(13.5) $$\mathscr{J}(\zeta(\rho)) = \mathscr{J}(s^*) + M\rho + o(\rho).$$

The definition of M and Δ imply that the mapping $M\rho$ maps γC into $\gamma\Delta$. Since the d^i are linearly independent, M^{-1} exists. The mapping defined for $y \in \gamma\Delta$ by

(13.6) $$\psi(y) = \mathscr{J}(\zeta(M^{-1}y)) - \mathscr{J}(s^*)$$

maps $\gamma\Delta$ into E^k. Since M^{-1} maps $\gamma\Delta$ into γC, Eq. (13.5) implies for $y \in \gamma\Delta$

(13.7) $$|\psi(y) - y| \leqq |o(M^{-1}y)|.$$

Let $a = (-\frac{1}{2}, 0, \ldots, 0)$. By the construction of Δ for some ε, $0 < \varepsilon < 1$, a spherical neighborhood $N(a, \varepsilon)$ of radius ε of a is interior to Δ. Choose an ε so this is true. Since

(13.8) $$\lim_{|M^{-1}y| \downarrow 0} \frac{|o(M^{-1}y)|}{|M^{-1}y|} = 0$$

and for $\eta > 0$, $M^{-1}y$ maps $\eta\Delta$ into ηC there is an η, $0 < \eta < \gamma$ so that if $y \in \eta\Delta$

(13.9) $$|o(M^{-1}y)| < \varepsilon\eta.$$

Choose such an η.

Let π denote the projection of E^k on E^{k-1} defined by

(13.10) $$\pi(y_1, \ldots, y_k) = (y_2, \ldots, y_k).$$

We have $\pi(a) = 0$. Since $N(a, \varepsilon)$ is contained in Δ, the $\eta\varepsilon$ neighborhood of zero, $N(0, \eta\varepsilon)$ in E^{k-1} is contained in $\pi(\eta\Delta)$. Define

$$\chi\colon \pi(\eta\Delta) \to E^{k-1}$$

by $\chi(y_2, \ldots, y_k) = (y_2, \ldots, y_k) - \pi(\psi(-\eta, y_2, \ldots, y_k))$. Now if $(y_2, \ldots, y_k) \in \pi(\eta\Delta)$, (13.7) and (13.9) imply

$$\begin{aligned} |\chi(y_2, \ldots, y_k)| &= |\pi(\psi(-\eta, y_2, \ldots, y_k)) - \pi(-\eta, y_2, \ldots, y_k)| \\ &\leqq |\psi(-\eta, y_2, \ldots, y_k) - (-\eta, y_2, \ldots, y_k)| < \eta\varepsilon. \end{aligned}$$

Hence since $\pi(\eta\Delta)$ contains the $\eta\varepsilon$ neighborhood of zero in E^{k-1} we have

$$\chi(\pi(\eta\Delta)) \subset \pi(\eta\Delta).$$

Since χ is clearly continuous, Brouwer's fixed point theorem, Hurewicz Wallman [1], implies that χ must have a fixed point on $\pi(\eta\Delta)$. Thus there exists a point

$(\bar{y}_2, \ldots, \bar{y}_k)$ such that

$$\chi(\bar{y}_2, \ldots, \bar{y}_k) - (\bar{y}_2, \ldots, \bar{y}_k) = -\pi(\psi(-\eta, \bar{y}_2, \ldots, \bar{y}_k)) = 0.$$

Since $|\psi(-\eta, \bar{y}_2, \ldots, \bar{y}_k) - (-\eta, \bar{y}_2, \ldots, \bar{y}_k)| < \eta\,\varepsilon < \eta$ and $\pi(\psi(-\eta, \bar{y}_2, \ldots, \bar{y}_k)) = 0$, we must have

$$\psi(-\eta, \bar{y}_2, \ldots, \bar{y}_k) = (y_1, 0, \ldots, 0) \quad \text{with } y_1 < 0.$$

But this and the definition of ψ imply

$$J_1(\zeta(M^{-1}(-\eta, \bar{y}_2, \ldots, \bar{y}_k))) - J_1(s^*) = y_1 < 0$$

while

$$J_i(\zeta(M^{-1}(-\eta, \bar{y}_2, \ldots, \bar{y}_k))) - J_i(s^*) = 0 \qquad i = 2, \ldots, k$$

which contradicts the optimality of s^*. □

§ 14. A Cone of Variations for the Problem of Optimal Control

Recall the set $\mathscr{G}$ and the mapping $\mathscr{J}$ defined in § 12.

Theorem 14.1. *The convex cone generated by the vectors* (i)–(iv) *listed below is a cone of variations of the mapping* $\mathscr{J}(x_0, u) = \phi(e)$ *at* (x_0, u).

(i) $$\phi_{x_0}(e)\, y + \phi_{x_1}(e)\, \delta x(t_1)$$

where y *is an element of* E^n *and* $\delta x(t)$ *is the solution*

(14.1) $$\delta \dot{x}(t) = f_x(t, x(t), u(t))\, \delta x(t)$$

with

(14.2) $$\delta x(t_0) = y.$$

(ii) $$\phi_{x_1}(e)\, \delta x(t_1)$$

where $\delta x(t)$ *is a solution of* (14.1) *with boundary condition given for some* $\tau \in (t_0, t_1]$ *and* $u \in U$ *by*

(14.3) $$\delta x(\tau) = f(\tau, x(\tau), u) - f(\tau, x(\tau), u(\tau)).$$

(iii) $$\pm[\phi_{t_0}(e) + \phi_{x_0}(e)\, f(t_0, x(t_0), u(t_0))].$$

(iv) $$\pm[\phi_{t_1}(e) + \phi_{x_1}(e)\, f(t_1, x(t_1), u(t_1))].$$

In the rest of this section we prove this theorem. To begin its proof we must construct mappings $\zeta\colon \eta\, C \to \mathscr{G}$ so that $\mathscr{J}(\zeta(\rho))$ have appropriate conical differentials. Consider a vector d in the convex cone generated by vectors of the form (i)–(iv). Each vector will be a positive linear combination of the vectors (i)–(iv). Thus each such vector will have a representation of the form (14.4) in terms of a vector of E^n, pairs of times and control values (τ_i, u^i), $i = 2, \ldots, j$, and numbers a_i, $i=0, \ldots, j+1$ such that a_i, $i=1, \ldots, j$ are nonnegative:

$$\begin{aligned} d = {} & a_0[\phi_{t_0}+\phi_{x_0} f(t_0, x(t_0), u(t_0))] + a_1[\phi_{x_0} y + \phi_{x_1} \delta x(t_1)^1] \\ & + \sum_{i=2}^{j} a_i \phi_{x_1} \delta x(t_1)^i + a_{j+1}[\phi_{t_1}+\phi_{x_1} f(t_1, x(t_1), u(t_1))]. \end{aligned} \tag{14.4}$$

In (14.4), $\delta x(t)^1$ is the solution of (14.1) with boundary condition (14.2) and $\delta x(t)^i$, $i=2, \ldots, j$, is the solution of (14.1) with boundary condition of (14.3) corresponding to τ_i and u^i. In (14.4), (τ_i, u^i) and (τ_m, u^m) are distinct pairs in the representation if $\tau_i=\tau_m$ but $u^i \neq u^m$.

Notice that given k such vectors $d^1, \ldots, d^k$, we may assume each of the vectors is represented in terms of the same set of quantities $\delta x(t_1)^i$, corresponding to (τ_i, u^i) $i=2, \ldots, j$. If this were not the case it could be accomplished by considering the union of the set of these quantities appearing in the representation of all the vectors. Then a given vector could be represented in terms of this set using a coefficient $a_i=0$ if a particular $\delta x(t_1)^i$ was not involved in its previous representation. Thus given k such vectors we may arrange the numbers appearing in their representation into a $(j+1)\times k$ matrix A whose columns are the numbers a_i in the representation of each vector. Let a^i denote the row matrix which is i-th row of this matrix.

Let the times τ_i, $i=2, \ldots, j$ in the representation (14.4) be given in nondecreasing order. Let s_i be the largest integer so that $\tau_i=\tau_m$ if $i \leqq m \leqq s_i$. Define for $\rho \in C$ the function $l_i(\rho)$ by

$$l_i(\rho) = \sum_{m=i}^{s_i} a^m \rho. \tag{14.5}$$

If $\rho \in \eta C$ and η is small enough the intervals

$$\tau_i - l_i(\rho) < t \leqq \tau_i - l_{i+1}(\rho) \qquad \text{if } i \neq s_i \tag{14.6}$$

$$\tau_i - l_i(\rho) < t \leqq \tau_i \qquad \text{if } i = s_i \tag{14.7}$$

and the two intervals which have respective endpoints

$$t_0, t_0 + a^0 \rho \tag{14.8}$$

and

$$t_1, t_1 + a^{j+1} \rho \tag{14.9}$$

are nonoverlapping. For a given control u defined on $[t_0, t_1]$ and $\eta>0$ such that the intervals (14.6)–(14.9) are nonoverlapping define for $\rho \in \eta C$ a control u^ρ on $[t_0+a^0\rho, t_1+a^{j+1}\rho]$ by

$$u^\rho(t) = \begin{cases} u(t_0) & \text{if } t_0 + a^0 \rho \leqq t \leqq t_0 \\ u^i & \text{if } \tau_i - l_i(\rho) < t \leqq \tau_i - l_{i+1}(\rho) \\ u(t_1) & \text{if } t_1 < t \leqq t_1 + a^{j+1} \rho \\ u(t) & \text{otherwise.} \end{cases} \tag{14.10}$$

The vectors $a_{11} y^1, \ldots, a_{1k} y^k$ which appear in the representation of $d^1, \ldots, d^k$ can be arranged into a matrix N whose columns are these vectors. Define the mapping

(14.11) $$\zeta(\rho) = (x_0 + N\rho, u^\rho).$$

We must show that this mapping maps ηC, for η small enough, into $\mathscr{G}$. The control u^ρ is of the form $\tilde{u}$ described in § 3 and hence the assumption made there implies $u^\rho \in \mathscr{U}$. The existence theorem for differential equations implies that Eq. (3.1) corresponding to u^ρ has a solution in some small interval adjacent to a point at which a boundary condition is specified. The theorem for continuous dependence of solutions of differential equations on initial conditions will imply that there is a solution of (3.1) for both u^ρ and u on an interval $\tau_i \leqq t \leqq b$ on which $u^\rho(t) \equiv u(t)$ provided that $x^\rho(\tau_i)$ is close enough to $x(\tau_i)$. Hence by using these two theorems successively, it can be seen that if η is small enough and $\rho \in \eta C$ that there is a solution of (3.1) on the interval $[t_0 + a^0\rho, t_1 + a^{j+1}\rho]$ (or on $[t_0, t_1 + a^{j+1}\rho]$ if $a^0\rho > 0$) which satisfies the boundary condition

(14.12) $$x^\rho(t_0) = x_0 + N\rho.$$

Thus for small enough η, ζ does map ηC into the set $\mathscr{G}$. Let us choose such an $\eta > 0$.

The theorem on continuous dependence of solutions on initial conditions can be used in a manner similar to that sketched above to show that the mapping

(14.13) $$\rho \to \left(x^\rho(t_0 + a^0\rho), x^\rho(t_1 + a^{j+1}\rho)\right)$$

is a continuous mapping of ηC into E^{2n}. It can be used further to show that on $[t_0, t_1]$, $x^\rho(t)$ converges uniformly to $x(t)$ as $|\rho|$ approaches zero.

From the preceeding remarks it follows that

(14.14) $$\mathscr{J}(x_0 + N\rho, u^\rho) = \phi\left(t_0 + a^0\rho, t_1 + a^{j+1}\rho, x^\rho(t_0 + a^0\rho), x^\rho(t_1 + a^{j+1}\rho)\right)$$

is a continuous function of ρ on ηC. Therefore, we may conclude that D is a cone of variations if we can conclude that $\mathscr{J}(x_0, u)$ has the appropriate conical differential at (x_0, u) along ζ. The next two lemmas will enable us to establish this.

Lemma 14.1. *Let the times τ_i, $i = 2, \ldots, j$ defined in the representation of the vectors $d^1, \ldots, d^k$ be ordered by magnitude. Define $\tau_1 = t_0$ and $\tau_{j+1} = t_1$. Then if $\tau_i \leqq t \leqq \tau_{i+1}$*

(14.15) $$x^\rho(t) = x(t) + \delta x(t)\rho + o(t, \rho)$$

where

(14.16) $$\delta x(t) = \sum_{m=1}^{i} \delta x(t)^m$$

where $\delta x(t)^1$ is the $n \times n$ matrix which is the solution of

(14.17) $$\delta\dot{x}(t) = f_x\left(t, x(t), u(t)\right)\delta x(t)$$

with initial condition

(14.18) $$\delta x(t_0) = N$$

and $\delta(t)^m$ for $m \geqq 1$ is the $n \times n$ matrix which is the solution of (14.17) *with boundary condition*

$$\delta x(\tau_m) = [f(\tau_m, x(\tau_m), u^m) - f(\tau_m, x(\tau_m), u(\tau_m))]\, a^m. \tag{14.19}$$

Proof. We will prove Lemma 14.1 by induction on the number of distinct times of $\tau_1, \ldots, \tau_{j+1}$ which are $\leqq t$. If there is one such time, $\tau_1 = t_0$, we have by the definition of the initial condition $x^\rho(t_0) = x_0 + N\rho$, hence the theorem follows from Theorem 10.4. Suppose the lemma has been established when there are s distinct times $\leqq t$ and we wish to prove it when there are $s+1$. Let τ_i be the largest element of $\tau_1, \ldots, \tau_{j+1}$ which is $\leqq t$. Suppose q is the smallest integer so that $\tau_q = \tau_i$. Now[1]

$$x^\rho(\tau_i) = x^\rho(\tau_i - l_q(\rho)) + \sum_{m=q}^{i} \int_{\tau_i - l_m(\rho)}^{\tau_i - l_{m+1}(\rho)} f(t, x^\rho(t), u^m)\, dt \tag{14.20}$$

and

$$x(\tau_i) = x(\tau_i - l_q(\rho)) + \sum_{m=q}^{i} \int_{\tau_i - l_m(\rho)}^{\tau_i - l_{m+1}(\rho)} f(t, x(t), u(t))\, dt. \tag{14.21}$$

There are fewer than s distinct times of the set of τ_k's which are $\leqq \tau_i - l_q(\rho)$, hence by the induction hypothesis

$$x^\rho(\tau_i - l_q(\rho)) = x(\tau_i - l_q(\rho)) + \delta x(\tau_i - l_q(\rho))\, \rho + o(\tau_i - l_q(\rho), \rho) \tag{14.22}$$

where

$$\delta x(\tau_i - l_q(\rho)) = \sum_{j=1}^{q-1} \delta x(\tau_i - l_q(\rho))^j. \tag{14.23}$$

Since the $\delta x(t)^j$ are continuous and $l_q(\rho)$ approaches zero as $|\rho|$ approaches zero

$$\sum_{j=1}^{q-1} \delta x(\tau_i - l_q(\rho))^j \rho + o(\tau_i - l_q(\rho), \rho) = \sum_{j=1}^{q-1} \delta x(\tau_i)^j \rho + o(\tau_j, \rho). \tag{14.24}$$

Since $x^\rho(t) = x(t) + O(t, \rho)$ and the integrands in (14.20) and (14.21) are left continuous,

$$\begin{aligned} &\sum_{m=q+1}^{i} \int_{\tau_i - l_m(\rho)}^{\tau_i - l_{m+1}(\rho)} [f(t, x^\rho(t), u^m) - f(t, x(t), u(t))]\, dt \\ &\quad = \sum_{m=q}^{i} [f(\tau_m, x(\tau_m), u^m) - f(\tau_m, x(\tau_m)\, u(\tau_m))]\, a^m \rho + o(\tau, \rho). \end{aligned} \tag{14.25}$$

Hence combining (14.22)–(14.25) gives

$$\begin{aligned} x^\rho(\tau_i) &= x(\tau_i) + \sum_{j=1}^{q-1} \delta x(\tau_i)^j \rho \\ &\quad + \sum_{m=q}^{i} [f(\tau_i, x(\tau_i), u^m) - f(\tau_i, x(\tau_i), u(\tau_i))]\, a^m \rho + o(\tau_i, \rho). \end{aligned} \tag{14.26}$$

---

[1] If $i = s_i$, interpret $l_{i+1}(\rho)$ as zero in (14.20), (14.21) and (14.25).

The lemma now follows from (14.26) by applying Theorem 10.4. □

Lemma 14.2. *If* $\delta x(t_1)$ *is as defined in Lemma* 14.1 *then*

$$x^{\rho}(t_0+a^0\rho)=x_0(t_0)+N\rho+f(t_0,x(t_0),u(t_0))\,a^0\rho+o(\rho) \tag{14.27}$$

and

$$x^{\rho}(t_1+a^{j+1}\rho)=x(t_1)+\delta x(t_1)\,\rho+f(t_1,x(t_1),u(t_1))\,a^{j+1}\rho+o(\rho). \tag{14.28}$$

The proof of Lemma 14.2 uses arguments similar to parts of Lemma 14.1 and is left for the reader.

It now follows, applying Lemma 14.2 and the chain rule for differentiation to the vector function $\phi(t_0+a^0\rho, t_1+a^{j+1}\rho, x^{\rho}(t_0+a^0\rho), x^{\rho}(t_1+a^{j+1}\rho))$, and using the representation (14.4) for d^i and formulas (14.16) through (14.19) that $\mathscr{J}(x_0,u)$ has a conical differential at (x_0,u) along ζ and that the matrix M in this differential is the matrix whose columns are the vectors $d^1,\dots,d^k$ of the type given in (14.4). This and the remarks preceding Lemma 14.1 establish Theorem 14.1.

§ 15. Verification of Pontryagin's Principle

We are now in a position to deduce the assertions of § 5, the Pontryagin Principle, from the results of § 13, and § 14. For brevity we write $u^*=u$, $x^*=x$. Theorems 13.1 and 14.1 imply the existence of a nonzero vector λ such that $\lambda_1\leqq 0$ and

$$\lambda' d\leqq 0 \tag{15.1}$$

for vectors d of the convex cone D generated by the vectors (i)–(iv) of Theorem 14.1. Let $P(t)$ be the solution of (5.1) with boundary condition (5.3). Applying formula (10.5) with $G(t)=0$, $y(t)=\delta x(t)$ which is the solution of (14.1) with boundary condition (14.3), gives

$$P(\tau)'[f(\tau,x(\tau),u)-f(\tau,x(\tau),u(\tau))]-\lambda'\phi_{x_1}(t_0,t_1,x(t_0),x(t_1))\,\delta x(t_1)=0. \tag{15.2}$$

Hence (15.1) for the vector (ii) of Theorem 14.1 and (15.2) imply (5.2).

Applying (10.5) with $y(t)=\delta x(t)$ which is the solution of (14.1) with boundary condition (14.2) gives

$$P(t_0)'y-\lambda'\phi_{x_1}(t_0,t_1,x(t_0),x(t_1))\,\delta x(t_1)=0. \tag{15.3}$$

Now (15.1) implies for the vector (i) of Theorem 14.1 that

$$\lambda'\phi_{x_0}(t_0,t_1,x(t_0),x(t_1))\,y+\lambda'\phi_{x_1}(t_0,t_1,x(t_0),x(t_1))\,\delta x(t_1)\leqq 0. \tag{15.4}$$

Adding (15.3) and (15.4) gives

$$[P(t_0)'+\lambda'\phi_{x_0}(t_0,t_1,x(t_0),x(t_1))]\,y\leqq 0. \tag{15.5}$$

Since this must hold for each $y\in E^n$, the quantity in brackets must be zero which establishes (5.4).

For the vectors (iv) of Theorem 14.1 we must have

$$(15.6)\quad \pm\lambda'[\phi_{t_1}(t_0, t_1, x(t_0), x(t_1)) + \phi_{x_1}(t_0, t_1, x(t_0), x(t_1))\, f(t_1, x(t_1), u(t_1))] \leqq 0.$$

Hence (15.6) and (5.3) imply (5.5). A similar argument using (iii) of Theorem 14.1. and (5.4) establishes (5.6). Condition (5.7) is implied by (5.2) and (5.6). This result is asserted in Corollary 15.1. Lemma 15.1 is a general result of this type.

Lemma 15.1. *Let $h(t, u)$ be a continuous function defined on $E^1 \times E^m$ which has a continuous partial derivative with respect to t. Let U be a closed subset of E^m. Let $u(t)$ be a left continuous piecewise continuous function defined on $[t_0, t_1]$ with values in U. If*

$$(15.7)\qquad \max_{u \in U} h(t, u) = h(t, u(t))$$

for each $t \in [t_0, t_1]$ then $h(t, u(t))$ is piecewise continuously differentiable on $[t_0, t_1]$ and

$$(15.8)\qquad h(t, u(t)) = \int_{t_0}^{t} h_t(s, u(s))\, ds + h(t_0, u(t_0)).$$

Proof. First we will show that (15.7) implies $h(t, u(t))$ is continuous. Since it is piecewise continuous and left continuous, this will follow if we show it is right continuous at each interior point of $[t_0, t_1]$. For such a point (15.7) implies

$$h(t, u(t+\tau)) \leqq h(t, u(t)) \quad \text{and} \quad h(t+\tau, u(t)) \leqq h(t+\tau, u(t+\tau)).$$

Taking limits as τ decreases to zero implies

$$h(t, u(t)^+) \leqq h(t, u(t)) \leqq h(t, u(t)^+).$$

Thus $h(t, u(t))$ is continuous.

Let t be a point of continuity of $u(t)$ and consider the difference

$$d(\tau) = h(t+\tau, u(t+\tau)) - h(t, u(t)).$$

From (15.7)

$$h(t+\tau, u(t)) - h(t, u(t)) \leqq d(\tau) \leqq h(t+\tau, u(t+\tau)) - h(t, u(t+\tau)).$$

By the theorem of the mean there are θ_1, θ_2 between zero and one such that

$$h_t(t+\theta_1 \tau, u(t)) \leqq \frac{d(\tau)}{\tau} \leqq h_t(t+\theta_2 \tau, u(t+\tau)).$$

Since t is a point of continuity of $u(t)$ taking limits as τ approaches zero we have

$$h_t(t, u(t)) = \frac{d}{dt} h(t, u(t)).$$

Thus $h(t, u(t))$ has a piecewise continuous derivative $h_t(t, u(t))$. From this and the continuity of $h(t, u(t))$ we conclude that (15.8) holds. □

Corollary 15.1. *If $f(t, x, u)$ and $u(t)$ are as in Theorem 5.1, condition (5.7) of the statement of Theorem 5.1 is implied by conditions (5.1), (5.2) and (5.6).*

Proof. Apply Lemma 15.1 with

$$h(t, u) = P(t)' f(t, x(t), u).$$

Then

$$h_t(t, u(t)) = -P(t)' f_x(t, x(t), u(t)) f(t, x(t), u(t)) + P(t)' f_t(t, x(t), u(t))$$
$$+ P(t)' f_x(t, x(t), u(t)) f(t, x(t), u(t)) = P(t) f_t(t, x(t), u(t)).$$

Hence (5.7) follows directly from (15.8) and (5.6). □

Problems—Chapter II

(1) Formulate as an optimal control problem the problem of putting a satellite into a planar elliptical orbit about the earth by means of a rocket using a minimum amount of fuel. Use simplified point mass equations to represent the equations of motion of the rocket.

(2) Let the end conditions (3.3) in the statement of the optimal control problem be replaced by the requirements that t_0 be a fixed time, x_0 a fixed initial state, and that the terminal time t_1 and state $x(t_1)$ satisfy equations

$$\phi_i(t_1, x(t_1)) = 0 \qquad i = 2, \ldots, l.$$

Let the performance function ϕ_1 be a function $\phi_1(t_1, x_1)$ of only the terminal time and terminal state.

Show that for this optimal control problem necessary conditions for optimality are given by changing conditions (5.3), (5.5) and of Pontryagin's principle to the statement: There is a nonzero l-dimensional vector λ with $\lambda_1 \leqq 0$ such that

(5.3)′ $$P(t_1)' = \lambda' \phi_x(t_1, x^*(t_1))$$

(5.5)′ $$P(t_1)' f(t_1, x^*(t_1), u^*(t_1)) = -\lambda' \phi_t(t_1, x^*(t_1))$$

(5.7)′
$$P(t)' f(t, x^*(t), u^*(t)) = -\lambda' \phi_t(t_1, x^*(t_1)) - \int_t^{t_1} P(s)' f_t(s, x^*(s), u^*(s))\, ds$$

and omitting conditions (5.4) and (5.6).

(3) For the optimal control problem with equations of motion:

$$\dot{x} = f(t, x, u)$$

initial conditions:

$$x(t_0) = x_0,$$

in which t_0 and x_0 are fixed, terminal conditions:

$$\phi_i(t_1, x(t_1)) = 0 \qquad i = 2, \ldots, l$$

and performance criterion:

$$\int_{t_0}^{t_1} L(t, x(t), u(t))\, dt,$$

show that necessary conditions for $u^*(t)$ to be an optimal control are:

There is a nonzero l dimensional vector λ such that $\lambda_1 \leqq 0$ and a n-dimensional vector function $P(t)$ such that for $t \in [t_0, t_1]$

$$\dot{P}(t)' = -\lambda_1 L_x(t, x^*(t), u^*(t)) - P(t)' f_x(t, x^*(t), u^*(t)).$$

For $t \in [t_0, t_1]$ and $v \in U$

$$\lambda_1 [L(t, x^*(t), v) - L(t, x^*(t), u^*(t)] + P(t)' [f(t, x^*(t), v) - f(t, x(t), u^*(t))] \leqq 0$$

$$P'(t_1) = \tilde{\lambda}' \phi_x(t_1, x^*(t_1))$$

$$\lambda_1 L(t_1, x^*(t_1), u^*(t_1)) + P(t_1)' f(t_1, x^*(t_1), u^*(t_1)) = -\tilde{\lambda}' \phi_t(t_1, x^*(t_1))$$

where $\tilde{\lambda}' = (\lambda_2, \ldots, \lambda_l)$ and $\phi(t, x)' = (\phi_2(t, x), \ldots, \phi_l(t, x))$.

(4) Find an extremal control for the optimization problem with:
Equation of motion:

$$\dot{x} = x + u$$

in which x and u are scalars,
Control set:

$$U = \{u \colon -\infty < u < \infty\}$$

Initial condition:

$$x(0) = 1;$$

Terminal condition:

$$t_1 = 1$$

and performance index:

$$\tfrac{1}{2} \int_0^1 [x(t)^2 + u(t)^2] \, dt.$$

(5) The equations of motion in rectangular coordinates of a vehicle moving with an acceleration of magnitude one are:

$$\dot{x} = u$$

$$\dot{y} = v$$

$$\dot{u} = \cos \beta$$

$$\dot{v} = \sin \beta.$$

Consider the control problem of steering the vehicle from

$$x(0) = 0, \quad y(0) = 0, \quad u(0) = 0, \quad v(0) = 0$$

to

$$y(T) = 1, \quad v(T) = 0$$

in a fixed time T while maximizing $u(T)$. Compute extremal controls for this problem.

(6) Show that if the matrix

$$(\phi_{t_0}(t_0, t_1, x_0, x_1), -\phi_{x_0}(t_0, t_1, x_0, x_1), -\phi_{t_1}(t_0, t_1, x_0, x_1), \phi_{x_1}(t_0, t_1, x_0, x_1))$$

has rank k at $e=(t_0, t_1, x(t_0), x(t_1))$, then the condition $\lambda \neq 0$ of Theorem 5.1 is equivalent to $P(t) \neq 0$ for $t \in [t_0, t_1]$.

(7) An alternate model for a one sector economy to that given in Example 2.2 takes into account population growth. A production function $f(k)$ gives the rate of increase of capital per worker due to production. If the population has growth rate constant β, there is a rate of decrease $-\beta k$ of capital per worker due to population growth. A fraction u of new production is retained in the economy and the remaining fraction $1-u$ is consumed. Hence the equation

$$\dot{k} = u f(k) - \beta k$$

governs the capital per worker in the economy. Let $h(c)$ denote the utility to the economy of consuming at rate c. Consider the problem of choosing a savings plan $u(t)$ to increase the capital per worker from k_1 to k_2 in the fixed time T, while maximizing

$$\int_0^T h([1-u(t)]\, f(k(t)))\, dt.$$

For $f(k) = \alpha k$, $h(c) = c$ compute an extremal control law. Notice that $0 \leqq u(t) \leqq 1$ must be satisfied.

(8) Let $P_i(t)$ denote the probability distribution of a finite state Markov process $\xi(t)$. That is, $\xi(t)$ has states $\{1, \ldots, n\}$ and

$$P_i(t) = Pr\{\xi(t) = i\}.$$

For a finite state Markov process there is a matrix $(a_{ij}(t))$ such that

$$a_{ij}(t) \geqq 0, \quad i \neq j$$
$$a_{ii}(t) = -\sum_{j \neq i} a_{ij}(t)$$

and $P_j(t)$ is the solution of

$$\dot{P}_j(t) = \sum_{i=1}^{n} P_i(t)\, a_{ij}(t).$$

The quantity $a_{ij}(t)\,\Delta t$ is approximately the conditional rate at which jumps from i to j take place at time t.

Consider such a process which is controllable in the following sense. The components of the matrix are continuous functions $a_{ij}(t, u)$ of time and a control variable u. A class $\mathscr{U}$ of control functions

$$\mathscr{U} = \{u(t, i)\}$$

of time and the current state whose values lie in the closed control set U is given. The controlled process then satisfies the equation

$$\dot{P}_j(t) = \sum_{i=1}^{n} P_i(t)\, a_{ij}(t, u(t, i)).$$

Let f_j be a function defined on $\{1, \ldots, n\}$, T a fixed time, and $P_i(0)$ a given initial probability distribution. Consider the problem of choosing the control $u(t, i)$ in $\mathscr{U}$

so that the expectation

$$E\{f_{\xi(T)}\}$$

is a minimum.

Apply Pontryagin's principle to this problem to show that necessary conditions for optimality are:

There are solutions $\phi_i(t)$ of

$$\dot{\phi}_i(t) = -\sum_{j=1}^{n} a_{ij}(t, u(t, i))\, \phi_j(t)$$

with $\phi_i(T) = -f_i$ and

$$\max_{v \in U} \sum_j P_i(t)\, a_{ij}(t, v)\, \phi_j(t) = \sum_j P_i(t)\, a_{ij}(t, u(t, i))\, \phi_j(t), \qquad i = 1, \ldots, n, \ t \in (0, T].$$

(9) Consider an optimal control problem in which u is a scalar control and

$$f(t, x, u) = a(x) + b(x)\, u$$

where $a(x)$ and $b(x)$ are C^2 vector functions. If

$$P(t)'\, b(x(t)) = 0$$

on a time interval $\alpha \leqq t \leqq \beta$, the Hamiltonion does not depend on u and the problem is *singular*. Show that under these conditions

$$P(t)'\, q(x(t)) = 0 \qquad \alpha \leqq t \leqq \beta$$

where $q(x) = b_x(x)\, a(x) - a_x(x)\, b(x)$. Show further that if

$$P(t)'\, [q_x(x(t))\, b(x(t)) - b_x(x(t))\, q(x(t))] \neq 0$$

that

$$u(t) = -\frac{P(t)'\, [q_x(x(t))\, a(x(t)) - a_x(x(t))\, q(x(t))]}{P(t)'\, [q_x(x(t))\, b(x(t)) - b_x(x(t))\, q(x(t))]}.$$

(10) The equations of motion of a vehicle with velocity of magnitude one for which the angular rate of the velocity is controlled are

$$\dot{x} = \cos\theta$$

$$\dot{y} = \sin\theta$$

$$\dot{\theta} = u.$$

Assume

$$|u| \leqq 1.$$

Consider the problem of transfering the vehicle from initial conditions

$$x(0) = 4, \qquad y(0) = 0, \qquad \theta(0) = \frac{\pi}{2}$$

to

$$x(t_1) = 0, \qquad y(t_1) = 0$$

in minimum time t_1. Compute an extremal control law for this problem.

(11) Let $H(u)$ be a function defined on a closed convex subset C of E^n. Let $H(u)$ have a tangent plane at u_0. Show that a necessary condition that $H(u)$ have

a maximum at u_0 is that

$$H_u(u_0)\, v \leqq 0$$

for each $v \in E^n$ such that $u_0 + v \in C$. Apply this result to compare Theorem 11.2 and Theorem 11.5.

(12) A concept analogous to a cone of variations is the concept of *derived set.* Let Y be a subset of E^p. A set $D \subset E^p$ will be said to be a derived set for Y at y_0 if for every finite subset $d_1, \ldots, d_n$ of D, for some $\delta > 0$, there is a continuous mapping $y(\varepsilon)$ of $\{\varepsilon\colon 0 \leqq \varepsilon_i \leqq \delta;\ i = 1, \ldots, n\} \subset E^n$ into Y such that

$$y(\varepsilon) = y_0 + \sum_{i=1}^{n} \varepsilon_i d_i + o(\varepsilon).$$

Show that the convex cone generated by a derived set for Y at y_0 is a derived set for Y at y_0.

(13) Carry out explicitly the computation of the conical differential of the performance $J(x_0, u)$ discussed at the end of § 14.

(14) In the notation of Chap. I, let

$$J(x) = \int_{t_0}^{t_1} L(t, x(t), \dot{x}(t))\, dt, \qquad I(x) = \int_{t_0}^{t_1} M(t, x(t), \dot{x}(t))\, dt.$$

Given a real number a, consider the problem of minimizing $J(x)$ among all piecewise C^1 functions $x(\cdot)$ on $[t_0, t_1]$ which satisfy $x(t_0) = x(t_1)$ and $I(x) = a$. Suppose that x^* minimizes in this problem, and that x^* is not an extremal for $I(x)$ according to the definition in Sect. I.3.

(a) Show that there exists a scalar μ such that x^* is an extremal for $J + \mu I$ (you can use Theorem 13.1 or give a direct proof).

(b) Apply (a) to the classical isoperimetric problem in the plane E^2, in which

$$J(x) = \int_{t_0}^{t_1} \sqrt{\dot{x}_1^2 + \dot{x}_2^2}\, dt$$

$$I(x) = \tfrac{1}{2} \int_{t_0}^{t_1} (x_1 \dot{x}_2 - x_2 \dot{x}_1)\, dt$$

correspond respectively to length, area. Show that any minimizing x^* describes a circle. *Hint.* Introduce arc length as a parameter.

(15) Consider the characteristic value problem

$$\frac{d^2 X}{ds^2} + \lambda\, U(s) X = 0, \qquad 0 \leqq s \leqq 1,$$

$$X(0) = X(1) = 0,$$

where U is piecewise continuous, $0 \leqq U(s) \leqq M$. Let $\lambda_1 = \lambda_1(U)$ be the smallest characteristic value.

(a) From Pontryagin's principle show that $\lambda_1(U)$ is minimum when $U(s) \equiv M$ *Hint.* The substitution $t = \sqrt{\lambda}\, s$ reduces this to the problem of minimum time t

such that $x(t_1)=0$, where

$$\ddot{x}+ux=0, \quad x(0)=0, \quad \dot{x}(0)=1, \quad 0 \leqq u(t) \leqq M.$$

(b) What can you say about the higher characteristic values?

(c) In part (a) add the constraint $c=\int_0^1 U(s)\,ds$ where $0 \leqq c \leqq M$. Find the optimal U.

(16) Let x be a vector of variables in E^n and A be an $n \times n$ symmetric matrix. Show that the gradient of the quadratic form $x'Ax$ with respect to the variables x is given by $2x'A$.

Chapter III. Existence and Continuity Properties of Optimal Controls

§ 1. The Existence Problem

In the two previous chapters we found necessary conditions for a minimum in problems of calculus of variations and optimal control. We turn now to a different question: does the performance index J in fact attain its minimum value? This is the existence problem.

From a practical viewpoint, an existence theorem insures that the problem has a solution before attempting to calculate an optimal control. We shall see from examples that reasonable appearing minimum problems do not always have solutions. It turns out that a certain convexity condition is crucial in establishing the existence of a minimum (Theorems 2.1, 4.1). If the control set U is unbounded, a certain growth condition as $|u| \to \infty$ is also needed in Theorem 4.1.

Occasionally existence theorems can be obtained by methods of differential equations. This happens, for instance, if those methods give an extremal and the convexity conditions in Theorem I.2.3 hold. However, the existence of a minimum must usually be proved directly using methods of real analysis. For this purpose it is unsatisfactory to deal merely with piecewise continuous controls u and piecewise C^1 trajectories x as we did in Chap. I and II. The reason is that the spaces of such controls and such trajectories are not complete in their "natural" metrics. Therefore, in the present chapter we admit control functions which are Lebesgue integrable. The corresponding trajectories x, satisfying (2.1), are then absolutely continuous functions.

There remains the question whether an optimal control function u^* is necessarily a rather nicely behaved function, even though control functions which are merely Lebesgue integrable have been admitted. For instance, one may ask whether u^* is (say) continuous or at least piecewise continuous. A sufficient condition for u^* to be continuous is proved in § 6. The problem of giving general sufficient conditions for piecewise continuity of u^* is difficult. We state in Theorem 6.3 a partial result (without proof) and some counterexamples about this problem.

To indicate the need for growth and convexity conditions in the existence theorems, let us give two examples. Both examples are of the simplest type in calculus of variations (one dimensional): find the minimum of $J(x)$ in I.(3.1) subject to fixed end conditions I.(3.2).

Example 1.1. Minimize

$$J(x)=\int_0^1 [x(t)]^2\,dt$$

with end conditions $x(0)=1$, $x(1)=0$. As in Chap. I, let $\mathscr{X}_e$ denote the set of piecewise C^1 functions x satisfying the end conditions. Clearly, $J(x)\geqq 0$; and $J(x)=0$ only when $x(t)\equiv 0$. Hence $J(x)>0$ for all $x\in\mathscr{X}_e$. For $r=1,2,\ldots$ let

$$x^r(t)=1-rt, \qquad 0\leqq t\leqq r^{-1},$$
$$x^r(t)=0, \qquad r^{-1}\leqq t\leqq 1.$$

Then $J(x^r)=(3r)^{-1}$, which tends to 0 as $r\to\infty$. Hence

$$\inf_{\mathscr{X}_e} J(x)=0,$$

but $J(x)$ does not have a minimum on $\mathscr{X}_e$. {Instead of $\mathscr{X}_e$ we could in this example equally well have taken the set $\mathscr{X}_e'$ of absolutely continuous x satisfying the end conditions; see notation in § 3 below.}

The sequence x^r, $r=1,2,\ldots$, is called a minimizing sequence (see definition in § 3). Although the problem of minimizing $J(x)$ on $\mathscr{X}_e$ has no solution, we may regard x^r as an approximate solution for large r.

More generally, if $L=M(t,x)+N(t,x)\dot{x}$ is an integrand linear in $\dot{x}$, then the Euler equation is an equation $M_x=N_t$ not involving $\dot{x}$; see Chap. I.4. One may expect J to have a minimum on $\mathscr{X}_e$ only when the end conditions happen to satisfy the equation $M_x=N_t$.

Example 1.2. Let $L(x,\dot{x})=x^2+(1-\dot{x}^2)^2$,

$$J(x)=\int_0^1 L[x(t),\dot{x}(t)]\,dt$$

with end conditions $x(0)=x(1)=0$. Again $J(x)\geqq 0$. Define x^r for $r=1,2,\ldots$ by requiring $x^r(0)=0$ and

$$\dot{x}^r(t)=1, \qquad \frac{2i-2}{2r}<t<\frac{2i-1}{2r}$$
$$\dot{x}^r(t)=-1, \qquad \frac{2i-1}{2r}<t<\frac{2i}{2r}, \qquad i=1,\ldots,r.$$

The graph of x^r is of "sawtooth" shape. Moreover,

$$0\leqq x^r(t)\leqq 1/2r, \qquad x^r(1)=0$$
$$J(x^r)=r^{-2}/12.$$

The sequence x^r, $r=1,2,\ldots$, is minimizing. However, no x can satisfy both $J(x)=0$ and the end conditions. Hence J has no minimum on $\mathscr{X}_e$.

The simplest problem in calculus of variations becomes a control problem by taking as control function $u(t)=\dot{x}(t)$ and imposing no control constraint ($U=E^1$). Since $\dot{x}(t)$ and the initial state $x_0=x(t_0)$ determine $x(t)$, $t_0\leqq t\leqq t_1$, we may use the notation $J(x_0,u)$ of Chap. II instead of the notation $J(x)$ of Chap. I.

Among the existence theorems which we prove in this chapter, the one which applies to the simplest problem in calculus of variations is Theorem 4.1. According to that theorem, $J(x)$ has a minimum on $\mathscr{X}_e$ provided: (1) L is continuous, (2) $L(t, x, \dot{x})$ is convex as a function of $\dot{x}$ for each (t, x); and (3) $L(t, x, \dot{x}) \geqq g(\dot{x})$ where $|\dot{x}|^{-1} g(\dot{x}) \to \infty$ as $|\dot{x}| \to \infty$. In Example 1.1, $L = x^2$ does not satisfy (3). More generally, condition (3) fails if $L = M + N\dot{x}$ is linear in $\dot{x}$, although (2) holds. In Example 1.2, L satisfies (3) with $g(\dot{x}) = (1 - \dot{x}^2)^2$, but L does not have the convexity property (2).

In Example 1.1, $|\dot{x}(t)|$ must be large for certain t if $J(x)$ is to be near its infimum. If an additional constraint $|\dot{x}(t)| \leqq B$ is imposed, then there is a minimum (which in Example 1.1 can be found by inspection). More generally, suppose that the simplest problem in calculus of variations is modified by imposing a constraint $|\dot{x}(t)| \leqq B$. This is a control problem with $\dot{x} = u$, $U = \{|u| \leqq B\}$. Condition (3) is irrelevant for this problem since $\dot{x}$ is bounded. According to Theorem 4.1 there is a minimum if (1), (2) hold, and if $|x_1 - x_0| \leqq B(t_1 - t_0)$. The latter condition insures that the set of functions x satisfying $|\dot{x}(t)| \leqq B$ and the end conditions is not empty. Actually, Theorem 4.1 asserts that there is a minimizing x^* such that $u^* = \dot{x}^*$ is Lebesgue measurable and $|u^*(t)| \leqq B$. If L is regular (see I.3), then it will be shown in § 6 that u^* is continuous. In that case x^* is C^1. Regularity means $L_{\dot{x}\dot{x}} > 0$, while for L of class C^2 the convexity condition (2) means $L_{\dot{x}\dot{x}} \geqq 0$. It seems to be an open question what set of general assumptions on L guarantees that x^* is piecewise C^1 when (2) holds but L is not regular.

§ 2. An Existence Theorem (Mayer Problem, *U* Compact)

Let us consider a control problem with system equations

$$\dot{x} = f(t, x(t), u(t)), \qquad t_0 \leqq t \leqq t_1, \tag{2.1}$$

where as in Chap. II the vector $x(t) \in E^n$ denotes the system state and $u(t) \in U$ the control applied at time t. End conditions are imposed of the type $e \in S$, where

$$e = (t_0, t_1, x(t_0), x(t_1)) \tag{2.2}$$

and S is a given subset of E^{2n+2}. In Chap. II the set S is determined by the equations $\phi_j(e) = 0$, $j = 2, \ldots, k$; see II(3.3). As performance index let us take an expression of Mayer type

$$J(x_0, u) = \phi(e). \tag{2.3}$$

In the notation of Chap. II, $\phi = \phi_1$. (Recall that $x(t_1)$ is determined through (2.1) by the control function u, the initial data (t_0, x_0), $x_0 = x(t_0)$, and the final time t_1.) In § 4 we state existence theorems in which J has the more general Bolza form.

Throughout the present chapter the following assumptions about f are made:

(2.4) *f is continuous; moreover, there exist positive constants C_1, C_2 such that*

(a) $|f(t, x, u)| \leqq C_1(1 + |x| + |u|)$,

(b) $|f(t, x', u) - f(t, x, u)| \leqq C_2 |x' - x| (1 + |u|)$

for all $t \in E^1$, $x, x' \in E^n$, and $u \in U$.

If f is C^1, then (2.4) is implied by suitable bounds on partial derivatives of f and on $f(t,0,0)$. See Problem 7. If the control set U is bounded, then the term $|u|$ on the right-hand side of (2.4a, b) can be omitted, with C_1, C_2 replaced by different constants.

In the present chapter we admit control functions $u=u(\cdot)$ which are Lebesgue integrable on $[t_0, t_1]$, instead of admitting only piecewise continuous controls as in Chap. II. By solution of (2.1) we now mean a function $x=x(\cdot)$ satisfying

$$(2.1') \qquad x(t)=x(t_0)+\int_{t_0}^{t} f(s,x(s),u(s))\,ds, \qquad t_0 \leqq t \leqq t_1.$$

The solution x of (2.1′) exists, and is absolutely continuous on $[t_0, t_1]$. It is uniquely determined by the control function u and the initial data $(t_0, x(t_0))$. See McShane [5, Chap. 9].

Let $m(t)=|x(t)-x(t_0)|$. From (2.1′), (2.4)(a) and $|x(s)|\leqq m(s)+|x(t_0)|$, we have

$$0 \leqq m(t) \leqq C_1 \int_{t_0}^{t} m(s)\,ds + C_1 \int_{t_0}^{t} (1+|x(t_0)|+|u(s)|)\,ds.$$

Using Gronwall's inequality (Appendix A) we get the useful inequality

$$(2.5) \qquad |x(t)-x(t_0)| \leqq C \int_{t_0}^{t} (1+|x(t_0)|+|u(s)|)\,ds$$

with

$$C=e^{C_1(T_1-T_0)}C_1$$

provided $T_0 \leqq t_0 \leqq t \leqq T_1$.

Let $\mathscr{F}'$ denote the class of all (x_0, u) such that u is a Lebesgue-integrable function on an interval $[t_0, t_1]$ with values in U and the solution of (2.1′) satisfies the end conditions $e \in S$. Here $x_0 = x(t_0)$ and e is as in (2.2). The interval $[t_0, t_1]$ on which $u(t)$ is defined may vary with the control function u. Clearly $\mathscr{F} \subset \mathscr{F}'$, where $\mathscr{F}$ is the corresponding class in Chap. II with u required to be piecewise continuous. In this chapter we call any $(x_0, u) \in \mathscr{F}'$ a feasible pair.

In stating an existence theorem for the Mayer problem the following sets play an important role. For each $(t,x) \in E^{n+1}$ let

$$(2.6) \qquad F(t,x)=\{f(t,x,u)\colon u \in U\}.$$

Thus $F(t,x)$ is the image in E^n of the control set U under the function $f(t,x,\cdot)$. Since f is continuous, $F(t,x)$ is a compact set if U is compact.

Theorem 2.1. *Suppose that assumptions* (2.4) *hold, and moreover that:*

(2.7) (a) *$\mathscr{F}'$ is not empty;*

(b) *U is compact;*

(c) *S is compact and ϕ is continuous on S;*

(d) *$F(t,x)$ is convex for each $(t,x) \in E^{n+1}$.*

Then there exist (x_0^*, u^*) minimizing $J(x_0, u)$ on $\mathscr{F}'$.

Theorem (2.1) and its proof are essentially due to Filippov [1]. We defer the proof to § 3. Obviously assumption (a) is needed if the minimum problem is to make sense. In many applications (b) appears naturally in the form of bounds on the physical positions of control devices, rates of expenditure or investment, etc.

If f is linear in u and U is convex, then the image $F(t,x)$ of U under f is also convex; see Appendix C. Thus:

Corollary 2.1. *In Theorem* 2.1, *assumption* (d) *can be replaced by:*

(d′) *U is convex and $f(t,x,u)=\alpha(t,x)+\beta(t,x)u$.*

Note. When (d) does not hold, a kind of "generalized solution" to the problem can be obtained by broadening the concept of control. Such generalized solutions were invented by L.C. Young to treat nonregular problems in calculus variations. He called them generalized curves. In control theory they reappeared with the names relaxed, or chattering, controls. A relaxed control assigns at each t not merely a point $u(t)\in U$ but rather a probability measure $\nu(t)$ on U. It turns out, since $U\subset E^m$, that it suffices to consider atomic probability measures $\nu(t)$ concentrated on no more than $m+1$ points of U. The relaxed control problem can then be stated as one to which Theorem 2.1 applies. See Lee-Markus [1, p. 266], Young [1].

The compactness assumption on S in Theorem 2.1 can often be weakened. In fact, it suffices to know that S is a closed set and that e is bounded for those $(x_0,u)\in\mathscr{F}'$ such that $J(x_0,u)$ is near its infimum.

Corollary 2.2. *Let $\mu=\inf J(x_0,u)$ and $\mu_1>\mu$. In Theorem* 2.1 *assumption* (c) *can be replaced by:*

(c′) *ϕ is continuous on S; there exists a compact $S'\subset S$ such that $e\in S$ and $J(x_0,u)\leqq\mu_1$ imply $e\in S'$.*

To obtain Corollary 2.2 we merely replace S by S' in Theorem 2.1.

Fixed Initial Data. In many problems the initial data $t_0, x_0=x(t_0)$ are given and it is required that $(t_1, x(t_1))\in S_1$, where S_1 is the "terminal set". Thus

$$S=\{e=(t_0,t_1,x_0,x_1)\colon (t_1,x_1)\in S_1\}.$$

The pair (x_0,u) is in $\mathscr{F}'$ provided u steers to S_1 in finite time, i.e. $(t_1,x(t_1))\in S_1$ for some $t_1>t_0$. As a consequence of Corollary 2.2, let us state an existence theorem for the problem of time optimal control.

Corollary 2.3. (Time optimal control, fixed initial data (t_0,x_0)). *If* (2.7)(a)(b)(d) *hold and S_1 is closed, then there exists a control u^* which steers to S_1 in minimum time.*

Proof. Let $\phi(e)=t_1-t_0$, and suppose that u^1 is a control steering to S_1 in time T_1. Let $\mu_1=T_1-t_0=J(x_0,u^1)$. From (2.5) and (2.7)(b), there exists a constant M_1 (depending on x_0 but not on the control u) such that $|x(t)|\leqq M_1$ for $t_0\leqq t\leqq T_1$. In Corollary 2.2 we take

$$S'=\{(t_0,x_0,t_1,x_1)\colon t_0\leqq t_1\leqq T_1, |x_1|\leqq M_1, (t_1,x_1)\in S_1\}. \quad \square$$

We next give two counterexamples to show that there may be no minimum when some of the above assumptions are violated.

Example 2.1. Let $n=1$, $\dot{x}=u$, $U=[-1,1]$, with $(t_0,x_0)=(0,0)$, $S_1=\{(t,x): x=t^{-1}, t>0\}$. Let $J(x_0,u)=x(t_1)$. Clearly $J(x_0,u)>0$. However, $\mu=0$; and hence no (x_0,u) minimizes J on $\mathscr{F}'$. To see that $\mu=0$, let $u(t)=0$ for $0\leqq t\leqq t_2$, then $u(t)=1$ until the first time t_1 when $x(t_1)=t_1^{-1}$. Then $J(x_0,u)=t_1^{-1}<t_2^{-1}$. By taking t_2 sufficiently large, $J(x_0,u)$ can be made arbitrarily near 0. Note that the compact set S' required in Corollary 2.2 does not exist in this example.

Example 2.2. Let $n=2$. The system equations are $\dot{x}_1=-(x_2)^2+u^2$, $\dot{x}_2=u$, with the control constraints $|u(t)|\leqq 1$. We take initial data $(t_0,x_{01},x_{02})=(0,-1,0)$. The problem is to steer $x(t)=(x_1(t),x_2(t))$ to the circle $(x_1)^2+(x_2)^2=a^2$, $a<1$, in minimum time t_1. In the notation above, $S_1=\{(t,x): t\geqq 0, (x_1)^2+(x_2)^2=a^2\}$. Clearly $t_1\geqq 1-a$ since $\dot{x}_1\leqq 1$ for any control function u. For $r=1,2,\dots$ let

$$u^r(t)=1, \qquad \frac{2i-2}{2r}<t<\frac{2i-1}{2r}$$

$$u^r(t)=-1, \qquad \frac{2i-1}{2r}<t<\frac{2i}{2r}, \qquad i=1,2,\dots.$$

As in Example 1.2, $0\leqq x_2^r(t)\leqq 1/2r$. It can then be seen that $(t_1^r, x^r(t_1^r))\in S_1$ with $J(x_0,u^r)=t_1^r\to 1-a$ as $r\to\infty$. Hence, $1-a=\mu$ is the infimum of times to reach the circle from x_0. However, there is no control u^* with $J(x_0,u^*)=1-a$. In this example, $F(t,x_1,x_2)$ is a segment of a parabola – not a convex set.

§ 3. Proof of Theorem 2.1

Since the set S to which the end conditions e belong is compact, there is a fixed interval $[T_0,T_1]$ such that $T_0\leqq t_0\leqq t_1\leqq T_1$ for all $e\in S$. It is convenient to extend control functions u to $[T_0,T_1]$ by setting $u(t)=u_1$ for $t\in[T_0,t_0)$ and $t\in(t_1,T]$ where u_1 is some (fixed) element of U. The function x is then also defined on $[T_0,T_1]$ by (2.1′).

Let $\mathscr{X}_e'$ denote the set of all solutions of (2.1′) on $[T_0,T_1]$ corresponding to $(x_0,u)\in\mathscr{F}'$.

Lemma 3.1. *There exist positive constants M_1, M_2 such that, for all $x\in\mathscr{X}_e'$,*

$$(3.1)\qquad \begin{aligned}&\text{(a)}\ |x(t)|\leqq M_1, \qquad T_0\leqq t\leqq T_1,\\ &\text{(b)}\ |x(t)-x(s)|\leqq M_2(t-s), \qquad T_0\leqq s\leqq t\leqq T_1.\end{aligned}$$

Proof. Since S is compact, $|x(t_0)|\leqq M_0$ for all $x\in\mathscr{X}_e'$. Since U is compact, $|u(t)|\leqq B$ for all controls u. Here M_0, B are certain constants. By (2.5),

$$|x(t)-x(t_0)|\leqq C(T_1-T_0)(1+M_0+B).$$

In (a) we may take $M_1=M_0+C(T_1-T_0)(1+M_0+B)$. By (2.1′) and (2.4)(a),

$$|x(t)-x(s)|\leqq C_1\int_s^t (1+|x(r)|+|u(r)|)\,dr\leqq M_2(t-s),$$

where $M_2=C_1(1+M_1+B)$. □

A set $\mathscr{X}$ of functions is called *equicontinuous* on $[T_0, T_1]$ if given $\varepsilon>0$ there exists $\delta=\delta(\varepsilon)>0$ such that $|s-t|<\delta$ implies $|x(t)-x(s)|\leqq\varepsilon$. The set $\mathscr{X}$ is *uniformly bounded* if there exists M_1 such that $|x(t)|\leqq M_1$ for all $x\in\mathscr{X}$ and $T_0\leqq t\leqq T_1$. We recall Graves [1, p. 122]:

Ascoli's Theorem. *Let $\mathscr{X}$ be an equicontinuous, uniformly bounded set of functions on $[T_0, T_1]$. Then any sequence in $\mathscr{X}$ has a subsequence converging uniformly on $[T_0, T_1]$.*

In particular, $\mathscr{X}_e'$ is an equicontinuous, uniformly bounded set by Lemma 3.1 (take $\delta=\varepsilon M_2^{-1}$).

Definition. Let $\mu=\inf\limits_{\mathscr{F}'} J(x_0, u)$. A sequence $(x_0^r; u^r)$, $r=1, 2, \ldots$, in $\mathscr{F}'$ is *minimizing* if

$$\mu=\lim_{r\to\infty} J(x_0^r; u^r).$$

Many minimizing sequences exist; for the proof of Theorem 2.1 it makes no difference which minimizing sequence we take. We observe that any subsequence of a minimizing sequence is also a minimizing sequence.

Methods of constructing particular minimizing sequences are important for the practical approximate solution of optimal control problems. For discussion of such methods we refer to Bryson-Ho [1, Chap. 7], Beltrami [1, Chap. 4], Falb-DeJong [1], Zadeh-Neustadt-Balakrishnan [1], Polak [1].

If (x_0^r, u^r), $r=1, 2, \ldots$, is a sequence in $\mathscr{F}'$, the control u^r is initially defined on an interval $[t_0^r, t_1^r]$. We extend $u^r(t)$ to $[T_0, T_1]$ as indicated above. Let x^r be the corresponding solution on $[T_0, T_1]$ of (2.1′). The corresponding end conditions are

$$e^r=(t_0^r, t_1^r, x_0^r, x_1^r), \qquad x_i^r=x^r(t_i^r), \qquad i=0, 1.$$

Lemma 3.2. *There exists a minimizing sequence (x_0^r, u^r), $r=1, 2, \ldots$, such that x^r tends uniformly on $[T_0, T_1]$ to a limit x^* as $r\to\infty$. Moreover, x^* satisfies inequalities* (3.1) *and e^r tends to*

$$e^*=(t_0^*, t_1^*, x_0^*, x_1^*), \qquad x_i^*=x^*(t_i^*), \quad i=0, 1.$$

Proof. Let $(\bar{x}_0^j; \bar{u}^j)$, $j=1, 2, \ldots$, be some minimizing sequence and $\bar{x}^j$ the corresponding solution of (2.1′). Since S is compact, $\bar{e}^j$ is bounded. Using Ascoli's Theorem with $\mathscr{X}=\mathscr{X}_e'$, there is a subsequence $j_1, j_2, \ldots$ such that $x^r=\bar{x}^{j_r}$ tends uniformly on $[T_0, T_1]$ to x^* and $e^r=\bar{e}^{j_r}$ tends to a limit e^* as $r\to\infty$. We also take $u^r=\bar{u}^{j_r}$. Since x^r satisfies (3.1) for each $r=1, 2, \ldots, x^*$ also satisfies (3.1). Since e^r tends to e^* and x^r tends to x^* uniformly,

$$\lim_{r\to\infty} t_i^r=t_i^*, \qquad \lim_{r\to\infty} x^r(t_i^r)=x^*(t_i^*), \quad i=0, 1. \quad \square$$

The next two lemmas will establish that there is a control u^* such that x^* satisfies (2.1′) with $u=u^*$. For this purpose the convexity (2.7)(d) of the sets $F(t, x)$ is needed.

Since x^* satisfies (3.1)(b), x^* is Lipschitz. Hence, x^* is absolutely continuous. The derivative $\dot{x}^*(t)$ exists almost everywhere, and the fundamental theorem of calculus holds.

Lemma 3.3. *If $\dot{x}^*(t)$ exists, then $\dot{x}^*(t) \in F(t, x^*(t))$.*

Proof. Suppose that $T_0 < t < T_1$ (obvious changes using left and right hand derivatives are made at endpoints.) For brevity let $F(t, x^*(t)) = F$. For $a > 0$ consider the a-neighborhood of F:

$$\mathcal{O}_a = \{z \in E^n \colon \operatorname{dist}(z, F) < a\}.$$

Since F is convex, $\mathcal{O}_a$ is also convex. See Appendix C. The continuous function f is uniformly continuous on the compact set $[T_0, T_1] \times \{|x| \leqq M_1\} \times U$. Hence, there exists positive η_a tending to 0 as $a \to 0$ such that

$$F(s, x') \subset \mathcal{O}_a \quad \text{whenever} \quad |s-t| + |x' - x^*(t)| < \eta_a.$$

In particular, let $x' = x^r(s)$, $u = u^r(s)$, and use the inequality

$$|x^r(s) - x^*(t)| \leqq |x^r(s) - x^*(s)| + |x^*(s) - x^*(t)|.$$

Let $\delta_a \leqq \eta_a (3M_2)^{-1}$, and small enough that $\delta_a < \eta_a/3$, $\delta_a \leqq |T_i - t|$, $i = 0, 1$. Choose r_a such that

$$|x^r(s) - x^*(s)| < \tfrac{1}{3}\eta_a \quad \text{for } r \geqq r_a, \quad T_0 \leqq s \leqq T_1.$$

Let $z^r(s) = f(s, x^r(s), u^r(s))$. Then $z^r(s) \in F(s, x^r(s))$, and hence $z^r(s) \in \mathcal{O}_a$ whenever $|t-s| < \delta_a$, $r \geqq r_a$. By (2.1′)

$$\frac{x^r(t+h) - x^r(t)}{h} = \frac{1}{h} \int_t^{t+h} z^r(s)\, ds, \qquad |h| < \delta_a.$$

Since $\mathcal{O}_a$ is convex and $z^r(s) \in \mathcal{O}_a$, the right side is in $\bar{\mathcal{O}}_a$ (Appendix C). Let $r \to \infty$. Since $\bar{\mathcal{O}}_a$ is closed,

$$\frac{x^*(t+h) - x^*(t)}{h} \in \bar{\mathcal{O}}_a, \qquad |h| < \delta_a.$$

We let $h \to 0$, obtaining $\dot{x}^*(t) \in \bar{\mathcal{O}}_a$. But $a > 0$ is arbitrary. Hence $\dot{x}^*(t) \in F$. □

Lemma 3.4. *There exists u^* such that $(x_0^*, u^*) \in \mathscr{F}'$ and*

(2.1°) $$\dot{x}^* = f(t, x^*(t), u^*(t)),$$

almost everywhere in $[T_0, T_1]$.

By Lemma 3.3 and the definition of $F(t, x)$ it is immediate that $u^*(t)$ exists such that (2.1°) holds. However, in order that $(x_0^*, u^*) \in \mathscr{F}'$ the function u^* must be Lebesgue measurable. A selection lemma in Appendix B guarantees that such a measurable selection of $u^*(t)$ is possible. □

Proof of Theorem 2.1. Let u^r, e^r, x^r, x^*, e^* be as in Lemma 3.2 and u^* as in Lemma 3.4. Since S is compact and each $e^r \in S$, we have $e^* \in S$. Moreover, x^* is the solution of (2.1′) corresponding to the pair $(x_0^*, u^*) \in \mathscr{F}'$. Since ϕ is continuous, $\phi(e^r) \to \phi(e^*)$ as $r \to \infty$. However, $J(x_0^r, u^r) = \phi(e^r)$ tends to μ, since the sequence is minimizing. Therefore, $J(x_0^*, u^*) = \phi(e^*) = \mu$. Thus, (x_0^*, u^*) minimizes $J(x_0, u)$ on $\mathscr{F}'$. □

§ 4. More Existence Theorems

In this section Theorem 2.1 will be extended in two ways. First, instead of the Mayer form (2.3) we suppose that

$$(4.1) \qquad J(x_0, u) = \int_{t_0}^{t_1} L(t, x(t), u(t))\, dt + \phi(e).$$

Such a performance index J is said to be of *Bolza* type. Second, we allow the control set U to be not compact. The present section is based on Cesari [1].

Instead of the sets $F(t,x)$ in (2.6) let us consider the following subsets $\tilde{F}(t,x)$ of E^{n+1}. Let us write a vector $\tilde{z} \in E^{n+1}$ as

$$\tilde{z} = (z, z_{n+1}), \quad \text{where} \quad z = (z_1, \ldots, z_n) \in E^n.$$

For each $(t,x) \in E^{n+1}$ let

$$(4.2) \qquad \tilde{F}(t,x) = \{\tilde{z} : z = f(t,x,u),\ z_{n+1} \geqq L(t,x,u),\ u \in U\}.$$

Let us show that the sets $\tilde{F}(t,x)$ are closed if (4.3)(b)(e) below hold. Suppose that $\tilde{z}^j \in \tilde{F}(t,x)$ for $j = 1, 2, \ldots$ and that $\tilde{z}^j \to \tilde{z}$ as $j \to \infty$. Then

$$z^j = f(t,x,u^j), \qquad z^j_{n+1} \geqq L(t,x,u^j)$$

where $u^j \in U$. The sequence $\tilde{z}^j$ is convergent, hence bounded. Hence, $g(u^j) \leqq z^j_{n+1}$ is bounded. By (e), the sequence u^j is also bounded. By taking subsequences we may assume that u^j tends to a limit u as $j \to \infty$. Since U is closed, $u \in U$. Moreover, by continuity of f and L,

$$z = f(t,x,u), \qquad z_{n+1} \geqq L(t,x,u).$$

Thus $\tilde{z} \in \tilde{F}(t,x)$. This proves that $\tilde{F}(t,x)$ is closed.

Theorem 4.1. *Suppose that assumptions* (2.4) *hold, that* L *is continuous, and moreover that:*

(4.3) (a) $\mathscr{F}'$ *is not empty;*

(b) U *is closed;*

(c) S *is compact and* ϕ *is continuous on* S;

(d) $\tilde{F}(t,x)$ *is convex for each* $(t,x) \in E^{n+1}$;

(e) $L(t,x,u) \geqq g(u)$, *where* g *is continuous and* $|u|^{-1} g(u) \to +\infty$ *as* $|u| \to \infty$, $u \in U$.

Then there exist (x_0^*, u^*) *minimizing* $J(x_0, u)$ *on* $\mathscr{F}'$.

We postpone the proof of Theorem 4.1 to § 5.

Some of the assumptions in the theorem can be replaced by others, more easily verified. If U is compact, then (e) holds vacuously.

Corollary 4.1. *In Theorem* 4.1 *any of the assumptions* (c), (d), (e) *can be replaced by the corresponding* (c′), (d′), (e′) *where:*

(c′) *Same as in Corollary* 2.2.

(d′) U *is convex*, $f(t,x,u)=\alpha(t,x)+\beta(t,x)\,u$, $L(t,x,\cdot)$ *is convex on* U.

(e′) $L(t,x,u)\geqq c_1\,|u|^\beta-c_2,\ c_1>0,\ \beta>1$.

Proof. The argument that (c) can be replaced by (c′) is the same as for Corollary 2.2. In Appendix C it is shown that (d′) implies (d). Finally, (e′) implies (e) with $g(u)=c_1\,|u|^\beta-c_2$. □

Note. (e′) implies $u^*(t)$ is integrable to power β.

Example 4.1. Let $\dot{x}=u$, $U=E^n$. When the end conditions are fixed this is the n-dimensional simplest problem in calculus of variations, discussed in I.6, also III.1. In this case

$$\tilde{F}(t,x)=\{\tilde{z}\colon z_{n+1}\geqq L(t,x,z),\ z\in E^n\}.$$

In this example, convexity of $\tilde{F}(t,x)$ is equivalent to convexity of the function $L(t,x,\cdot)$.

If L is linear in u, then (e) cannot hold. For such L we may not expect J to have a minimum on $\mathscr{F}'$ when U is not compact. See remarks in III.1 about the case $L=M(t,x)+N(t,x)\,\dot{x}$.

Note. Minimum problems with linear L and unbounded U often have solutions in terms of controls involving instantaneous impulses of the Dirac "delta function" type. Such an impulse causes the state $x(t)$ to jump instantaneously from one value to another. Impulsive controls should properly be regarded as measures on $[t_0,t_1]$, and not integrable functions. For results in terms of impulsive controls we refer to Lee-Markus [1, p. 279], Rishel [1].

Example 4.2. In the simplest problem of calculus of variations ($n=1$) let $L(t,\dot{x})=t\,\dot{x}^2$, with endpoints (0, 1), (1, 0). The growth condition (e) fails; the difficulty occurs near $t=0$. (If $t_0>0$, then one could take $g(\dot{x})=t_0\,\dot{x}^2$.) $J(x)>0$ for every $x\in\mathscr{X}_e'$. Let $x^r(t)=1-t^{1/r}$, $0\leqq t\leqq 1$, $r=1,2,\ldots$. Since $x^r(0)=1$, $x^r(1)=0$, we have $x^r\in\mathscr{X}_e'$. Moreover, $J(x^r)=(2r)^{-1}$, which tends to 0 as $r\to\infty$. Thus, x^r, $r=1,2,\ldots$, is a minimizing sequence, and $\inf_{\mathscr{X}_e'} J(x)=0$. However, no $x^*\in\mathscr{X}_e'$ with $J(x^*)=0$ exists.

§ 5. Proof of Theorem 4.1

The proof we give is a modification, due to Cesari [1], of the proof of Theorem 2.1. The first step is to find x_0^r, u^r, e^r, x^r as in Lemma 3.2. Since the control set U need no longer be compact, the growth condition (4.3)(e) is used to establish uniform convergence of x^r to x^* and absolute continuity of x^*.

As in § 3 we arrange that all functions u, x are defined on a fixed interval $[T_0,T_1]$, setting $u(t)=u_1$ outside $[t_0,t_1]$. Let $|A|$ denote Lebesgue measure of a set A.

Lemma 5.1. *Given* $v>0$, $\eta>0$ *there exists* $\delta>0$ *such that* $|A|<\delta$ *and* $\int_{T_0}^{T_1} g(u(t))\,dt\leqq v$ *imply* $\int_A |u(t)|\,dt<\eta$.

Proof. By (4.3)(e) there exists θ_B, tending to 0 as $B \to \infty$, such that $|u| \leqq \theta_B\, g(u)$ whenever $|u| > B$, $u \in U$. Choose B large enough that $\theta_B\, v < \frac{1}{2}\eta$, and then choose δ small enough that $B\,\delta < \frac{1}{2}\eta$. Let $A_1 = \{t \in A : |u(t)| \leqq B\}$. Then

$$\int_A |u(t)|\,dt = \int_{A_1} |u(t)|\,dt + \int_{A - A_1} |u(t)|\,dt$$

$$\leqq B\,\delta + \theta_B\, v < \eta. \quad \square$$

A set $\mathscr{X}$ of functions is called *equiabsolutely continuous* on $[T_0, T_1]$ if given $\varepsilon > 0$ there exists $\delta = \delta(\varepsilon) > 0$ with the following property: if $A \subset [T_0, T_1]$ is a finite disjoint union of intervals, namely $A = \bigcup_{i=1}^{l} [s_i, t_i]$ with $t_i < s_{i+1}$ for $i = 1, \ldots, l-1$, and $|A| < \delta$, then

$$\sum_{i=1}^{l} |x(t_i) - x(s_i)| \leqq \varepsilon. \tag{5.1}$$

If $\mathscr{X}$ is an equiabsolutely continuous set and x^r, $r = 1, 2, \ldots$, is a sequence in $\mathscr{X}$ tending uniformly on $[T_0, T_1]$ to a limit x^*, then x^* is absolutely continuous. This follows from the fact that (5.1) holds with $x = x^r$, $r = 1, 2, \ldots$, hence also with $x = x^*$.

Any equiabsolutely continuous set $\mathscr{X}$ is in particular equicontinuous (take $l = 1$).

Lemma 5.2. *Same as Lemma* 3.2 *except read* "x^* *is absolutely continuous*" *instead of* "x^* *satisfies inequalities* (3.1)".

Proof. Since ϕ is continuous on the compact set S, $|\phi(e)| \leqq N$ for suitable N. Let $(\bar{x}_0^j; \bar{u}^j)$, $j = 1, 2, \ldots$ be some minimizing sequence. Then $J(\bar{x}_0^j; \bar{u}^j) \leqq \gamma$ for suitable γ. Since

$$J(\bar{x}_0^j; \bar{u}^j) = \int_{\bar{t}_0^j}^{\bar{t}_1^j} L(t, \bar{x}^j(t), \bar{u}^j(t))\,dt + \phi(\bar{e}^j),$$

$$\bar{e}^j = (\bar{t}_0^j, \bar{t}_1^j, \bar{x}^j(\bar{t}_0^j), \bar{x}^j(\bar{t}_1^j))$$

we have from (4.3)(e) and the fact that $\bar{u}^j(t) = u_1$ outside $[\bar{t}_0^j, \bar{t}_1^j]$,

$$\int_{\bar{t}_0^j}^{\bar{t}_1^j} g[\bar{u}^j(t)]\,dt - N < \gamma,$$

$$\int_{T_0}^{T_1} g[\bar{u}^j(t)]\,dt \leqq v, \quad \text{where} \tag{5.2}$$

$$v = \gamma + N + (T_1 - T_0)\,|g(u_1)|.$$

Let

$$\mathscr{X}_e^v = \left\{ x \in \mathscr{X}_e' : \int_{T_0}^{T_1} g[u(t)]\,dt \leqq v \right\}.$$

Let us show that $\mathscr{X}_e^v$ is an equiabsolutely continuous set. Assumption (4.3)(e) implies that $\int_{T_0}^{T_1} |u(t)|\,dt$ is uniformly bounded whenever $\int_{T_0}^{T_1} g[u(t)]\,dt \leqq v$. By (2.5) and (4.3)(c) there exists M_1 such that $|x(t)| \leqq M_1$ for all $x \in \mathscr{X}_e^v$ and $T_0 \leqq t \leqq T_1$. Thus $\mathscr{X}_e^v$ is uniformly bounded. If A is the union of disjoint intervals $[s_i, t_i]$,

$i=1, \ldots, l$, then by applying (2.5) on each interval $[s_i, t_i]$

$$\sum_{i=1}^{l} |x(t_i)-x(s_i)| \leqq \tilde{C}[|A| + \int_A |u(s)|\, ds],$$

where $\tilde{C}=C(1+M_1)$. Given $\varepsilon>0$ let $\eta=\frac{1}{2}\varepsilon\tilde{C}^{-1}$, and take δ as in Lemma 5.1 small enough that $\tilde{C}\delta \leqq \frac{1}{2}\varepsilon$. Thus $\mathscr{X}_e^v$ is equiabsolutely continuous. The proof of Lemma 5.2 is then completed using Ascoli's theorem in the same way as for its counterpart in § 3. □

The function g in (4.3)(e) is continuous and $g(u)\to+\infty$ as $|u|\to\infty$, $u\in U$. Hence $g(u)\geqq -c$ for some c. Since $L\geqq g$, $L+c\geqq 0$.

Lemma 5.3. *Let*

$$Z^r(t)=\int_{T_0}^{t} L(s, x^r(s), u^r(s))\, ds, \qquad r=1,2,\ldots.$$

The sequences in Lemma 5.2 can be chosen such that $Z^r(t)$ tends to a limit $Z^*(t)$ as $r\to\infty$ for every $t\in[T_0, T_1]$. Moreover, $Z^*(t)+ct$ is a monotone function on $[T_0, T_1]$.

Proof. By the proof of Lemma 5.2,

$$\int_{t_0^r}^{t_1^r} [L(t, x^r(t), u^r(t))+c]\, dt$$

is bounded above since $J(x_0^r, u^r)$ is bounded above. Since $|x^r(t)|\leqq M_1$ for some M_1, $L(t, x^r(t), u_1)$ is bounded. Moreover $L+c\geqq 0$. Therefore, there exists v such that

$$0\leqq \int_{T_0}^{t} [L(s, x^r(s), u^r(s))+c]\, ds \leqq v$$

for $r=1,2,\ldots$, $T_0\leqq t\leqq T_1$. Let

$$\psi^r(t)=Z^r(t)+c(t-T_0).$$

Then ψ^r is monotone on $[T_0, T]$ for each $r=1,2,\ldots$, and

$$0\leqq \psi^r(t)\leqq v.$$

By a theorem of Helly there is a subsequence which converges pointwise to a limit $\psi^*(t)$. See Feller [1, p. 261]. For notational simplicity we denote this subsequence again by ψ^r, and the corresponding sequences of trajectories, controls, and end conditions again by x^r, u^r, e^r. Then $Z^*(t)=\psi^*(t)-ct$ is the required limit. □

For noncompact U the $(n+1)$-dimensional analogues of the sets $\mathcal{O}_a$ in § 3 are no longer convenient. Instead, for fixed (t,x) and $\eta>0$, let

$$\tilde{\mathcal{O}}_\eta = \bigcup_{|s-t|+|x'-x|<\eta} \tilde{F}(s, x'),$$

$$C_\eta = co\, \tilde{\mathcal{O}}_\eta,$$

where co denotes convex hull.

Lemma 5.4. $\tilde{F}(t,x)=\bigcap_{\eta>0}\bar{C}_\eta$.

Proof. Clearly $\tilde{F}(t,x)$ is contained in the right side. Let us show that $\tilde{z}\in\bar{C}_\eta$ for every $\eta>0$ implies $\tilde{z}\in\tilde{F}(t,x)$. Consider such a $\tilde{z}$, and take $\eta=j^{-1}$, $j=1,2,\ldots$. Then

$$\tilde{z}=\lim_{j\to\infty}\tilde{z}^j,\qquad \tilde{z}^j\in C_{j^{-1}}.$$

Since $\tilde{\mathcal{O}}_\eta\subset E^{n+1}$, every point of its convex hull C_η is a convex combination of $n+2$ (or fewer) points of $\tilde{\mathcal{O}}_\eta$. See Fleming [1, p. 20]. Thus

$$\tilde{z}^j=\sum_{i=0}^{n+1}\lambda_i^j\tilde{z}_i^j,\qquad \text{with}$$

$$0\leqq\lambda_i^j,\qquad \sum_{i=0}^{n+1}\lambda_i^j=1,\qquad \tilde{z}_i^j\in\tilde{F}(t_i^j,x_i^j),$$

where $|t_i^j-t|+|x_i^j-x|<j^{-1}$. Thus $\tilde{z}_i^j=(z_i^j,z_{n+1,i}^j)$ where

$$z_i^j=f(t_i^j,x_i^j,u_i^j)$$
$$z_{i,n+1}^j\geqq L(t_i^j,x_i^j,u_i^j),\qquad u_i^j\in U.$$

By taking subsequences, we may assume that λ_i^j tends to a limit λ_i as $j\to\infty$. Clearly,

$$(5.3)\qquad 0\leqq\lambda_i,\qquad \sum_{i=0}^{n+1}\lambda_i=1.$$

We may also assume, by taking further subsequences, that for each $i=0,1,\ldots,n+1$, either u_i^j tends to a limit u_i or that $|u_i^j|\to+\infty$ as $j\to\infty$. We may suppose that the first occurs for $i=0,1,\ldots,l$, and the second for $i=l+1,\ldots,n+1$. Since U is closed, $u_i\in U$ for $i=0,1,\ldots,l$. Let us show that $\tilde{z}=(z,z_{n+1})$, where

$$(5.4)\qquad \text{(a)}\quad z=\sum_{i=0}^{l}\lambda_i f(t,x,u_i),$$

$$\text{(b)}\quad z_{n+1}\geqq\sum_{i=0}^{l}\lambda_i L(t,x,u_i).$$

Now

$$(5.5)\qquad z_{n+1}^j\geqq\sum_{i=0}^{n+1}\lambda_i^j L(t_i^j,x_i^j,u_i^j)\geqq\sum_{i=0}^{n+1}\lambda_i^j g(u_i^j),$$

and z_{n+1}^j tends to the finite limit z_{n+1} as $j\to\infty$. Therefore λ_i^j must tend to 0 as $j\to\infty$ for those i such that $g(u_i^j)\to+\infty$.

By (4.3)(e), $\lambda_i=0$ for $l+1,\ldots,n+1$. Also, for such i we have $g(u_i^j)>0$ for all large j. Hence, from (5.5), we get (5.4)(b). Now

$$z^j=\sum_{i=0}^{n+1}\lambda_i^j f(t_i^j,x_i^j,u_i^j)$$

tends to z as $j\to\infty$. For $i>l$ and j large enough that $g(u_i^j)>0$, we have by (2.4)

$$0\leqq\frac{|f(t_i^j,x_i^j,u_i^j)|}{L(t_i^j,x_i^j,u_i^j)}\leqq\frac{C_1(1+|x_i^j|+|u_i^j|)}{g(u_i^j)}.$$

The right side tends to 0 as $j\to\infty$, for $i=l+1,\dots,n+1$, by (4.3)(e). By (5.5) $\lambda_i^j g(u_i^j)$ is bounded. Hence

$$\lim_{j\to\infty}\lambda_i^j f(t^j,x^j,u_i^j)=0,\qquad i=l+1,\dots,n+1.$$

This proves (5.4)(a). Choose $v_0,\dots,v_l$ such that each $v_i\geqq 0$ and

$$z_{n+1}=\sum_{i=0}^{l}\lambda_i[L(t,x,u_i)+v_i].$$

Then $\tilde z_i=(f(t,x,u_i),L(t,x,u_i)+v_i)$ is in $\tilde F(t,x)$. By (5.3) and (5.4),

$$\tilde z=\sum_{i=0}^{l}\lambda_i\tilde z_i$$

is a convex combination of points of $\tilde F(t,x)$. Since $\tilde F(t,x)$ is convex by (4.3), $\tilde z\in\tilde F(t,x)$. □

Lemma 5.5. *Let $\tilde x^r=(x^r,Z^r)$, $r=1,2,\dots$ and $\tilde x^*=(x^*,Z^*)$. Then $\dot{\tilde x}^*(t)\in\tilde F(t,x^*(t))$ for each $t\in[T_0,T_1]$ where $\dot{\tilde x}^*(t)$ exists.*

Proof. The proof is almost the same as for Lemma 3.3. Given $\eta>0$ there exist $\delta_\eta>0$ and r_η such that

$$\tilde z^r(s)=(f(s,x^r(s),u^r(s)),L(s,x^r(s),u^r(s)))\in\tilde{\mathcal{O}}_\eta$$

whenever $|s-t|<\delta_\eta$ and $r\geqq r_\eta$. Then

$$\frac{\tilde x^r(t+h)-\tilde x^r(t)}{h}=\frac{1}{h}\int_t^{t+h}\tilde z^r(s)\,ds\in\bar C_\eta$$

for $r\geqq r_\eta$, $|h|<\delta_\eta$. Let $r\to\infty$, then let $h\to 0$. We get $\dot{\tilde x}^*(t)\in\bar C_\eta$, for each $\eta>0$. By Lemma 5.4, $\dot{\tilde x}^*(t)\in\tilde F(t,x)$. □

Since $x^*(t)$ is absolutely continuous and $\psi^*(t)=Z^*(t)+ct$ is monotone, their derivatives $\dot x^*(t)$ and $\dot Z^*(t)+c$ exist for almost all $t\in[T_0,T_1]$.

Lemma 5.6. *There exist an integrable function u^* and a measurable function v^* such that*

(5.6)
(a) $\dot x^*(t)=f(t,x^*(t),u^*(t))$
(b) $\dot Z^*(t)=L(t,x^*(t),u^*(t))+v^*(t),\qquad v^*(t)\geqq 0,$

almost everywhere in $[T_0,T_1]$.

Proof. Let $\tilde U=U\times[0,\infty)$, and for $\tilde u=(u,v)\in\tilde U$ let

$$\tilde f(t,x,\tilde u)=(f(t,x,u),L(t,x,u)+v).$$

Then $\tilde{F}(t,x)$ is the image of $\tilde{U}$ under the function $\tilde{f}(t,x,\cdot,\cdot)$. By a selection lemma in Appendix B and Lemma 5.5 there exist measurable functions u^*, v^* satisfying (5.6) almost everywhere, with $u^*(t)\in U$, $v^*(t)\geqq 0$. Since ψ^* is monotone,

$$\psi^*(T_1)-\psi^*(T_0)\geqq \int_{T_0}^{T_1}\dot{\psi}^*(t)\,dt.$$

By (4.3)(e) and (5.6)(b),

$$\dot{\psi}^*(t)\geqq g[u^*(t)]+c\geqq 0.$$

Thus $g[u^*(t)]$ is (Lebesgue) integrable on $[T_0,T_1]$. This implies that u^* is also integrable there. □

Proof of Theorem 4.1. Recall the notation in the proof of Theorem 2.1. We have

$$\mu=\lim_{r\to\infty} J(x_0^r,u^r)=\lim_{r\to\infty}[Z^r(t_1^r)-Z^r(t_0^r)+\phi(e^r)].$$

Moreover, $(x_0^*,u^*)\in\mathscr{F}'$. Since $\mu\leqq J(x_0,u)$ for all $(x_0,u)\in\mathscr{F}'$, in particular for (x_0^*,u^*), the proof that (x_0^*,u^*) minimizes will be completed by showing that $J(x_0^*,u^*)\leqq\mu$. Since $e^r\to e^*$, $\phi(e^r)\to\phi(e^*)$ as $r\to\infty$. Consider t_0,t_1 with $t_0^*<t_0<t_1<t_1^*$. For large r, $t_0^r<t_0$ and $t_1<t_1^r$. Since $\psi^r=Z^r+ct$ is monotone, for large r

$$\psi^r(t_1^r)-\psi^r(t_0^r)\geqq\psi^r(t_1)-\psi^r(t_0).$$

After subtracting $c(t_1^r-t_0^r)$ from both sides we get since $t_i^r\to t_i^*$ and $\psi^r(t_i)\to\psi^*(t_i)$, $i=0,1$,

$$\lim_{r\to\infty}[Z^r(t_1^r)-Z^r(t_0^r)]\geqq\psi^*(t_1)-\psi^*(t_0)-c(t_1^*-t_0^*).$$

However, using (5.6)(b)

$$\psi^*(t_1)-\psi^*(t_0)\geqq\int_{t_0}^{t_1}\dot{\psi}^*(t)\,dt\geqq\int_{t_0}^{t_1}L(t,x^*(t),u^*(t))\,dt+c(t_1-t_0),$$

$$\lim_{r\to\infty}[Z^r(t_1^r)-Z^r(t_0^r)]\geqq\int_{t_0}^{t_1}L(t,x^*(t),u^*(t))\,dt-c[(t_1^*-t_1)+(t_0-t_0^*)].$$

Now let $t_i\to t_i^*$, $i=1,2$. We get

$$\mu\geqq\int_{t_0^*}^{t_1^*}L(t,x(t),u^*(t))\,dt+\phi(e^*)=J(x_0^*,u^*).\quad\square$$

§ 6. Continuity Properties of Optimal Controls

To obtain Theorems 2.1 and 4.1 about existence of an optimal control, we had to enlarge the class of piecewise continuous control functions to include Lebesgue-integrable u. There remains the question whether under reasonable general conditions on the problem there is a piecewise continuous optimal control u^*. It is rather easy to find conditions which guarantee that an optimal u^*

is continuous (Theorem 6.2). These conditions are satisfied in many problems. However, they do not hold for many others of interest in which an optimal u^* is in fact usually discontinuous. An example is the problem of linear time-optimal control. In the second part of the section we state, but do not prove, a result (Theorem 6.3) about existence of piecewise continuous controls. This theorem concerns problems in Mayer form with f linear in the variables x.

The results in the present section are derived from the necessary condition II(5.2) in Pontryagin's principle. Thus we deal with u^* satisfying this necessary condition, but not necessarily minimizing.

Let us assume that f, L are of class C^1. Suppose that $(x_0^*, u^*)\in\mathscr{F}'$, u^* being defined on an interval $[t_0^*, t_1^*]$. Moreover, we suppose that there exists an absolutely continuous vector function $P=(P_1, \dots, P_n)$ on $[t_0^*, t_1^*]$ such that, for almost all $t\in[t_0^*, t_1^*]$,

$$\dot{P}(t)' = -P(t)' f_x(t, x^*(t), u^*(t)) + L_x(t, x^*(t), u^*(t)), \tag{6.1}$$

$$h(t, u) \leqq h(t, u^*(t)) \quad \text{for all } u\in U, \quad \text{where} \tag{6.2}$$
$$h(t, u) = P(t)' f(t, x^*(t), u) - L(t, x^*(t), u).$$

As usual, x^* is the solution of (2.1) corresponding to (x_0^*, u^*). It is assumed that $L_x(t, x^*(t), u^*(t))$ is integrable on $[t_0^*, t_1^*]$. The integrability of $f_x(t, x^*(t), u^*(t))$ follows from (2.4)(b), which for f of class C^1 is equivalent to $|f_x| \leqq C_2(1+|u|)$, and integrability of u^*.

If the minimum problem is rephrased in Mayer form by introducing a new system component x_{n+1}, obeying $\dot{x}_{n+1} = L(t, x(t), u(t))$ as explained in II.4, then in the Pontryagin necessary conditions II.5 we have taken $P_{n+1} = -1$. This is equivalent to assuming $P_{n+1} < 0$, since $P(t)$ is determined only up to a positive scalar multiple. From II (5.3), $P_{n+1} = \lambda_1$, where λ_1 is the first multiplier component in the abstract multiplier rule from which Pontryagin's principle was derived in II.15.

The function u^* can be arbitrarily redefined on a set of measure 0. When we state that u^* is continuous on an interval I we mean that u^* agrees almost everywhere with a function continuous on I.

Theorem 6.1. *Let* (6.1), (6.2) *hold, and let U be compact. Let I be an interval such that $u^*(t)$ is the unique point $u\in U$ at which $h(t, u)$ is maximum for all $t\in I$. Then u^* is continuous on I.*

Proof. Given $t\in I$, let t_j, $j=1, 2, \dots$, be any sequence in I tending to t. Since U is compact, for a subsequence $u^*(t_j)$ tends to a limit u'. For notational simplicity, we denote this sequence again by t_j, $j=1, 2, \dots$. It suffices to show that $u' = u^*(t)$. The function h is continuous on $I\times U$. Moreover,

$$h(t_j, u) \leqq h(t_j, u^*(t_j)) \quad \text{for all } u\in U.$$

Upon taking limits we get

$$h(t, u) \leqq h(t, u') \quad \text{for all } u\in U.$$

By assumption, $h(t, \cdot)$ is maximum on U only at $u^*(t)$. Thus $u' = u^*(t)$. □

Example 6.1. Suppose that $L=0$ (the Mayer form) and $f=A(t)x+B(t)u$. When $\varphi(e)=t_1-t_0$ this is the linear time-optimal control problem. Equations (6.1) become $\dot{P}'=-P'A(t)$, which are coupled with the state equations $\dot{x}=f$ only through the transversality conditions. Condition (6.2) is equivalent to the statement that the linear function $P'(t)B(t)u$ has a maximum on U at $u^*(t)$. Consider the particular control set $U=\{u: |u_i|\leqq 1$ for $i=1,\dots,m\}$, an m-dimensional cube. Let $b_i(t)$ denote the i-th column vector of the matrix $B(t)$. Then

$$P'(t)B(t)u=\sum_{i=1}^{m} P'(t)b_i(t)u_i.$$

If $P'(t)b_i(t)\neq 0$, then $u_i^*(t)=\pm 1$. Therefore, if for each $i=1,\dots,m$, $P'(t)b_i(t)=0$ for only finitely many t, then u^* is piecewise constant on $[t_0^*, t_1^*]$. This example is discussed further at the end of the section.

Let us next state a theorem giving conditions under which u^* is continuous at all points of $[t_0^*, t_1^*]$. It is no longer assumed that U is compact. For that reason condition (1) in the theorem is imposed.

Theorem 6.2. *Let* (6.1), (6.2) *hold, and also:*

(1) L *satisfies* (4.3)(e);

(2) $-h(t,\cdot)$ *is a strictly convex function on* U.

Then u^* *is continuous on* $[t_0^*, t_1^*]$.

Proof. The functions x^*, P are continuous on $[t_0^*, t_1^*]$, hence bounded there. By (2.4)(a) and (4.3)(e),

$$|P(t)'f(t,x^*(t),u)|\leqq C(1+|u|),$$
$$-h(t,u)\geqq g(u)-C(1+|u|)$$

for some constant C. Then $-|u|^{-1}h(t,u)\to\infty$ as $|u|\to\infty$, $u\in U$. This together with (2) imply that $h(t,\cdot)$ has a maximum on U at a unique $u(t)$, for every $t\in[t_0^*, t_1^*]$. By (6.2), $u(t)=u^*(t)$ almost everywhere in $[t_0^*, t_1^*]$; we may redefine $u^*(t)$ to be $u(t)$ on the remaining null set in $[t_0^*, t_1^*]$.

Fix some $u_1\in U$. Then, for some constant M, $|h(t,u_1)|\leqq M$. Hence

$$-h(t,u^*(t))\leqq -h(t,u_1)\leqq M,$$
$$g[u^*(t)]-C(1+|u^*(t)|)\leqq M.$$

Since $u^{-1}g(u)\to\infty$ as $|u|\to\infty$, this implies

$$|u^*(t)|\leqq B, \quad t_0^*\leqq t\leqq t_1^*,$$

for some constant B. Let $U_B=U\cap\{|u|\leqq B\}$. By Theorem 6.1, with U replaced by U_B, u^* is continuous on $[t_0^*, t_1^*]$. □

We note that (2) holds if $f(t,x,u)=\alpha(t,x)+\beta(t,x)u$ is linear in u and $L(t,x,\cdot)$ is strictly convex on U. A sufficient condition for strict convexity is L_{uu} positive definite. The control u^* in Theorem 6.2 is not generally C^1. However, if there are no control constraints, then more can be said:

Corollary 6.1. *Suppose that* $U = E^m$ *and that* L, f *are of class* C^r, $r \geqq 2$. *Moreover, instead of* (2) *assume in Theorem* 6.2:

(2') *The matrix* $L_{uu}(t, x^*(t), u)$ *is positive definite for all* $u \in U$, *and* $f(t, x^*(t), \cdot)$ *is linear for all* $t \in [t_0^*, t_1^*]$.
Then u^* *is of class* C^{r-1} *and* x^* *of class* C^r *on* $[t_0^*, t_1^*]$.

The proof is entirely similar to that for Corollary I.3.3, and hence is omitted.

Let us give some conditions under which (6.1) and (6.2) hold when (x_0^*, u^*) minimizes $J(x_0, u)$.

A) U is compact. Rewrite the problem in Mayer form, as in II.4. Pontryagin's principle was proved for piecewise continuous controls in Chap. II. Using nearly the same proof, the existence of P satisfying the Pontryagin necessary conditions II (5.1), II (5.2) almost everywhere in $[t_0^*, t_1^*]$ can be proved for a bounded measurable control u^* such that $J(x_0^*, u^*)$ minimizes $J(x_0, u)$. If $\lambda_1 \neq 0$, these conditions imply (6.1), (6.2). In fact, (6.2) holds at points where u^* is approximately continuous.

Note. These necessary conditions have been extended to unbounded control sets U. See McShane [1]. They do not include the transversality conditions II. (5.3)–(5.6), which must be rephrased when u^* is measurable but not piecewise continuous. Moreover, a different proof of transversality is needed.

When A) holds one can sometimes apply Theorem 6.1 to get continuity of u^* on certain subintervals of $[t_0^*, t_1^*]$ as was done for Example 6.1. Let us turn to two cases when Theorem 6.2 applies.

B) U compact, t_0, x_0, t_1 fixed, $x(t_1)$ free. Suppose that $f(t, x, u) = \alpha(t, x) + \beta(t, x) u$ and $L(t, x, \cdot)$ is strictly convex on U. Then u^* is continuous on $[t_0^*, t_1^*]$. A slight adaptation of the proof of Pontryagin's principle in the special case II.11 gives (6.1), (6.2). Condition (1) in Theorem 6.2 holds vacuously since U is compact, while our assumptions on f, L imply (2).

Example 6.2. (Saturated linear regulator). In the linear regulator problem Example II.2.3 impose control constraints $|u(t)| \leqq 1$. Unlike the linear regulator, there is no longer an explicit solution. However, the problem is of the type B). The optimal control u^* is continuous, but not generally C^1. In this problem J is a strictly convex function of u (Problem 4). By Theorem I.2.4, the optimal u^* is unique. Existence of an optimal u^* follows from Theorem 4.1.

C) The simplest problem in calculus of variations, in n dimensions. Suppose that $L(t, x, \dot{x})$ is regular and satisfies

$$L(t, x, \dot{x}) \geqq g(\dot{x}), \quad \text{where} \quad |u|^{-1} g(u) \to \infty \quad \text{as } |u| \to \infty,$$

$$|L_x| + |L_{\dot{x}}| \leqq C(1 + L) \quad \text{for some constant } C.$$

By Theorem 4.1 there exists an absolutely continuous function x^* minimizing $J(x)$ subject to given (fixed) end conditions.

A straightforward extension of Lemma I.3.2 shows that $P(t) = \int_{t_0}^{t} L_x \, d\tau + c$ satisfies, for suitable c, $-P(t) + L_{\dot{x}}(t, x^*(t), \dot{x}^*(t)) = 0$ almost everywhere in $[t_0, t_1]$. For L regular this is equivalent to (6.1), (6.2). By Corollary 6.1, x^* is C^r if L is C^r.

Conditions for Piecewise Continuity of u^*. Unfortunately there are few general results available insuring piecewise continuity of optimal controls. However, a good deal is known when the problem is in Mayer form ($L=0$) and

$$f(t,x,u)=A(t)\,x+B(t)\,u \tag{6.3}$$

is linear. Even in this special class difficulties arise. The Pontryagin minimum condition (6.2) does not always effectively determine u^*. There may be many optimal controls, not all of which can be piecewise continuous. (See for instance Problem 5b). In such cases one would like to show that some optimal control exists with u^* piecewise continuous, not that all such u^* are piecewise continuous. Unfortunately, even if A, B are C^∞ functions and the optimal u^* is unique, it still does not generally follow that u^* is piecewise continuous. See Problem 5d. One is led therefore to make the stronger assumption that A, B are analytic functions.

It is sometimes rather easy under further assumptions to establish piecewise continuity of u^*. To illustrate this let us return to Example 6.1. If A and B are analytic, then $P'B$ is also analytic since P satisfies the differential equation $\dot{P}'=-P'A$. We showed that u^* is piecewise constant if:

$$(*) \qquad P'(t)\,b_i(t)=0 \quad \text{for only finitely many } t, \text{ for each } i=1,\dots,m.$$

Since $P'B$ is analytic, either $(*)$ holds or $P'(t)\,b_i(t)\equiv 0$ for some i. In particular, suppose that A, B are constants (independent of t). If $P'(t)b_i\equiv 0$, then

$$0=\frac{d^j}{dt^j}(P'b_i)=(-1)^j P'A^j b_i, \qquad j=0,1,2,\dots.$$

In particular, the vector P' annihilates $b_i, Ab_i, \dots, A^{n-1}b_i$, which implies that these n vectors are linearly dependent. Thus, for constant A, B, the following is a sufficient condition for $(*)$:

$$(**) \qquad b_i, Ab_i, \dots, A^{n-1}b_i$$

are linearly independent vectors, for each $i=1,\dots,m$. For a discussion of the relationship of $(**)$ to the idea of controllability and to uniqueness of optimal controls we refer the reader to Hermes-LaSalle [1, §'s 15–17].

The following deeper result was proved by Halkin [1] and also by Levinson [1]. Suppose that A, B are analytic and U is a compact polyhedron. Then an optimal $(x_0^*, u^*)\in\mathscr{F}$ exists with $u^*(t)$ piecewise constant, with vertices of U as values of $u^*(t)$. This result was generalized by Halkin and Hendricks [1].

Problems—Chapter III

(1) Let $n=2$, $U=[-1,1]$, $\dot{x}_1=\cos\pi u$, $\dot{x}_2=\sin\pi u$,

$$J(u)=\int_0^1 [x_1(t)^2+x_2(t)^2]\,dt,$$

with $x(0)=0$ and $x(1)$ free. Show that the infimum of $J(u)$ is 0, but no minimizing u^* exists. Which assumption in Theorem 2.1 fails to hold?[1]

(2) Consider the following problem of linear regulator type: $\dot{x}=Ax+Bu$, with $x(\cdot)$ and $u(\cdot)$ scalar valued functions on $[0, 1]$, $U=E^1$, and

$$J(u)=\int_0^1 t[x(t)^2+u(t)^2]\,dt.$$

Here $x(0)=x_0 \neq 0$, with x_0 fixed, and $x(1)$ is free. Show that no Lebesgue integrable minimizing u^* exists. Which assumption in Theorem 4.1 fails to hold? *Hints.* Use the necessary conditions II(7.8), II(7.9). An optimal $x^*(t)$ cannot change sign.

(3) Let $\dot{x}=u$, $J(u)=\int_0^1 [1+u(t)^2]^{1/4}\,dt$, $U=E^1$, and fixed end conditions $x(0)=x(1)=0$. By inspection $u^*(t)\equiv 0$ is optimal. Show that there is a minimizing sequence $u^1, u^2, \ldots$ such that the corresponding $x^r(t)=\int_0^t u^r(v)\,dv$ does not converge to $x^*(t)\equiv 0$ as $r\to\infty$.

(4) In Example 6.2, prove that $J(u)$ is a strictly convex function of u.

(5) Let $n=1$, $\dot{x}=B(t)u$, $U=[-1, 1]$, $0\leqq t\leqq 1$, $x(0)=x_0$, $x(1)$ free and $J(u)=x(1)^2$. Show that:

(a) The set of possible values of $x(1)$ is a compact interval I.

(b) If 0 is an interior point of I, then there are infinitely many optimal u^* some of which are not piecewise continuous.

(c) If I is contained in the open half line $x<0$, then any optimal control satisfies $u^*(t)=|B(t)|^{-1}B(t)$ for almost all $t\in[0, 1]$ such that $B(t)\neq 0$.

(d) Let $B(t)=\sin(t^{-1})\exp(-t^{-1})$ for $t>0$, and $B(0)=0$. Then $B(t)$ is C^∞ on $[0, 1]$. However, the unique optimal u^* is not piecewise continuous, if I is as in (c).

(6) In Problem I.7 suppose that $g''(x)\geqq c_1>0$ and $g(x)\geqq -c_2$. Use Corollary 4.1 and Corollary 6.1 to show that the two point boundary problem $\ddot{x}=g'(x)$ with $x(t_0)=x_0$, $x(t_1)=x_1$ has a solution.

(7) Suppose that $f(t,x,u)$ is of class C^1 and that $|f(t,0,0)|\leqq C$, $|f_x(t,x,u)|\leqq C(1+|u|)$, $|f_u(t,x,u)|\leqq C$ for some constant C. Show that (2.4) holds for $C_1=C_2=C$. *Hint.* Apply the mean value theorem to $f(t,x,u)-f(t,x,0)$, $f(t,x,0)-f(t,0,0)$, and to $f(t,x',u)-f(t,x,u)$.

[1] When x_0 is fixed, we write $J(x_0,u)=J(u)$.

Chapter IV. Dynamic Programming

§ 1. Introduction

In Chap. II optimality problems were studied through differential properties of mappings into the space of controls. The method of Dynamic Programming takes a different approach. In Dynamic Programming a family of fixed initial point control problems is considered. The minimum value of the performance criterion is considered as a function of this initial point. This function is called the value function. Whenever the value function is differentiable it satisfies a first order partial differential equation called the partial differential equation of dynamic programming.

A sufficient condition for optimality can be phrased in terms of a continuously differentiable solution of the partial differential equation of dynamic programming. The linear regulator problem is solved by this method in § 5.

The method of Dynamic Programming encounters the difficulty that for many problems the value function is not differentiable everywhere. In § 6 we consider feedback controls whose discontinuities have a special form. Feedback controls of this type are called admissible.

This concept leads to Theorem 7.2 which asserts roughly that a necessary and sufficient condition for the existence of an optimal admissible feedback control is that the performance function of this control be a solution of the partial differential equation of dynamic programming except on a lower dimensional set.

The "method of characteristics" is a method for solving first order partial differential equations. In § 8 an analogue of this method is investigated for the partial differential equation of dynamic programming. The characteristic ordinary differential equations for the partial differential equation of dynamic programming are the equations of Pontryagin's principle. When there is an admissible feedback control which satisfies Pontryagin's principle, these "characteristic equations" can be integrated to show that the performance function of this control is a solution of the partial differential equation of dynamic programming except on a lower dimensional set. This gives the connection between Dynamic Programming and Pontryagin's principle.

A corresponding approach to Dynamic Programming concerned with the simplest problem of Calculus of Variations is called Hamilton-Jacobi theory. This is extensively discussed in Caratheodory [1]. Controls in feedback form when specialized to the simplest problem of Calculus of Variations also correspond to a classical concept. They correspond to the notion of a field as discussed for instance in Bliss [1][2].

§ 2. The Problem

Consider an optimization problem similar to the type considered in Chap. II and III. However now let the initial time t_0 and state x_0 be fixed and the other end conditions involve only the final time t_1 and state x_1. Let these terminal conditions be specified by the requirement

$$(t_1, x_1) \in M \tag{2.1}$$

where M is a closed subset of E^{n+1}. Call M the *terminal set*. (In Chap. III.2, M was denoted by S_1.) Later in § 6–8 we shall require that M be a smooth manifold.

Given initial data (t_0, x_0), let $\mathscr{F}_{t_0 x_0}$ be the class of piecewise continuous controls, defined on some interval $[t_0, t_1]$, beginning at the fixed time t_0, which are feasible for x_0. As before feasibility asserts that there is a corresponding solution of the differential Eqs. II (3.1) on $[t_0, t_1]$ with $x(t_0) = x_0$ and $(t_1, x(t_1)) \in M$. Hence the notation $\mathscr{F}_{t_0 x_0}$ is consistent with the usage of the symbol $\mathscr{F}$ in Chaps. II and III. Assume a function $\phi_1(t_1, x(t_1))$ of the terminal time and state is given as the performance index. The optimization problem is to minimize the performance index over all controls $u = u(\cdot)$ in $\mathscr{F}_{t_0 x_0}$.

§ 3. The Value Function

In this chapter it will be often convenient to denote the initial time and state of the optimization problem by (s, y) rather than (t_0, x_0). For the optimization problem (s, y) is considered fixed. However if we consider a family of optimization problems with different initial conditions (s, y), we can consider the dependence of the value of these optimization problems on their initial conditions. Thus define a value function by

$$V(s, y) = \inf_{u \in \mathscr{F}_{sy}} \phi_1(t_1, x(t_1)). \tag{3.1}$$

In (3.1) we shall follow the convention that $V(s, y) = \infty$ if $\mathscr{F}_{sy}$ is empty. With this convention the value function is an extended real valued function which may take on both the values plus and minus infinity. The method of Dynamic Programming, initiated by Bellman [2], studies the properties of this function.

As a first property of the value function we state:

Theorem 3.1. *The value function evaluated along any trajectory corresponding to a control feasible for its initial state is a nondecreasing function of time.*

Proof. Let u defined on $[t_0, t_1]$ be feasible for x_0 and x the corresponding solution II (3.1) We must show, for $t_0 \leqq \tau_1 \leqq \tau_2 \leqq t_1$, that

$$V(\tau_1, x(\tau_1)) \leqq V(\tau_2, x(\tau_2)). \tag{3.2}$$

For any t such that $t_0 \leqq t \leqq t_1$ the set $\mathscr{F}_{t\,x(t)}$ is not empty since the restriction of u to the interval $[t, t_1]$ is feasible for $x(t)$. Let $\tilde{u}$ be any control in $\mathscr{F}_{\tau_2\,x(\tau_2)}$. Define the control $u^\dagger$ on $[\tau_1, \tilde{t}_1]$ by

$$u^\dagger(r) = \begin{cases} u(r) & \text{if } \tau_1 \leqq r \leqq \tau_2 \\ \tilde{u}(r) & \text{if } \tau_2 \leqq r \leqq \tilde{t}_1. \end{cases}$$

Then $u^\dagger \in \mathscr{F}_{\tau_1\,x(\tau_1)}$. Hence

$$V(\tau_1, x(\tau_1)) \leqq \phi_1(\tilde{t}_1, x^\dagger(\tilde{t}_1)).$$

Since $\tilde{u}$ was any control in $\mathscr{F}_{\tau_2\,x(\tau_2)}$ taking the infimum over the controls in $\mathscr{F}_{\tau_2\,x(\tau_2)}$ gives

$$V(\tau_1, x(\tau_1)) \leqq V(\tau_2, x(\tau_2)). \quad \square$$

A corresponding result to Theorem 3.1 for optimal controls is:

Theorem 3.2. *The value function evaluated along any optimal trajectory is constant.*

Proof. Let u^* defined on $[t_0, t_1]$ be an optimal control for the problem starting from x_0. By Theorem 3.1 $V(t, x^*(t))$ is nondecreasing. Since for $t_0 \leqq t \leqq t_1$, u^* restricted to $[t, t_1]$ is a feasible control for $x^*(t)$

$$V(t, x^*(t)) \leqq \phi_1(t_1, x^*(t_1)).$$

The optimality of u^* for the problem with initial conditions t_0, x_0 implies

$$V(t_0, x_0) = \phi_1(t_1, x^*(t_1)).$$

Since $V(t, x^*(t))$ is nondecreasing we must have $V(t, x^*(t)) \equiv \phi_1(t_1, x^*(t_1))$ on $t_0 \leqq t \leqq t_1$. $\square$

Notice that the properties of the value function described in Theorems 3.1 and 3.2 are necessary conditions for optimality. It is interesting to ask if these properties are sufficient conditions for optimality. Theorem 3.3 asserts that this is so.

Theorem 3.3. *Let $W(s, y)$ be an extended real valued function defined on E^{n+1} such that $W(s, y) = \phi_1(s, y)$ if $(s, y) \in M$. Let t_0, x_0 be given initial conditions, and suppose, for each trajectory x corresponding to a control u in $\mathscr{F}_{t_0 x_0}$, that $W(t, x(t))$ is finite and nondecreasing on $[t_0, t_1]$. If u^* is a control in $\mathscr{F}_{t_0 x_0}$ such that for the corresponding trajectory x^*, $W(t, x^*(t))$ is constant then u^* is an optimal control and $W(t_0, x_0) = V(t_0, x_0)$.*

Proof. For any control u in $\mathscr{F}_{t_0 x_0}$

$$W(t_0, x_0) \leqq \phi_1(t_1, x(t_1)).$$

For the control u^*,

$$W(t_0, x_0) = \phi_1(t_1, x^*(t_1^*))$$

so u^* is optimal and $W(t_0, x_0) = V(t_0, x_0)$. $\square$

Corollary 3.1. *Let u^* be an optimal control in $\mathscr{F}_{t_0 x_0}$ and x^* the corresponding trajectory of* II (3.1), *then the restriction of u^* to $[t, t_1]$ is an optimal control for each control problem with initial conditions $(t, x^*(t))$ when $t_0 \leqq t \leqq t_1$.*

Proof. The value function $V(s, y)$ satisfies the conditions of Theorem 3.3 for each control problem with initial conditions $(t, x^*(t))$ for $t_0 \leqq t \leqq t_1$. □

So far by these simple remarks we have established that a necessary and sufficient condition for a control to be optimal is that there exist a function $W(s, y)$ defined on E^{n+1} such that: $W(s, y) = \phi_1(s, y)$ on the terminal set M, $W(s, y)$ is constant on the given trajectory, and $W(s, y)$ is nondecreasing on any other trajectory.

Immediately one asks "How practical is it to check whether a function is nondecreasing on each of a given set of trajectories? Can this requirement for checking be reduced to checking a smaller set of conditions?" Another question is "Can a systematic method be found for constructing the functions $W(s, y)$ which are the candidates for value functions?" The main results of this chapter will try to answer these questions.

§ 4. The Partial Differential Equation of Dynamic Programming

To begin to answer these questions, the next theorem asserts, at points at which the value function is "well behaved", that partial differential inequalities or equations hold. To formulate this theorem consider the following concept. Define

$$Q_0 = \{(s, y) \in E^{n+1}: \mathscr{F}_{sy} \neq \emptyset\}. \tag{4.1}$$

That is Q_0 is the set of points (s, y) from which it is possible to reach the terminal set M with some trajectory corresponding to a piecewise continuous control. Let us call Q_0 the *reachable set.*

We call a function $V(s, y)$ *differentiable* at an interior point (s_0, y_0) of its domain if there exist a scalar a and a vector b such that

$$\lim_{(s, y) \to (s_0, y_0)} (|s - s_0| + |y - y_0|)^{-1} \, |V(s, y) - V(s_0, y_0) - a(s - s_0) - b'(y - y_0)| = 0.$$

Differentiability of V is equivalent to the existence of a tangent plane at (s_0, y_0). The composite of two differentiable functions is differentiable. A sufficient condition for differentiability is that V have continuous first order partial derivatives. For a discussion of these facts see for instance Fleming [1].

Theorem 4.1. *Let (s, y) be any interior point of the reachable set Q_0 at which the function $V(s, y)$ is differentiable. Then $V(s, y)$ satisfies the partial differential inequality*

$$V_s + V_y f(s, y, v) \geqq 0, \tag{4.2}$$

for all $v \in U$.

If there is an optimal control u^* in $\mathscr{F}_{s,y}$, then the partial differential equation

$$\min_{v \in U} \{V_s + V_y f(s, y, v)\} = 0 \tag{4.3}$$

is satisfied. The minimum in (4.3) *is achieved by the right limit* $u^*(s)^+$ *of the optimal control at* s.

Proof. Since (s, y) is an interior point of Q_0, if an arbitrary constant control $v \in U$ is used over an interval $[s, s+k]$, and k is small enough, the corresponding trajectory $x(t)$ will lie in Q_0 for $s \leqq t \leqq s+k$. If $\tilde{u}$ is a control defined on $[s+k, t_1]$ feasible for $x(s+k)$, the control u_k defined on $[s, t_1]$ by

$$u_k(t) = \begin{cases} v & s \leqq t \leqq s+k \\ \tilde{u}(t) & s+k \leqq t \leqq t_1 \end{cases}$$

is in $\mathscr{F}_{sy}$. Let x_k denote the solution of II(3.1) with initial condition $x(s)=y$ corresponding to u_k. Let $D^+ g(t)$ denote the right derivative of a function $g(t)$. Since u_k is piecewise continuous

$$D^+ x_k(t) = f(t, x_k(t), u_k(t)^+)$$

where $u_k(t)^+$ is the limit from the right of u_k at t. From Theorem 3.1 $V(t, x_k(t))$ is nondecreasing, hence $D^+ V(t, x_k(t)) \geqq 0$ for any value of t for which this derivative exists. Hence computing this derivative at $t=s$ using the chain rule for differentiation we have

$$V_s + V_y f(s, y, v) \geqq 0. \tag{4.4}$$

If there is an optimal control u^* in $\mathscr{F}_{sy}$, with corresponding trajectory x^*, Theorem 3.2 implies that

$$V(t, x^*(t)) = \phi_1(t_1, x^*(t_1))$$

for $s \leqq t \leqq t_1$. Hence since $V(s, y)$ is differentiable at (s, y),

$$D^+ V(s, x^*(s)) = V_s + V_y f(s, y, u^*(s)^+) = 0. \tag{4.5}$$

Thus (4.4) and (4.5) give (4.3) and that the right limit of the optimal control $u^*(s)^+$ achieves the minimum in (4.3). □

The partial differential Eq. (4.3) is called the *partial differential equation of dynamic programming*. For problems for which optimal controls exist, Theorem 4.1 asserts that the value function must satisfy the partial differential equation of dynamic programming at each interior point of the reachable set at which it is differentiable.

In Theorem 4.2 we show, for a class of fixed time free endpoint problems that V is differentiable except at a set of $(n+1)$-dimensional Lebesgue measure zero A different kind of result will be that V is of class C^1 except on a lower dimensional set in case there is an optimal feedback control of the special kind described in § 6. Unfortunately for many problems V is not C^1 on all of Q_0.

In the proof of Theorem 4.2, and elsewhere in this chapter, the dependence of solutions of II(3.1) on their initial conditions will play an important role. When we wish to make this dependence clear we shall use the notation $x(t; s, y)$ to denote a solution of II(3.1) with initial conditions $x(s)=y$. Similarly we will use both $x_y(t)$ and $x_y(t; s, y)$, $x_s(t)$ and $x_s(t; s, y)$ to denote the respective matrix or vector of partial derivatives of $x(t; s, y)$ with respect to the components of y or of s. For a given control u and initial conditions (s, y), let $J(s, y, u)$ denote the corresponding performance index

$$J(s, y, u) = \phi_1(t_1, x(t_1; s, y)). \tag{4.6}$$

Call $J(s, y, u)$ the *performance function* of the control u. For a matrix A let

$$|A| = \sup_{|x|=1} |A x|.$$

Let $\| \; \|$ denote the supremum norm of a vector or matrix-valued function. In particular, if we restrict t to an interval $[T_0, T]$ we have

$$\|f_x\| = \sup \{|f_x(t, x, u)| : t \in [T_0, T], x \in E^n, u \in U\}. \tag{4.7}$$

A function $g(s, y)$ is called *Lipschitzian on a subset R of E^{n+1}* if there exists a constant K such that

$$|g(s_1, y_1) - g(s_2, y_2)| \leq K [|s_1 - s_2|^2 + |y_1 - y_2|^2]^{1/2} \tag{4.8}$$

for all $(s_1, y_1), (s_2, y_2) \in R$. Equivalently, an inequality of the form (4.8) is called a *Lipschitz condition* on g. We say that g is *locally Lipschitzian* on R if g restricted to any compact set $R_1 \subset R$ is Lipschitzian on R_1. By a theorem of Rademacher, Federer [2] p. 216, a locally Lipschitzian function is differentiable at almost every interior point of R (almost everywhere in the sense of Lebesgue $(n+1)$-dimensional measure).

If R is convex and g is of class C^1 on R, then a sufficient condition for (4.8) to hold on R is that

$$[|g_y|^2 + |g_s|^2]^{\frac{1}{2}} \leq K. \tag{4.9}$$

This follows from the mean value theorem.

Theorem 4.1 leads naturally to the question, "When will the value function $V(s, y)$ be differentiable?" Let $[T_0, T]$ be a fixed interval and consider a family of fixed time free endpoint problems of the type discussed in II.11. Let the problems be the class of problems which have fixed initial conditions (s, y) and fixed terminal time T, where $T_0 \leq s \leq T$.

Theorem 4.2. *For the above family of fixed terminal time free endpoint control problems, let the control set U be compact, $\|f_x\| < \infty$, and $\phi_1(t, x)$ locally Lipschitzian. Then the value function $V(s, y)$ is locally Lipschitzian on $[T_0, T] \times E^n$. Thus by Rademacher's theorem it is differentiable except on a set of Lebesgue measure zero.*

Proof. Since the infimum of a family of Lipschitzian functions with a common Lipschitz constant is Lipschitzian with this same Lipschitz constant, the theorem will follow if it can be shown for any control u, that $J(s, y, u)$ is locally Lipschitzian with constant independent of the control. First we shall show that the hypotheses of the theorem imply that solutions $x(t; s, y)$ of II(3.1) are locally Lipschitzian with constants which are independent of the control.

For a real number $a > 0$, let

$$\mathscr{S}_a = \{y \in E^n : |y| \leq a\}$$

and

$$C_1 = \sup\{|f(t,x,u)|: (t,x,u) \in [T_0, T] \times \mathscr{S}_a \times U\}.$$

From the theorem of the mean

$$|f(t, x(t; s, y), u(t)) - f(t, y, u(t))| \leqq \|f_x\| \, |x(t; s, y) - y|.$$

Taking norms in II(3.1) in integrated form, if $|y| \leqq a$,

$$|x(t; s, y) - y| \leqq C_1(t-s) + \int_s^t \|f_x\| \, |x(\tau; s, y) - y| \, d\tau.$$

Hence the Gronwall-Bellman inequality implies

$$(4.10) \qquad |x(t; s, y) - y| \leqq C_1[(t-s) + \int_s^t \|f_x\| (\tau - s) \, e^{\|f_x\|(t-\tau)} \, d\tau] \leqq C_2 |t-s|$$

where

$$C_2 = \max\{C_1, \|f_x\| [T - T_0] \, e^{\|f_x\|(T - T_0)}\}.$$

Again using II(3.1) in integrated form, if $|y| < a$, $|y'| < a$, and $s < s'$,

$$(4.11) \; |x(T; s, y) - x(T, s', y')| \leqq |y - y'| + |y - x(s'; s, y)| + \int_{s'}^T \|f_x\| \, |x(\tau; s, y) - x(\tau; s', y')| \, d\tau.$$

Hence the Corollary of the Gronwall-Bellman inequality and (4.10) imply if $|y| < a$, $|y'| < a$ that

$$(4.12) \qquad |x(T; s, y) - x(T; s', y')| \leqq K[|y - y'|^2 + (s - s')^2]^{1/2}$$

for a constant K which depends only on $T - T_0$, C_1 and $\|f_x\|$ and not on the control u.

Since $\phi_1(t, x)$ is locally Lipschitzian (4.12) implies that the performance function

$$J(s, y, u) = \phi_1(T, x(T, s, y))$$

is Lipschitzian on $[T_0, T] \times \mathscr{S}_a$ with a Lipschitz constant independent of the control. □

In II.9 the concept of a feedback control was mentioned. At this time we shall more formally define a feedback control to be a function

$$\mathbf{u} = \mathbf{u}(t, x)$$

from a subset Q of E^{n+1} into U such that for each (s, y) in Q there is a unique solution $x(t; s, y)$ of the differential equation

$$(4.13) \qquad \dot{x} = f(t, x, \mathbf{u}(t, x))$$

on an interval $s \leqq t \leqq t_1(s, y)$ with $x(s; s, y) = y$, such that $(t, x(t; s, y)) \in Q$ for $s \leqq t \leqq t_1(s, y)$ and $(t_1(s, y), x(t_1(s, y); s, y)) \in M$. Notice that this definition implies that Q is contained in the reachable set Q_0 for the system of differential Eqs. (4.13).

In Chap. II for the moon landing example problem it was possible to obtain an optimal control in feedback form. This feedback control simultaneously solved the optimal control problems formulated for any set of initial conditions in the reachable set. For many control problems it will be possible to find optimal controls in feedback form. Suppose for the problem formulated in § 2 it was possible to obtain a feedback control $\mathbf{u}(t, x)$ which simultaneously for each set of initial conditions (s, y) in the reachable set solved the optimization problem with these initial conditions. That is, if $x(t; s, y)$ is the solution of (4.13), then

$$u(t)=\mathbf{u}(t, x(t; s, y))$$

would be an optimal control for the control problem formulated with initial conditions (s, y). Call such a feedback control $\mathbf{u}(t, x)$ an *optimal feedback control.* When there is an optimal feedback control the equality

$$V(s, y)=J(s, y, \mathbf{u})=\phi_1(t_1(s, y), x(t_1(s, y); s, y))$$

holds between the performance function defined in (4.6) and the value function. Thus since ϕ_1 is C^1 the question of differentiability of the value function reduces to, "When are the terminal time and state $(t_1, x(t_1))$ differentiable with respect to the initial conditions of the trajectory?" We shall formalize this by stating Theorem 4.3.

Theorem 4.3. *If there is an optimal feedback control* $\mathbf{u}^*(t, x)$ *and* $t_1(s, y)$ *and* $x(t_1(s, y); s, y)$ *are the terminal time and terminal state for the trajectories of*

$$\dot{x}=f(t, x, \mathbf{u}^*(t, x)) \tag{4.13}$$

with initial conditions (s, y), *then the value function* $V(s, y)$ *is differentiable at each point at which* $t_1(s, y)$ *and* $x(t_1(s, y); s, y)$ *are differentiable with respect to* (s, y).

As we have seen in the moon landing problem, optimal feedback controls may be discontinuous. For a discontinuous feedback control $\mathbf{u}(t, x)$ we will have to consider carefully the meaning of a solution of (4.13) and the differentiability properties of these solutions with respect to initial conditions. These topics will be discussed in § 6.

Next let us turn to the question, "When does a solution $W(s, y)$ of the partial differential equation of dynamic programming satisfy the sufficient conditions for optimality in Theorem 3.3?" For the present consider only C^1 solutions $W(s, y)$. In § 7 this smoothness assumption will be relaxed.

Theorem 4.4. *Let* $W(s, y)$ *be a* C^1 *solution of the partial differential equation of dynamic programming* (4.3) *which satisfies the boundary condition*

$$W(s, y)=\phi_1(s, y) \quad \text{for } (s, y)\in M. \tag{4.14}$$

Let (t_0, x_0) *be a point of* Q, u *a control in* $\mathscr{F}_{t_0 x_0}$ *and* x *the corresponding solution of* II(3.1). *Then* $W(t, x(t))$ *is a nondecreasing function of* t. *If* u^* *is a control in* $\mathscr{F}_{t_0 x_0}$ *defined on* $[t_0, t_1^*]$ *with corresponding solution* x^* *of* II(3.1) *such that for* $t\in[t_0, t_1^*]$

$$W_s(t, x^*(t))+W_y(t, x^*(t))\, f(t, x^*(t), u^*(t))=0 \tag{4.15}$$

then u^* *is an optimal control in* $\mathscr{F}_{t_0 x_0}$ *and* $W(s, y)=V(s, y)$ *where* $V(s, y)$ *is the value function.*

Proof. From the partial differential equation of dynamic programming (4.3),

$$\frac{d}{dt} W(t, x(t)) = W_t + W_y f(t, x(t), u(t)) \geqq 0.$$

Hence $W(t, x(t))$ is nondecreasing.

Eq. (4.15) implies that

$$\frac{d}{dt} W(t, x^*(t)) = 0.$$

and since $W(t, x^*(t))$ is piecewise continuously differentiable that $W(t, x^*(t))$ is constant. Thus by Theorem 3.3 u^* is an optimal control and $W = V$. □

We call Theorem 4.4 a *verification theorem.* To verify that a control u is optimal, three conditions are to be verified, that there is a solution of (4.3), that the boundary condition (4.14), and the equality (4.15) are satisfied. Notice that (4.15) asserts that $u^*(t)$ achieves the minimum in (4.3) when $x = x^*(t)$. Often (4.15) is replaced by the equivalent statement, that along the optimal trajectory, the minimum in (4.3) is achieved by the value of the optimal control.

In § 5 we will apply Theorem 4.4 to compute the optimal control for the linear regulator problem.

§ 5. The Linear Regulator Problem

Recall the statement of the linear regulator problem in Example II.2.3. Convert this problem to an optimal control problem in Mayer form as was done in § II.7. Let

$$\tilde{y} = (y_1, \ldots, y_n, y_{n+1}) = (y, y_{n+1}).$$

Then Eq. (4.3) becomes

$$\text{(5.1)} \qquad \min_{u \in U} \{V_s + V_y [A y + B u] + V_{y_{n+1}} [y' M(s) y + u' N(s) u]\} = 0.$$

Since $x_{n+1}(t)$ represents the contribution to the performance index

$$\text{(5.2)} \qquad \int_{t_0}^{t_1} [x'(t) M(t) x(t) + u'(t) N(t) u(t)] \, dt + x(t_1)' D x(t_1)$$

made up to time t and the problem is linear with quadratic criteria, we might be led to suspect that the value function is linear in y_{n+1} and quadratic in y. Thus guess that the value function has the form

$$\text{(5.3)} \qquad W(s, \tilde{y}) = y_{n+1} + y' K(s) y$$

where $K(s)$ is a C^1 symmetric matrix with $K(t_1) = D$ and see if $W(s, \tilde{y})$ can be a solution of (5.1). For the function (5.3) the expression within the brackets of (5.1) is

$$y' \dot{K}(s) y + y' A' K(s) y + u' B'(s) K(s) y + y' K(s) A(s) y$$
$$+ y' K(s) B(s) u + y' M(s) y + u' N(s) u.$$

Taking the gradient with respect to u of this expression and setting it equal to zero, we see the minimum of this expression is achieved when

$$u = -N(s)^{-1} B(s)' K(s) y.$$

The value of the minimum is given by

$$y' \dot{K}(s) y + y'(A(s)' K(s) + K(s) A(s)) y - y' K(s) B(s) N(s)^{-1} B(s)' K(s) y + y' M(s) y.$$

A sufficient condition for this quantity to be equal to zero is that $K(s)$ satisfy the *matrix Riccati equation*

$$(5.4) \qquad \dot{K}(s) = -A(s)' K(s) - K(s) A(s) + K(s) B(s) N(s)^{-1} B(s)' K(s) - M(s).$$

Thus the conditions of Theorem 4.4 are satisfied and we have established the following theorem.

Theorem 5.1. *If there is a C^1 solution $K(s)$ of the matrix Riccati equation on $[t_0, t_1]$ with terminal condition $K(t_1) = D$, then $W(s, \tilde{y})$ defined by* (5.3) *is a C^1 solution on $[t_0, t_1] \times E^n$ of the partial differential equation of dynamic programming* (5.1) *for the linear regulator problem. The optimal control for this problem is given by*

$$(5.5) \qquad u(t) = -N(t)^{-1} B(t)' K(t) x(t).$$

Next in Theorem 5.2 an existence theorem is given for the matrix Riccati equation. For an example of nonexistence when the hypotheses of Theorem 5.2 are violated see Problem [5]. See Problem [4] for the correspondence between the treatment of the linear regulator problem in Theorem 5.1 and that given in § II.7.

Theorem 5.2. *Let the matrices $M(s)$ and D be nonnegative definite and $N(s)$ be positive definite. Then the matrix Riccati equation* (5.4) *with terminal condition $K(t_1) = D$ has a solution on $(-\infty, t_1]$.*

Proof. Theorems from differential equations imply that (5.4) with terminal condition $K(t_1) = D$ has a solution backward in time from t_1 until the solution becomes unbounded in E^{n^2}. The theorem will follow if it can be shown that $K(t)$ remains in a bounded subset of E^{n^2} for any finite interval $[\tau, t_1]$.

On any interval $[t_0, t_1]$ on which $K(t)$ exists it follows from Theorem 5.1 and Theorem 4.4 that (5.3) is the optimal performance for the linear regulator problem with initial conditions $\tilde{y}$. For the choice $y_{n+1} = 0$, (5.3) becomes

$$(5.6) \qquad y' K(s) y.$$

This must give the infimum of (5.2) over the various controls. The nonnegative definiteness of $M(t)$, $N(t)$ and D imply this infimum of (5.2) is nonnegative. Thus zero is a lower bound for (5.6).

The performance of the linear regulator $J(s, y, u)$ when u is the identically zero control gives an upper bound for (5.2) and hence an upper bound for the quadratic form (5.6). Since the quadratic form (5.6) is bounded above and below on each finite interval, $K(t)$ remains in a bounded subset of E^{n^2} on each finite interval and hence the solution of (5.4) must exist on the entire interval $(-\infty, t_1]$. □

§ 6. Equations of Motion with Discontinuous Feedback Controls

In Theorem 4.4 we gave sufficient conditions for optimality. However, those results presume the existence of a smooth solution W to the dynamic programming equation. Unfortunately, this assumption on W is quite restrictive. In the remainder of the chapter we justify the dynamic programming method under weaker conditions, which hold in many control problems of interest in applications. Our treatment is a modification of Boltyanskii [1][2] and Berkovitz [1][2][3].

Theorem 4.3 asserted that if there was an optimal feedback control that the value function was differentiable whenever the terminal time and state of the solution of (4.15) were differentiable with respect to the initial conditions. We begin to elaborate further on this idea by studying feedback controls with discontinuities which occur on smooth lower dimensional manifolds and the differentiability of solutions of (4.13) with respect to initial conditions for these types of feedback controls.

Let Q denote a subset of E^{n+1} with non empty interior. Suppose that $f(t,x,u)$ is defined on $Q \times U$ and $\mathbf{u}(t,x)$ defined on Q. The differentiability assumptions assumed in II.3 on $f(t,x,u)$ will be understood to hold. The terminal set M will assumed to be a smooth manifold defined by C^1 functions $\phi_2(t,x), \ldots, \phi_l(t,x)$ similar to that in II.3. That is $M = \{(t,x)\colon \varphi_i(t,x)=0,\ i=2,\ldots,l\}$.

To define a solution of

$$\dot{x} = f(t, x, \mathbf{u}(t,x)) \tag{4.13}$$

when a feedback control $\mathbf{u}(t,x)$ is discontinuous we shall assume that there is an appropriate smooth decomposition of the set Q and that the discontinuities of $\mathbf{u}(t,x)$ are located on the lower dimensional skeleton of this decomposition. Assumptions will be made and a procedure described so that a solution of (4.13) can be defined on each cell of the decomposition and the solution extended from cell to cell until the terminal set M is attained. The conditions A, B, C given below are assumptions which will be made on the control $\mathbf{u}(t,x)$. When they are satisfied $\mathbf{u}(t,x)$ will be said to have an *admissible set of discontinuities* or to be an *admissible feedback control*. The set $\mathscr{S}_i$ described below is to be interpreted as the i-dimensional skeleton of Q. The discontinuities of $\mathbf{u}(t,x)$ relative to the set $\mathscr{S}_i$ are contained in $\mathscr{S}_{i-1}$.

A) There are sets

$$\mathscr{S}_0 \subset \mathscr{S}_1 \cdots \subset \mathscr{S}_n \subset Q.$$

Some sets $\mathscr{S}_i$, $i<n$ may be empty. Each set $\mathscr{S}_i$ is closed in the relative topology of Q. Call the components of $Q-\mathscr{S}_n$ or $\mathscr{S}_i - \mathscr{S}_{i-1}$, $i=1,\ldots,n$, cells. Each cell of $Q-\mathscr{S}_n$ is an open subset of E^{n+1}. For each nonempty cell of $\mathscr{S}_i - \mathscr{S}_{i-1}$ there is a C^1 mapping defined on a neighborhood $\tilde{Q}$ of Q

$$\theta\colon \tilde{Q} \to E^{n+1-i}$$

such that the cell is a relatively open subset of

$$\{(t,x)\colon \theta(t,x)=0\}.$$

We shall call this function θ the *determining function of the cell*. For each cell C there is a C^1 mapping $\mathbf{u}_c(t,x)$ defined on a neighborhood of the closure of the cell such that $\mathbf{u}_c(t,x)=\mathbf{u}(t,x)$ on the cell C. Only a finite number of cells intersect any compact subset of Q. The terminal set M is contained in $\mathscr{S}_n$ and is the union of a finite number of cells. The components of the determining functions θ of cells of M include the functions $\phi_2(t,x), \ldots, \phi_l(t,x)$ which define M.

B) Cells can be classified into three catagories called type I, type II or type III. Each cell of $Q-\mathscr{S}_n$ is required to be of type I. If C is a cell of type I, there is a unique corresponding cell $\pi(C)$ whose associated function θ has one more component than that of C (has one component if C is a cell of $Q-\mathscr{S}_n$). For each $(s,y)\in C$ there is a necessarily unique trajectory of (4.13) starting at y at time s, which after a finite time leaves[1] C by reaching $\pi(C)$. Each trajectory upon leaving C strikes $\pi(C)$ at a nonzero angle. That is, if θ_i is the additional component of the function θ associated with $\pi(C)$ over that of the function θ associated with C and the trajectory exits from C at $(t,x)\in\pi(C)$, then

$$\theta_{it}(t,x)+\theta_{ix}(t,x)\,f(t,x,\mathbf{u}_c(t,x))\neq 0. \tag{6.1}$$

If C is a cell of type II, there is a cell $\Sigma(C)$ of type I whose associated function θ has one less component than that of C. From each $(s,y)\in C$ there starts a unique trajectory of (4.13) going into $\Sigma(C)$ which has only the point (s,y) in common with C. The function $\mathbf{u}(t,x)$ is of class C^1 on $C\cup\Sigma(C)$.

Cells of M are of type III. A trajectory of (4.13) may be continued from cell to cell until it hits M as follows: from C to $\pi(C)$ if $\pi(C)$ is of type I and from C to $\Sigma(\pi(C))$ if $\pi(C)$ is of type II. Furthermore it is assumed that:

C) (i) Every trajectory remains in Q and reaches M in a finite time.

(ii) Every trajectory goes through a finite number of cells to reach M.

The following are consequences of assumptions A–C. For any $(s,y)\in Q$ there will be a unique trajectory $x(t;s,y)$ of (4.13) passing through a finite number of type I cells $C_1, C_2, \ldots, C_q$ to reach M. For any i, $1<i<q$ if $\pi(C_i)$ is of type I then $C_{i+1}=\pi(C_i)$ and the trajectory goes directly from C_i to C_{i+1}. If $\pi(C_i)$ is of type II then $C_{i+1}=\Sigma(\pi(C_i))$ and the trajectory goes from C_i to C_{i+1} by crossing $\pi(C_i)$ in a single point.

Let $\tau_i(s,y)$, $i=1,2,\ldots,q$ denote the times at which the trajectory reaches $\pi(C_i)$. That is $\tau_i(s,y)$ $i=1,\ldots,q-1$ are the times the trajectory crosses from cell to cell and $\tau_q(s,y)$ is the time the trajectory reaches M. Let

$$x_i(s,y)=x(\tau_i(s,y);s,y). \tag{6.2}$$

Then $(\tau_i(s,y),\ x_i(s,y))\in\pi(C_i)$ and $(t,x(t;s,y))\in C_{i+1}$ for $\tau_i(s,y)\leqq t<\tau_{i+1}(s,y)$ if $\pi(C_i)$ is of type I or for $\tau_i(s,y)<t<\tau_{i+1}(s,y)$ if $\pi(C_i)$ is of type II.

Theorem 6.1. *Let C be a cell of $Q-\mathscr{S}_n$ and consider a trajectory $x(t;s,y)$ of* (4.13) *as a function of its initial conditions (s,y) on C. Then:*

A) *The times $\tau_j(s,y)$ and positions $x_j(s,y)$ at which $(t,x(t;s,y))$ leaves the respective cells C_j are of class C^1 on C.*

[1] A trajectory $x(t;s\,y)$ of (4.13) will be said to be in a cell C at time t if the point $(t,x(t;s,y))$ is in C.

B) $x(t; s, y)$ *is of class* C^1 *in* (t, s, y) *on*

$$\{(t, s, y): (s, y) \in C, \tau_{j-1}(s, y) < t < \tau_j(s, y)\}$$

for each j. *For fixed* $(s, y) \in C$, $x_y(t; s, y)$ *satisfies the matrix differential equation*

$$\begin{aligned}\dot{x}_y(t; s, y) = \big[&f_x(t, x(t; s, y), \mathbf{u}(t, x(t; s, y))\\ &+ f_u(t, x(t; s, y), \mathbf{u}(t, x(t; s, y)))\,\mathbf{u}_{c_j x}(t, x(t; s, y))\big]\,x_y(t; s, y)\end{aligned} \tag{6.3}$$

on

$$\{t: \tau_{j-1}(s, y) < t < \tau_j(s, y)\}$$

A similar differential equation holds for the vector of partial derivatives $x_s(t; s, y)$.

C) *At the times* $\tau_{j-1} = \tau_{j-1}(s, y)$ *and* $\tau_j = \tau_j(s, y)$, $x_y(t; s, y)$ *has the right and left hand limits*

$$x_y(\tau_{j-1}^+) = -f\big(\tau_{j-1}, x_{j-1}, \mathbf{u}_{c_j}(\tau_{j-1}, x_{j-1})\big)\,\tau_{(j-1)y} + x_{(j-1)y} \tag{6.4}$$

and

$$x_y(\tau_j^-) = -f\big(\tau_j, x_j, \mathbf{u}_{c_j}(\tau_j, x_j)\big)\,\tau_{jy} + x_{jy}. \tag{6.5}$$

Statements (6.4) *and* (6.5) *hold for* $x_s(t; s, y)$ *if* x_y *is replaced by* x_s *and* τ_{jy} *and* x_{jy} *replaced by* τ_{js} *and* x_{js}.

Proof of Theorem 6.1. Let C be a cell of $Q - \mathcal{S}_n$. Let $C_1 = C$ and $C_1, \ldots, C_q$ denote the succession of type I cells through which a trajectory $x(t; s, y)$ beginning at $(s, y) \in C$ passes to reach M. The proof will be by induction on the number j.

As a convention to begin the induction define $\tau_0(s, y)$ and $x_0(s, y)$ on C by

$$\tau_0(s, y) = s \qquad x_0(s, y) = y.$$

Then as an induction hypothesis suppose for an integer $j-1$ it has been shown that the conclusion of B is valid provided that $s \leqq t \leqq \tau_{j-1}(s, y)$, and that conditions A and C are valid for integers $\leqq j-1$. With this hypothesis up to $j-1$ it is desired to show the induction hypothesis is satisfied up to j.

To begin this argument recall that for $\tau_{j-1}(s, y) < t < \tau_j(s, y)$, $x(t; s, y)$ is in C_j. The function $\mathbf{u}_{c_j}(t, x)$ is defined and C^1 on a neighborhood of C_j and agrees with $\mathbf{u}(t, x)$ on C_j.

Theorems on continuous dependence of solutions of differential equations on initial conditions and theorems on continuation of solutions imply that for (w, z) in a neighborhood A_{j-1} of $(\tau_{j-1}(s, y), x_{j-1}(s, y))$ and $\varepsilon_j > 0$, there are solutions of

$$\dot{x} = f(t, x, \mathbf{u}_{c_j}(t, x)) \tag{6.6}$$

with initial conditions $x(w) = z$, defined on an interval

$$\tau_{j-1}(s, y) - \varepsilon_j \leqq t \leqq \tau_j(s, y) + \varepsilon_j.$$

(For consistency we always assume $|w - \tau_{j-1}(s, y)| < \varepsilon_j$ if $(w, z) \in A_{j-1}$.) Denote these solutions by $x_j(t; w, z)$.

Theorems on differentiability of solutions of differential equations with respect to initial conditions imply that it may be assumed that the mapping

$$(t, w, z) \to x_j(t; w, z)$$

is C^1 on

$$\{t \colon \tau_{j-1}(s, y) - \varepsilon_j \leqq t \leqq \tau_j(s, y) + \varepsilon\} \times A_j.$$

These theorems imply that the matrix of partial derivatives $x_{jz}(t; w, z)$ satisfies the matrix differential equation

$$\begin{aligned}\dot{x}_{jz}(t) = \big[&f_x(t, x_j(t; w, z), \mathbf{u}_{c_j}(t, x_j(t; w, z))) \\ &+ f_u(t, x_j(t; w, z), \mathbf{u}_{c_j}(t, x_j(t; w, z)))\, \mathbf{u}_{c_j x}(t, x_j(t; w, z))\big]\, x_{jz}(t)\end{aligned}$$

with initial condition $x_{jz}(w) = I$ where I is the identity matrix.

A similar equation holds for $x_{jw}(t)$. The initial condition $x_{jw}(w) = -f(w, z, \mathbf{u}_{c_j}(w, z))$ holds in this case.

Since solutions of (6.6) are unique and $\mathbf{u}_{c_j}(t, x) = \mathbf{u}(t, x)$ for $(t, x) \in C_j$ we must have

$$x_j(t; \tau_{j-1}(s, y), x_{j-1}(s, y)) = x(t; s, y) \tag{6.7}$$

if $\tau_{j-1}(s, y) < t < \tau_j(s, y)$.

Let $\theta_j(t, x)$ denote the extra component the determining function $\theta(t, x)$ of $\pi(C_j)$ has in addition to the determining function of C_j. Consider the equation

$$\theta_j(t, x_j(t; w, z)) = 0. \tag{6.8}$$

We shall verify that the conditions of the implicit function theorem are satisfied by (6.8) and that it can be solved for t to give t as a C^1 function $t = \gamma(w, z)$ of (w, z) for (w, z) in a neighborhood of $(\tau_{j-1}(s, y), x_{j-1}(s, y))$ included in A_{j-1}. To verify the conditions of the implicit function theorem, notice that

$$\theta_j(t, x_j(t; w, z))$$

is C^1 on $\{t \colon \tau_{j-1}(s, y) - \varepsilon \leqq t \leqq \tau_j(s, y) + \varepsilon_j\} \times A_j$, that from (6.7) and the definition of $x_j(s, y)$

$$\theta_j(\tau_j(s, y), x_j(\tau_j(s, y); \tau_{j-1}(s, y), x_{j-1}(s, y)) = \theta_j(\tau_j(s, y), x(\tau_j(s, y); s, y))) = 0,$$

and that

$$\frac{\partial}{\partial t}\theta_j(t, x_j(t; w, z)) = \theta_{jt} + \theta_{jx} f\big[t, x_j(t; w, z), \mathbf{u}_{c_j}(t, x_j(t; w, z))\big]. \tag{6.9}$$

The point $(t, w, z) = (\tau_j(s, y), \tau_{j-1}(s, y), x_{j-1}(s, y))$ is in $\pi(C_j)$ and hence by (6.1), (6.9) is different from zero at this point. Thus the conditions of the implicit function theorem are satisfied and we may conclude that a C^1 function γ does solve Eq. (6.8).

The definition of $\tau_j(s, y)$ and $\gamma(w, z)$ imply

$$\gamma(\tau_{j-1}(s, y), x_{j-1}(s, y)) = \tau_j(s, y). \tag{6.10}$$

Thus by our induction hypothesis and the differentiability of γ we have shown that τ_j is a composite function of differentiable functions and hence is differentiable. Since differentiability is a local property we conclude τ_j is differentiable on C.

It now follows from the differentiability of $\tau_j(s, y)$ on C and that

$$x_j(s, y) = x(\tau_j(s, y); s, y) = x_j(\tau_j(s, y); \tau_{j-1}(s, y), x_{j-1}(s, y))$$

that $x_j(s, y)$ is of class C^1 on C.

The remaining assertions of Theorem 6.1 follow by taking partial derivatives of the relationship

$$x(t; s, y) = x_j(t; \tau_{j-1}(s, y), x_{j-1}(s, y))$$

which holds for

$$\tau_{j-1}(s, y) < t < \tau_j(s, y)$$

and the two relationships

$$x_{j-1}(s, y) = x_j(\tau_{j-1}(s, y); \tau_{j-1}(s, y), x_{j-1}(s, y))$$

and

$$x_j(s, y) = x_j(\tau_j(s, y); \tau_{j-1}(s, y), x_{j-1}(s, y)). \quad \square$$

From Theorem 6.1 and the definition of $J(s, y, \mathbf{u})$ in (4.6) we have:

Remark 6.1. *If there is a control* $\mathbf{u}(t, x)$ *in feedback form with an admissible set of discontinuities, the performance function* $J(s, y, \mathbf{u})$ *is* C^1 *in* (s, y) *on* $Q - \mathscr{S}_n$.

We shall give a simple example to illustrate a control with an admissible set of discontinuities. In this example an extremal control in feedback form with an admissible set of discontinuities can be obtained. The performance function of this example is C^1 on $Q - \mathscr{S}_n$, but its partial derivatives have an infinite discontinuity on $\mathscr{S}_n$.

Consider the problem with equations of motion

$$\text{(6.11)} \qquad \begin{aligned} \dot{x}_1 &= x_2 \\ \dot{x}_2 &= u; \end{aligned}$$

control set $U = \{u: -1 \leqq u \leqq 1\}$; initial conditions $t_0 = s, x_1(s) = y_1, x_2(s) = y_2$; terminal conditions $x_1(t_1) = 0$, $x_2(t_1) = 0$; and performance index

$$\phi_1(t_1, x_1) = t_1 - s.$$

If the equation of motion is thought of as $\ddot{x}_1 = u$ this is the problem of controlling the acceleration of a system so as to drive it to the origin of position-velocity (x_1, x_2) space, in minimum time. The acceleration is constrained to have absolute value not more than one.

Consider applying Pontryagin's principle to computing an extremal feedback control law for this problem. Applying the conditions of II.5 to this problem, Eq. II (5.1) are

$$\text{(6.12)} \qquad \dot{P}_1(t) = 0, \quad \dot{P}_2(t) = -P_1(t).$$

Eq. II(5.9) is

$$\text{(6.13)} \qquad \max_{-1<u<1} \{P_1(t)\, x_2(t) + P_2(t)\, u\} = P_1(t)\, x_2(t) + P_2(t)\, u(t).$$

The vector function ϕ is given by

$$\phi(t_0, t_1, x_0, x_1) = \begin{pmatrix} t_1 - s \\ t_0 - s \\ x_{01} - y_1 \\ x_{02} - y_2 \\ x_{11} \\ x_{12} \end{pmatrix} \tag{6.14}$$

Eq. II(5.3)–II(5.6) are

$$\begin{aligned} (P_1(t_1), P_2(t_1)) &= (\lambda_5, \lambda_6) \\ (P_1(t_0), P_2(t_0)) &= -(\lambda_3, \lambda_4) \\ H(t_1, x(t_1), u(t_1)) &= -\lambda_1 \\ H(t_0, x(t_0), u(t_0)) &= \lambda_2 . \end{aligned} \tag{6.15}$$

Eq. (5.7) is

$$P_1(t)\, x_2(t) + P_2(t)\, u(t) = \lambda_2 \qquad 0 \leqq t \leqq t_1 .$$

From (6.12) and (6.15)

$$P_1(t) = -\lambda_3 \qquad P_2(t) = \lambda_3(t-s) - \lambda_4 . \tag{6.16}$$

Not both of λ_3 and $\lambda_4 = 0$, for if they did, by (6.15) and (6.16) λ would be zero, contradicting $\lambda \neq 0$. From (6.13) and (6.16)

$$u(t) = \begin{cases} 1 & \text{if } \lambda_3(t-s) - \lambda_4 > 0 \\ -1 & \text{if } \lambda_3(t-s) - \lambda_4 < 0. \end{cases}$$

Hence the control may switch between 1 and -1, but does this at most once on the interval $[s, t_1]$. Integrating equations (6.11) we find that on an interval on which $u(t) = 1$

$$x_1(t) = \tfrac{1}{2}(x_2(t))^2 + c_1 \tag{6.17}$$

where c_1 is a constant of integration. Similarly on an interval on which $u(t) = -1$

$$x_1(t) = -\tfrac{1}{2}(x_2(t))^2 + c_2 \tag{6.18}$$

where c_2 is a constant of integration. Thus in either case the trajectories follow along parabolas. When $u = -1$, then $\dot{x}_2 = -1$, thus the trajectories move downward along parabolas of the form (6.18). Similarly the trajectories move upward along parabolas of the form (6.17). The parabolas of the family (6.17) are illustrated in Fig. IV.2 and those of (6.18) in Fig. IV.1. It can be seen that the only way a trajectory starting at a given point y_1, y_2 of (x_1, x_2) space can reach $(0, 0)$ traveling in the directions indicated and switching at must once is that it be one of the family of trajectories indicated in Fig. IV.3. The heavily marked curve in Fig. IV.3 has the equation $x_1 = -\tfrac{1}{2} x_2 |x_2|$. Call this curve the switching curve since as the trajectory reaches this curve the control must switch from -1 to 1 if the trajectory

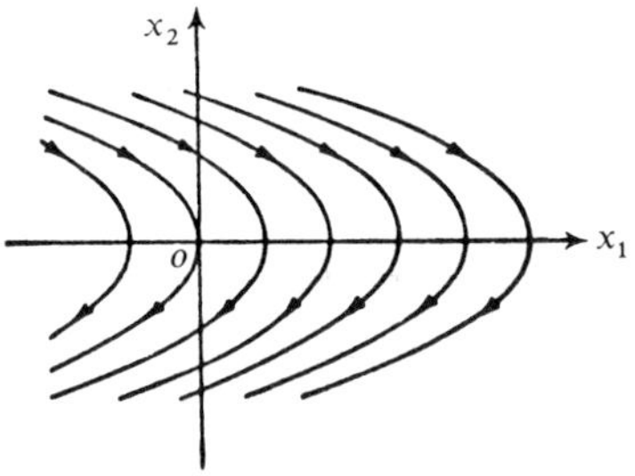

Fig. IV.1. Parabolas $x_1 = -{}^1/_2(x_2)^2 + c_1$; $u = -1$

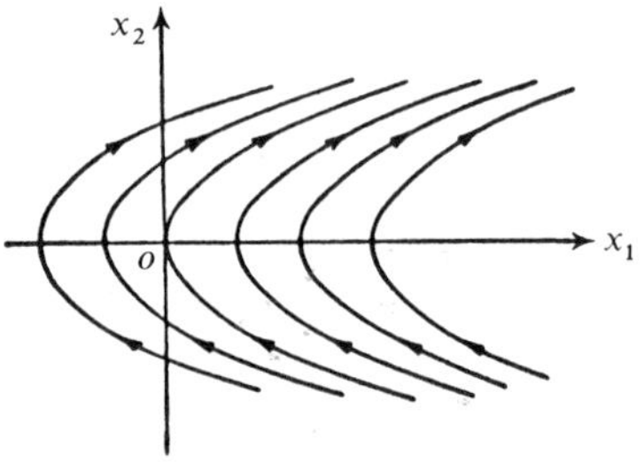

Fig. IV.2. Parabolas $x_1 = {}^1/_2(x_2)^2 + c_2$; $u = 1$

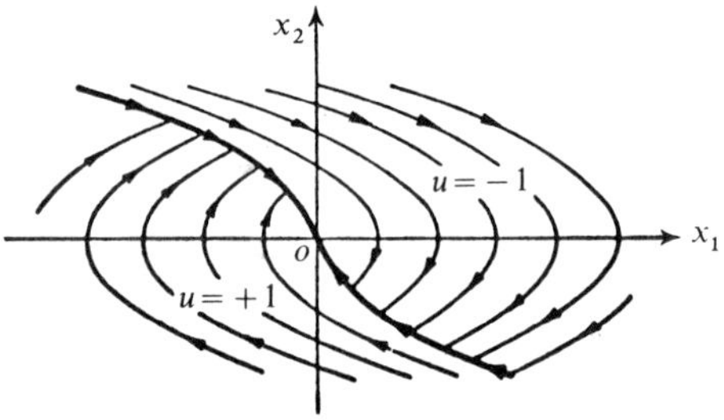

Fig. IV.3. Extremal trajectories

began above it or from 1 to -1 if it began below it. The control can be written in feedback form by

$$(6.19) \qquad \mathbf{u}(t, x_1, x_2) = \begin{cases} -1 & \text{if } x_1 > -\frac{1}{2} x_2 |x_2| \\ 1 & \text{if } x_1 > 0 \quad \text{and} \quad x_1 = -\frac{1}{2} x_2 |x_2| \\ 1 & \text{if } x_1 < -\frac{1}{2} x_2 |x_2| \\ -1 & \text{if } x_1 < 0 \quad \text{and} \quad x_1 = -\frac{1}{2} x_2 |x_2|. \end{cases}$$

While this control does not depend on t, we will denote it by $\mathbf{u}(t, x_1, x_2)$ and consider it as a function of t which is constant in this variable in order to agree with our previous notation for feedback controls.

The control (6.19) can be shown to have an admissible set of discontinuities. To do this let $Q = E^3$ and construct sets $\mathscr{S}_2$ and $\mathscr{S}_1$ by

$$\mathscr{S}_2 = \{(t, x_1, x_2) \colon x_1 = -\tfrac{1}{2} x_2 |x_2|\}$$

$$\mathscr{S}_1 = \{(t, x_1, x_2) \colon x_1 = 0, x_2 = 0\}.$$

The conditions, A, B, and C of § 6 can be shown to hold. The performance index corresponding to this control can be shown to be

$$J(s, y_1, y_2, \mathbf{u}) = \begin{cases} y_2 + 2(y_1 + \frac{1}{2} y_2^2)^{1/2} & \text{if } y_1 \geqq -\frac{1}{2} y_2 |y_2| \\ -y_2 + 2(-y_1 + \frac{1}{2} y_2^2)^{1/2} & \text{if } y_1 \leqq -\frac{1}{2} y_2 |y_2|. \end{cases}$$

This function is continuous and is C^1 on $E^3 - \mathscr{S}_2$. Its partial derivatives do approach $-\infty$ as $\mathscr{S}_2$ is approached from certain directions. However, the control $\mathbf{u}(t, x_1, x_2)$ does have an admissible set of discontinuities and the conclusion of Remark 6.1 is satisfied.

§ 7. Sufficient Conditions for Optimality

Remark 6.1 and Theorems 4.1 and 4.3 imply that if there is an optimal feedback control with an admissible set of discontinuities that the partial differential equation of dynamic programming must hold on $Q - \mathscr{S}_n$. We could ask, "Is there a sufficiency theorem like Theorem 4.4 when the partial differential equation of dynamic programming holds except on $Q - \mathscr{S}_n$?" Theorem 7.1 implies that this is the case.

Theorem 7.1. *Let Q denote a subset of E^{n+1} with nonempty interior and $W(s, y)$ a continuous function defined on Q such that $W(t, x) = \phi_1(t, x)$ for $(t, x) \in M \cap Q$. Let $\mathscr{S}$ denote a subset of Q such that $Q - \mathscr{S}$ is an open subset of E^{n+1} and the intersection of $\mathscr{S}$ with each compact subset of Q is contained in the union of a finite number of sets of the form*

$$\{(s, y): \theta(s, y) = 0\}$$

where θ is a C^1 mapping defined on a neighborhood $\tilde{Q}$ of Q, with

$$\operatorname{grad} \theta \neq 0 \qquad \text{on } \mathscr{S}.$$

Suppose that on $Q - \mathscr{S}$, $W(s, y)$ is of class C^1 and the dynamic programming equation

$$\min_{v \in U} \{W_s + W_y f(s, y, v)\} = 0$$

holds. Let (t_0, x_0) be a point of Q and $\tilde{\mathscr{F}}_{t_0 x_0}$ be the class of feasible controls for x_0 whose corresponding trajectories $(t, x(t))$ lie in Q for $t_0 \leqq t \leqq t_1$. Suppose $u^*(\cdot)$ is a control in $\tilde{\mathscr{F}}_{t_0 x_0}$ such that $W(t_0, x_0) = \phi_1(t_1^*, x^*(t_1^*))$. Then $u^*(\cdot)$ is an optimal control in the class $\tilde{\mathscr{F}}_{t_0 x_0}$.

The proof of Theorem 7.1 will try to use the simple argument of Theorem 4.4. However, it will be necessary to avoid the difficulty that when a trajectory follows along discontinuities of W_s or W_y

$$W_s(t, x(t)) + W_y(t, x(t)) f(t, x(t), u(t))$$

does not have a meaning.

The following lemma will enable this difficulty to be circumvented and will play a crucial role in the proof of Theorem 7.1.

Lemma 7.1. *Let (t_0, x_0) and $\mathcal{S}$ be as in the hypotheses of Theorem 7.1. Let $u(t)$ be a control in $\mathcal{F}_{t_0x_0}$ defined on $[t_0, t_1]$. Then in every neighborhood of x_0 there are vectors y such that the solution of* II(3.1) *with initial condition $x(t_0)=y$ exists on $[t_0, t_1]$ and meets $\mathcal{S}$ in at most a finite set of points.*

It follows directly from a general co-area formula of Federerer [1] p. 426, [2] p. 248 that for n-dimensional Lebesgue measure almost every y in a neighborhood of x_0 that the trajectory $(t, x(t; t_0, y))$ meets $\mathcal{S}$ in a finite set of points. A proof of this is given in Appendix F.

Proof of Theorem 7.1. Let $u(t)$ be an arbitrary control in $\tilde{\mathcal{F}}_{t_0x_0}$ defined on an interval $[t_0, t_1]$. From Lemma 7.1 and theorems on continuous dependence of solutions of II(3.1) on initial conditions, there is a sequence of points y_n approaching x_0 such that the solution of II(3.1) with initial condition $x(t_0)=y_n$ exists on $[t_0, t_1]$ and each solution meets $\mathcal{S}$ in at most a finite set of points.

For a particular y_n let $\tau_1, \ldots, \tau_k$ denote the set of times at which the solution $x(t; t_0, y_n)$ meets $\mathcal{S}$ and let $\tau_0 = t_0$, $\tau_{k+1} = t_1$. Then from (4.3), and the continuity of W_s and W_y at $(t, x(t; t_0, y_n))$ for $\tau_{i-1} < t < \tau_i$, and the continuity of $W(s, y)$

$$0 \leqq \int_{\tau_{i-1}}^{\tau_i} \left[W_s(t, x(t; t_0, y_n)) + W_y(t, x(t; t_0, y_n)) f(t, x(t; t_0, y_n)), u(t)\right] dt$$
$$= W(\tau_i, x(\tau_i; t_0, y_n)) - W(\tau_{i-1}, x(\tau_{i-1}; t_0, y_n)).$$

Adding these equations from $i=1$ to $k+1$ we have

$$0 \leqq W(t_1, x(t_1; t_0, y_n)) - W(t_0, y_n).$$

Now taking limits as y_n approaches x_0 using the continuity of $W(s, y)$ and that $x(t; t_0, y_n)$ converges to $x(t; t_0, x_0)$ gives

$$\lim_{n\to\infty} W(t_1, x(t_1; t_0, y_n)) = W(t_1, x(t_1; t_0, x_0)) = \phi_1(t_0, x(t_1; t_0, x_0))$$
$$= J(t_0, x_0, u) \geqq \lim_{n\to\infty} W(t_0, y_n) = W(t_0, x_0).$$

Thus the performance $J(t_0, x_0, u)$ of an arbitrary control in $\tilde{\mathcal{F}}_{t_0x_0}$ is greater than or equal to $W(t_0, x_0)$. If $u^*(t)$ is a control whose performance equals $W(t_0, x_0)$ it must be optimal. □

Theorem 7.1 was a sufficient condition for the optimality of a single control. By combining our previous results the following fundamental theorem gives necessary and sufficient conditions for the existence of an admissible optimal feedback control. We now take $Q = Q_0$, the reachable set.

Theorem 7.2. *Let $\mathbf{u}^*(t, x)$ be a feedback control with an admissible set of discontinuities and let $J(s, y, \mathbf{u}^*)$ be continuous on Q. Then a necessary and sufficient condition that $\mathbf{u}^*$ be optimal is that the following hold:*

(a) *$J(s, y, \mathbf{u}^*)$ is a C^1 solution of the dynamic programming equation* (4.3) *on $Q - \mathcal{S}_n$,*

(b) *For all* $(s, y) \in Q - \mathscr{S}_n$

$$\min_{v \in U} J_y(s, y, \mathbf{u}^*) f(s, y, v) = J_y(s, y, \mathbf{u}^*) f(s, y, \mathbf{u}^*(s, y)).$$

Proof. Necessity follows from Theorem 4.1, (with $V = W$) and Remark 6.1. Sufficiency follows by noting that (a) and the definition (4.6) of $J(s, y, \mathbf{u}^*)$ imply that, if for each (s, y) a control u^* is defined by

$$u^*(t) = \mathbf{u}^*(t, x^*(t; s, y)),$$

then the conditions of Theorem 7.1 are satisfied and hence u^* is an optimal control in $\tilde{\mathscr{F}}_{sy}$. Since this is true for each (s, y) $\mathbf{u}^*$ is an optimal feedback control law.

Remark 7.1. While the performance function of a feedback control with an admissible set of discontinuities must be C^1 on $Q - \mathscr{S}_n$ it is not difficult to give examples in which it need not be continuous on Q.

Remark 7.2. For the linear regulator problem solution of the partial differential equation of dynamic programming yielded the control (5.5). However, in general it is quite difficult to use Theorem 7.2 to determine a feedback control law. The reader may convince himself of this by trying to compute the feedback control law for the example of §6 from Theorem 7.2. One problem is that the set $\mathscr{S}_n$ and the value of the control on the sets $\mathscr{S}_n$ are not specified explicity from the partial differential equation of dynamic programming. However, since the condition is necessary and sufficient for the existence of an admissible optimal feedback control law, this information must be contained in an implicit manner in the equation.

§8. The Relationship between the Equation of Dynamic Programming and Pontryagin's Principle

Theorem 7.2 suggests that methods for constructing solutions of the partial differential equation of dynamic programming should be considered. This equation is a nonlinear first order partial differential equation. There is a classical method in the theory of partial differential equations for obtaining solutions of nonlinear first order partial differential equations. It is the "method of characteristics". See for instance Courant Hilbert [1], or John [1], for expositions of this method.

The method of characteristics consists of two parts. The first asserts that if there is a C^2 solution of the partial differential equation then a family of ordinary differential equations for curves called characteristic strips can be derived. The second shows that a solution of the partial differential equation can be constructed from a family of solutions of the ordinary differential equations for the characteristic strips.

The analogue for the partial differential equation of dynamic programming of the first part of the method of characteristics is carried out in Theorem 8.1. It is interesting to note that the differential equations for the characteristic strips are the same differential equations involved in Pontryagin's Principle. Thus

Theorem 8.1 could be considered as a derivation of Pontryagin's Principle via the method of dynamic programming. However, the twice continuous differentiability assumed in Theorem 8.1 for the value function is very restrictive. There are many examples in which the value function does not have continuous derivatives. Because of this Theorem 8.1 is more important as an indication of the correspondence between Pontryagin's principle and the method of characteristics applied to the partial differential equation of dynamic programming than as a way of deriving Pontryagin's Principle.

Theorem 8.2 is concerned with the second part of the method of characteristics, the construction of a solution of the partial differential equation of dynamic programming from solutions of the characteristic equations. It asserts that if there is a feedback control with an admissible set of discontinuities, satisfying Pontryagin's Principle (i.e., the characteristic equations for the partial differential equation of dynamic programming), then its performance function is a solution of the partial differential equation of dynamic programming except on a lower dimensional set. This theorem combined with Theorem 7.2 give quite general sufficient conditions for the optimality of such a control.

Theorem 8.1. *Let Q denote a subset of E^{n+1}. Let (t_0, x_0) be an interior point of $Q-M$. For the optimization problem of § 2 with initial conditions (t_0, x_0), let u^* be an optimal control. Suppose the corresponding trajectory $(t, x^*(t))$ except perhaps for the terminal point (t_1, x_1) lies in the interior of Q. Suppose the value function $V(s, y)$ defined by (3.1) is twice continuously differentiable on $Q-M$.*

Define $P(t)$ by $P(t)' = -V_x(t, x^*(t))$; then $P(t)$ satisfies the differential equation

$$\dot{P}(t)' = -P(t)' f_x(t, x^*(t), u^*(t)) \tag{8.1}$$

and the condition

$$\max_{v \in U} \{P(t)' f(t, x^*(t), v)\} = P(t)' f(t, x^*(t), u^*(t)) \tag{8.2}$$

holds for each $t \in (t_0, t_1)$.

Eqs. II(3.1) and (8.1) are the characteristic equations for the partial differential equation of dynamic programming. Notice that Eqs. (8.1) and (8.2) are the conclusion of Pontryagin's Principle without the transversality conditions II(5.3)–II(5.6).

Proof of Theorem 8.1. Consider the curve $(t, x^*(t))$ where

$$\dot{x}^*(t) = f(t, x^*(t), u^*(t)).$$

Consider the vector $V_y(s, y)$ along the curve $(t, x^*(t))$. Let t be a point of continuity of u^*. We must have

$$\dot{V}_y(t, x^*(t)) = V_{ys}(t, x^*(t)) + V_{yy}(t, x^*(t)) f(t, x^*(t), u^*(t)). \tag{8.3}$$

Notice from Theorem 4.1 that

$$V_s(s, y) + V_y(s, y) f(s, y, u) \geqq 0$$

for all (s, y) and $u \in U$. In particular this holds if we replace (s, u) by $(t, u^*(t))$. If moreover we replace y by $x^*(t)$, Theorem 4.1 implies

$$V_s(t, x^*(t)) + V_y(t, x^*(t)) f(t, x^*(t), u^*(t)) = 0.$$

Thus $x^*(t)$ minimizes the expression

$$V_s(t, y) + V_y(t, y) f(t, y, u^*(t)).$$

This is a differentiable function of y; hence its partial derivatives with respect to y must equal zero at $y = x^*(t)$. Setting these partial derivatives equal to zero and using that orders of partial differentiation can be interchanged gives

$$V_{ys}(t, x^*(t)) + V_{yy}(t, x^*(t)) f(t, x^*(t), u^*(t)) + V_y(t, x^*(t)) f_x(t, x^*(t), u^*(t)) = 0. \tag{8.4}$$

Combining (8.3) and (8.4) we obtain

$$\dot{V}_y(t, x^*(t)) = -V_y(t, x^*(t)) f_x(t, x^*(t), u^*(t)). \tag{8.5}$$

Recalling the definition of $P(t)$ we see that (8.5) is (8.1) and that (8.2) follows from (4.3). □

Next we shall wish to consider the converse problem of trying to construct a solution of the partial differential equation of dynamic programming from families of solutions of the equations of Pontryagin's Principle.

In Theorem 8.2 it is shown that if a control in feedback form with an admissible set of discontinuities satisfies Pontryagin's Principle then its performance is a solution of the dynamic programming partial differential equation on $Q - \mathscr{S}_n$. Theorem 8.2 combined with Theorem 7.2 gives the sufficiency Theorem 8.3 mentioned previously. Let $\phi = (\phi_1, \ldots, \phi_l)$ with ϕ_1 as in § 2 and $\phi_2, \ldots, \phi_l$ defining M as in § 6.

Theorem 8.2. *Let $\mathbf{u}(t, x)$ be a feedback control defined on the entire reachable set Q_0 with an admissible set of discontinuities. For each $(s, y) \in Q_0$ and each control*

$$u(t) = \mathbf{u}(t, x(t; s, y))$$

defined on $[s, t_1]$, let Pontryagin's Principle and the transversality conditions as given in Problem II.2 *be satisfied. Let the matrix $(\phi_x(t, x), \phi_t(t, x))$ have rank l on M. Then the performance function $J(s, y, \mathbf{u})$ defined in* (4.6) *is differentiable on each cell of $Q_0 - \mathscr{S}_n$. The partial derivatives of J and the adjoint variables of Pontryagin's Principle are related on $Q_0 - \mathscr{S}_n$ by*

$$\lambda_1 J_y(s, y, \mathbf{u}) = P(s)$$

and

$$\lambda_1 J_s(s, y, \mathbf{u}) = -H(s),$$

where

$$H(s) = P(s)' f(s, y, \mathbf{u}(s, y)).$$

The partial differential equation of dynamic programming

$$\min_{v \in U} \{J_s + J_y f(s, y, v)\} = 0$$

is satisfied at each point of $Q_0 - \mathscr{S}_n$. *The control* $\mathbf{u}(s, y)$ *achieves the minimum in this equation.*

A lemma will be needed in the proof of Theorem 8.2.

Lemma 8.1. *Let* $\mathbf{u}(t, x)$ *be a feedback control with an admissible set of discontinuities. Let* (s, y) *be a point of* $Q_0 - \mathscr{S}_n$. *Let* $\tau_j(s, y), j = 0 \dots q$, *be as defined in* § 6. *If on* $\tau_{j-1}(s, y) < t < \tau_j(s, y)$, $P(t)$ *is a vector function such that*

$$\max_{v \in U} P(t)' f(t, x(t; s, y), v) = P(t)' f(t, x(t; s, y), \mathbf{u}_{c_j}(t, x(t; s, y)))$$

then

$$P(t)' f_u(t, x(t; s, y)), \mathbf{u}(t, x(t; s, y)) \, \mathbf{u}_{c_j x}(t, x(t; s, y)) = 0. \tag{8.6}$$

Proof. For (w, z) in a neighborhood of the cell C_j, in which $x(t; s, y)$ lies for $\tau_{j-1}(s, y) < t < \tau_j(s, y)$,

$$P(t)' f(s, x(t; s, y), \mathbf{u}_{c_j}(w, z))$$

is a differentiable function of (w, z). It achieves it maximum at $(w, z) = (t, x(t; s, y))$. Hence, its matrix of partial derivatives with respect to z must equal zero. Noticing that $\mathbf{u}_{c_j z}$ and $\mathbf{u}_{c_j x}$ have the same meaning, we see that this matrix is

$$P(t)' f_u \, \mathbf{u}_{c_j x}. \quad \square \tag{8.7}$$

Proof of Theorem 8.2. From Theorem 6.1, $t_1(s, y)$ and $x(t_1(s, y); s, y)$ are differentiable functions of (s, y) on $Q_0 - \mathscr{S}_n$. Since $J(s, y, \mathbf{u}) = \phi_1(t_1(s, y), x(t_1(s, y); s, y))$,

$$J_y = \phi_{1t} t_{1y} + \phi_{1x} x(t_1)_y. \tag{8.8}$$

Let $P(t)$ be the solution II(5.1) with terminal condition $P(t_1) = \lambda' \phi_x(t_1, x(t_1))$. Multiplying (6.3) by $P(t)'$ we have

$$P(t)' \dot{x}_y(t) = P(t)' f_x x_y(t) + P(t)' f_u \mathbf{u}_{c_j x} x_y(t). \tag{8.9}$$

Multiplying II(5.1) by $x_y(t)$ we have

$$\dot{P}(t)' x_y(t) = -P(t)' f_x x_y(t). \tag{8.10}$$

Adding and using (8.6) we have that on the intervals $\tau_{j-1}(s, y) < t < \tau_j(s, y)$

$$\frac{d}{dt} P(t)' x_y(t) = 0.$$

Integrating from τ_{j-1} to τ_j gives

$$P(\tau_j)' x_y(\tau_j^-) - P(\tau_{j-1})' x_y(\tau_{j-1}^+) = 0; \tag{8.11}$$

adding formula (8.11) over the switching times τ_j gives

$$P(s)' x_y(s) = \sum_j P(\tau_j)' [x_y(\tau_j^-) - x_y(\tau_j^+)] + P(t_1)' x_y(t_1). \tag{8.12}$$

Since $(t_1, x(t_1))$ must satisfy

$$\phi_i(t_1, x(t_1))=0, \quad i=2, \ldots, l$$

it follows that

$$\phi_{it}\, t_{1y}+\phi_{ix}\, x(t_1)_y=0 \qquad i=2, \ldots, l. \tag{8.13}$$

We have from (6.5) with $\tau_q=t_1$, (8.13) and the transversality conditions that

$$\begin{aligned} P(t_1)'\, x_y(t_1) &= P(t_1)'\, x(t_1)_y - P(t_1)' f\, t_{1y} = \lambda'\, \phi_x\, x(t_1)_y + \lambda'\, \phi_t\, t_{1y} \\ &= \lambda_1\, \phi_{1x}\, x(t_1)_y + \lambda_1\, \phi_{1t}\, t_{1y} = \lambda_1\, J_y(s, y, \mathbf{u}). \end{aligned} \tag{8.14}$$

From (6.4), (6.5) and the continuity of the Hamiltonian

$$\begin{aligned} &\sum_k P(\tau_k)' [x_y(\tau_k^-) - x_y(\tau_k^+)] \\ &= \sum P(\tau_k)' \{ f[\tau_k, x(\tau_k), \mathbf{u}(\tau_k, x(\tau_k))^-] - f[\tau_k, x(\tau_k), \mathbf{u}(\tau_k, x(\tau_k, x(\tau_k)))^+] \}\, \tau_{ky} = 0. \end{aligned} \tag{8.15}$$

Hence since $x_y(s; s, y)=I$ we have from (8.12), (8.14) and (8.15) that

$$P(s)' = \lambda_1\, J_y. \tag{8.16}$$

To establish that $H(s) = -\lambda_1 J_s$ multiply the equation for $x_s(t; x, y)$ similar to (6.3) by $P(t)'$ and II(5.1) by $x_s(t)$. Then proceeding similarly to the derivation of (8.10)–(8.15) using the transversality conditions gives

$$-P(s)' f(s, y, \mathbf{u}(s, y)) = \sum_k P(\tau_k)' [f^- - f^+]\, \tau_{ks} + \lambda'\, \phi_x\, x(t_1)_s + \lambda'\, \phi_t\, t_{1s}.$$

Arguing similarly to (8.15) gives

$$H(s) = -\lambda_1\, J_s. \tag{8.17}$$

Notice that if λ_1 were zero (8.16) would imply $P(s)=0$ and hence through II(5.1) that $P(t)\equiv 0$ for $t \in [s, t_1]$. If $P(t)\equiv 0$, II(5.3) and II(5.4) would imply $\lambda'\, \phi_x(t_1, x(t_1))=0$ and $\lambda'\, \phi_t(t_1, x(t_1))=0$. Since (ϕ_x, ϕ_t) has rank l on M, this would imply $\lambda=0$, contradicting $\lambda \neq 0$, so we must have $\lambda_1<0$.

From Pontryagin's principle we must have that

$$\max_{v \in U} \{P(s)' f(s, y, v)\} = H(s) = P(s)' f(s, y, \mathbf{u}(s, y))$$

or since $\lambda_1<0$ from (8.16) and (8.17) that

$$\min_{v \in U} \{J_s(s, y) + J_y(s, y) f(s, y, v)\} = 0 = J_s(s, y, \mathbf{u}) + J_y(s, y, \mathbf{u}) f(s, y, \mathbf{u}(s, y)).$$

The above argument held for (s, y) an arbitrary point in any cell of $Q_0 - \mathscr{S}_n$; hence the conclusion of Theorem 8.2 holds. □

Theorem 8.3. *In addition to the hypotheses of Theorem 8.2, let the performance function $J(s, y, \mathbf{u})$ be a continuous function on Q_0. Then $\mathbf{u}^*$ is an optimal feedback control law.*

Proof. Apply Theorems 7.2 and 8.2.

Problems—Chapter IV

(1) If in the problem formulated in § 2

$$f(t, x, u) = f(x, u), \qquad \phi_1(t, x) = \phi_1(x)$$

and

$$M = \{(t, x): \ -\infty < t < \infty \ x \in \tilde{M} \subset E^n\}$$

this problem is called autonomous. Show for an autonomous control problem that the value function $V(t, x)$ is constant as a function of t and could consequently be written $V(x)$.

(2) For the simplest problem in calculus of variations (Chap. I) consider the minimum as a function of the initial endpoint $s = t_0$, $y = x_0$ with fixed final end point (t_1, x_1). Thus let χ_{sy} be the class of piecewise C^1 functions $x(t)$ on $[s, t_1]$ such that $x(s) = y$, $x(t_1) = x_1$ and let

$$V(s, y) = \inf_{\chi_{s,y}} \int_s^{t_1} L(x(t), \dot{x}(t))\, dt.$$

Show that the partial differential equation of dynamic programming is

(a)
$$V_s(s, y) + \min_u \{L(y, u) + u\, V_y(s, y)\} = 0$$

and if $x(t)$ is optimal in $\chi_{s,y}$ that

(b)
$$V_s(t, x(t)) + L(x(t), \dot{x}(t)) + \dot{x}(t)\, V_y(t, x(t)) = 0.$$

Show that if V is C^2 then (a)(b) imply that the Euler equation

$$\frac{d}{dt} L_{\dot{x}}(x(t), \dot{x}(t)) = L_x(x(t), \dot{x}(t))$$

holds.

(3) If the optimal control problem of § 2 is formulated in Lagrange form with a performance index given by II(4.2) and a value function defined by

$$V(s, y) = \inf_{u \in \mathscr{F}_{sy}} \left\{ \int_s^{t_1} L(t, x(t), u(t))\, dt \right\}$$

show that the analogue of Theorem 3.1 is the statement

$$V(s, y) \leqq \int_s^{s+h} L(t, x(t), u(t))\, dt + V(s+h, x(s+h)).$$

This statement has been named the "Principle of Optimality" by Bellman. Derive this statement and use it to show that in this case the partial differential equation of dynamic programming is

$$\min_{u \in U} \{V_s + V_y f(s, y, u) + L(s, y, u)\} = 0.$$

(4) Show that a sufficient condition for there to be a solution $x(t)$, $\tilde{P}(t)$ of II.7.8 of the form $\tilde{P}(t) = -K(t)\, x(t)$ is that $K(t)$ be a solution of the matrix Riccati equation (5.4).

(5) The linear regulator problem of minimizing

$$\int_0^{\pi/2} (-x^2+u^2)\,dt$$

subject to $\dot{x}=u$ and $x(0)=0$ is related to Example I.4.1 and the principle of least action for the oscillating spring mentioned in I.6. Show for this problem that the Riccati equation (5.4) with $K(\pi/2)=0$ does not have a solution on $0 \leqq t \leqq \pi/2$. Show that the Eqs. II(7.8) with boundary conditions II(7.9) do not have a unique solution in this case.

(6) Let $\phi_2(t,x)$ be a C^1 real valued function defined on E^{n+1} such that grad $\phi_2 = (\phi_{2t}, \phi_{2x}) \neq 0$. Let

$$M=\{(t,x)\colon \phi_2(t,x)=0\}$$

and

$$Q=\{(t,x)\colon \phi_2(t,x)\geqq 0\}.$$

Let $\mathbf{u}(t,x)$ be a C^1 feedback control defined on Q such that

$$\phi_{2t}(t,x)+\phi_{2x}(t,x)\,f(t,x,\mathbf{u}(t,x)) < -\gamma < 0 \quad \text{on } Q.$$

Assume solutions of (4.13) corresponding to the control $\mathbf{u}(t,x)$ exist on arbitrary large time intervals. Show for each $(s,y)\in Q$ that there is a finite first time $t_1(s,y)$ at which the corresponding trajectory $x(t;s,y)$ of (4.13) reaches M. Give a direct proof to show $t_1(s,y)$ is C^1 on Q. What is the result analogous to property C) of Theorem 6.1 concerning the time $t_1(s,y)$?

(7) Under the hypotheses of Problem 6 state a special case of Theorem 8.2. Use the results of Problem 6 to give a simplified proof of Theorem 8.2 in this special case.

(8) Suppose that $\dot{x}=u$, $U=\{|u|\leqq 1\}$, and the problem is minimum time to reach a target $\tilde{M}\subset E^n$ starting at $x(0)=y$.

(a) Show that the autonomous dynamic programming equation has the form $|V_y(y)|=1$ where $V_y(y)$ is the gradient vector.

(b) Find the minimum time $V(y_1,y_2)$ when $n=2$, $y=(y_1,y_2)$, and $\tilde{M}$ is the unit circle $x_1^2+x_2^2=1$ in E^2. Where does $V(y_1,y_2)$ fail to be differentiable?

(c) Let $\tilde{M}$ consist of two parallel straight lines in E^2. Where does the minimum time $V(y_1,y_2)$ fail to be differentiable?

(9) (Bushaw's problem). Let $\dot{x}_1=x_2$, $\dot{x}_2=-x_1+u$, $U=\{u\colon -1\leqq u\leqq 1\}$. The problem is to reach a target $\tilde{M}$ consisting of the single point $(0,0)$ in E^2 in minimum time, starting at (y_1,y_2). Find an optimal feedback control. *Hint.* Compare with the worked example with system equations (6.11).

(10) (1 dimensional saturated linear regulator) In Example III.6.2 let $n=1$ and $B>0$. Show that an optimal feedback control is

$$\mathbf{u}^*(s,y)=\begin{cases} -1 & \text{if } y>G(s)\\ \mathbf{u}(s,y) & \text{if } -G(s)\leqq y\leqq G(s)\\ 1 & \text{if } y<-G(s)\end{cases}$$

where $\mathbf{u}(s,y)$ is the optimal feedback control for the linear regulator in § 5 and $G(s)$ is determined by the equation $\mathbf{u}(s,y)=-1$.

Chapter V. Stochastic Differential Equations and Markov Diffusion Processes

§ 1. Introduction

In this chapter we review a part of the theory of continuous parameter stochastic processes which is of interest for the study of Markov models of systems which arise in applications. With one exception the present chapter involves no ideas from control theory, and may be read independently of the rest of the book. (The exception is Theorem 9.2 about the Kalman-Bucy filter; the proof we give depends on the solution to the linear regulator problem.) In the chapter we emphasize material needed to discuss in a mathematically correct way optimal control of diffusion processes in Chap. VI.

Systems arising in applications are often modelled by differential equations

$$\frac{dx}{dt} = b(t, x(t)), \tag{1.1}$$

where the vector $x(t) = (x_1(t), \ldots, x_n(t))$ describes the system state at time t. (For the optimal control problem considered in previous chapters, $b(t, x) = f(t, x, u(t))$, where $u(t)$ represents the control applied at time t.) However, many systems are subject to imperfectly known disturbances, which may be taken as random. A stochastic model is now appropriate in which the system states evolve according to some vector-valued stochastic process $\xi(t) = (\xi_1(t), \ldots, \xi_n(t))$.

We begin in §2 by defining a class of processes of particularly simple probabilistic structure—the brownian motions. Then we consider processes ξ which are solutions of a stochastically perturbed form of (1.1). The perturbed equations can be written formally as

$$\frac{d\xi}{dt} = b(t, \xi(t)) + \sigma(t, \xi(t)) \frac{dw}{dt}, \tag{1.2}$$

with dw/dt the formal time derivative of a brownian motion w. (In engineering literature dw/dt is called a white noise.) Actually, the derivative dw/dt does not exist. In the rigorous treatment of (1.2), the equation is first rewritten in differential notation $d\xi = b\,dt + \sigma\,dw$ and then integrated from a fixed lower limit s to a variable upper limit t. The integral $\int_s^t \sigma\,dw$ is defined as a stochastic integral in the sense of K. Ito. See §3, 4. Other definitions of stochastic integral are due to Stratonovich [2] and McShane [2][7]. For solutions of (1.2) the ordinary chain rule of differentiation is not valid. It must be replaced by a formula ((3.9) or (3.14)) called the Ito stochastic differential rule. This formula and its consequences (5.8), (7.6) are crucial relationships in the development of the theory of stochastic differential equations.

We shall omit in this chapter proofs of many results about stochastic integrals and stochastic differential equations which are readily available in the literature. However, we try to include enough discussion and examples to give the flavor of the subject. We shall follow rather closely the development in Gikhman-Skorokhod [2], which contains a well organized exposition of results we need. Other books which the reader may find helpful are Gikhman-Skorokhod [1, especially Chap. VIII] and Wong [1]. Some concepts from probability theory which we use are reviewed in Appendix D.

Solutions of (1.2) have the Markov property. In fact, they belong to a class of Markov processes called diffusions (§5). The vector-valued function $b=(b_1, \ldots, b_n)$ and the matrix-valued function $a=(a_{ij})$, $i, j=1, \ldots, n$, defined by $a=\sigma\sigma'$, have an important role. For small time increment h, $b_i(t, \xi(t))h$ is the expected change $\Delta\xi_i$ of a component ξ_i given the state vector $\xi(t)$, up to a term $o(h)$ small compared to h. The covariance of increments $\Delta\xi_i$, $\Delta\xi_j$ given $\xi(t)$ will be $a_{ij}(t, \xi(t))h+o(h)$. For diffusions, the functions $b(t, x)$, $a(t, x)$ together with the state $\xi(s)$ at an initial time s are enough to specify the probability distribution of $\xi(t)$ at any time $t>s$.

There is a close connection between diffusion processes and second order partial differential equations of parabolic or elliptic type. In the second half of the chapter we develop this connection in some detail, focusing on results needed in Chap. VI.

The functions b, a associated with (1.2) determine a pair of second order partial differential operators. They are called the backward and forward operators, and are formal adjoints of each other. We first discuss backward partial differential equations, with either Cauchy data at a final time T (Sect. 6) or boundary data (Sect. 7). Under suitable assumptions the probability density of a diffusion satisfies a forward partial differential equation (Sect. 8).

We use repeatedly the method of probabilistic solutions for backward equations. Such solutions involve expectations of certain functionals of a family of diffusions with the given backward operator, considered as functions of the initial data. A feature of the method is that one sometimes has a probabilistic solution in cases when this is not guaranteed by the theory of partial differential equations of parabolic type. See for instance Theorem 6.1.

Linear stochastic differential equations and the corresponding gaussian diffusion processes are discussed in Sect. 9. We also derive the equations of the Kalman-Bucy filter there.

For the study of optimal control problems in Chap. VI, one needs to consider stochastic differential equations for which b may be a discontinuous function. In Sects. 10 and 11 we outline a way to deal with this possibility. The method is based on a formula of Girsanov for absolutely continuous substitution of probability measures (Theorem 10.1).

§2. Continuous Stochastic Processes; Brownian Motion Processes

Let Ω be a set, $\mathscr{F}$ a σ-algebra of subsets of Ω, and P a probability measure on Ω $(P(\Omega)=1)$. The triple $(\Omega, \mathscr{F}, P)$ is called a *probability space*. A random variable is a real valued function on a probability space which is $\mathscr{F}$-measurable. More generally, a function from a probability space into some separable complete metric space Σ will be called a *random vector*, if it is measurable with respect to $\mathscr{F}$ and the σ-algebra $\mathscr{B}(\Sigma)$ of Borel subsets of Σ. (In this book, usually $\Sigma=E^n$ for some finite n.) A *stochastic process* ξ assigns to each time t in some set $\mathscr{T}$ a random vector $\xi(t)$. Thus $\xi=\xi(\cdot,\cdot)$ is a function from the cartesian product $\mathscr{T}\times\Omega$ into Σ, where $\mathscr{T}\subset E^1$, such that $\xi(t,\cdot)$ satisfies the above measurability condition for each $t\in\mathscr{T}$. We call Σ the *state space* of the process, and $\xi(t,\omega)$ the *state* at time t. For brevity, we usually write (as is customary in probability theory) $\xi(t)$ to denote the random vector $\xi(t,\cdot)$. The more cumbersome notations $\xi(t,\cdot)$, $\xi(\cdot,\cdot)$ will be used where needed for clarity.

In the discussion to follow $\mathscr{T}$ is an interval; in fact, we usually take $\mathscr{T}=[s,T]$ a compact interval with initial time s and final time T. A stochastic process is called *continuous* if $\xi(\cdot,\omega)$ is a continuous function on $\mathscr{T}$ for P-almost every $\omega\in\Omega$. The function $\xi(\cdot,\omega)$ is called a *sample function* of the process ξ. Thus, continuity means that sample functions are continuous on $\mathscr{T}$ with probability 1. We deal in this chapter with diffusion processes, which are continuous.

A stochastic process ξ is *measurable* if $\xi(\cdot,\cdot)$ is measurable with respect to the σ-algebras $\mathscr{B}(\mathscr{T})\times\mathscr{F}$, $\mathscr{B}(\Sigma)$. Any continuous process is measurable (in Chap. VI we use control processes u which are measurable, but not necessarily continuous).

Nonanticipative Processes, Stopping Times. Suppose that, besides the σ-algebra $\mathscr{F}$, a σ-algebra $\mathscr{F}_t$ is given for each $t\in\mathscr{T}$ with

$$\mathscr{F}_r\subset\mathscr{F}_t\subset\mathscr{F} \qquad \text{if } r<t.$$

Such a family of σ-algebras is called *increasing*. Frequently, the σ-algebra $\mathscr{F}_t$ describes in a set theoretic way a certain past history available at time t. Let ξ be a measurable process on $\mathscr{T}$. Then ξ is *nonanticipative* with respect to the family $\{\mathscr{F}_t\}$ of σ-algebras if there is a process $\tilde{\xi}$ such that $\xi(\cdot,\omega)=\tilde{\xi}(\cdot,\omega)$ with probability 1 and the random vector $\tilde{\xi}(t)$ is $\mathscr{F}_t$-measurable for every $t\in\mathscr{T}$.

Example 2.1. If ξ is a measurable process, let $\mathscr{F}(\xi(v), v\leqq t)$ denote the least σ-algebra with respect to which the random vectors $\xi(v)$ are measurable for all $v\leqq t$, $v\in\mathscr{T}$. Then $\mathscr{F}(\xi(v), v\leqq t)$ is called the σ-algebra *generated* by the process ξ up to time t. Obviously ξ is nonanticipative with respect to the family of σ-algebras it generates.

Let τ be a random variable with values in $\mathscr{T}$. Then τ is called a *stopping time* if there exists Γ with $P(\Gamma)=0$ such that the event $\{\tau>t\}$ differs from an $\mathscr{F}_t$-measurable event by a subset of Γ, for each $t\in\mathscr{T}$.

Example 2.2. Let $\mathscr{T}=[s,T]$ be a compact interval, and ξ a continuous, nonanticipative process on $\mathscr{T}$. Let B be an open set, $B\subset\Sigma$, with $\xi(s)\in B$. Let τ denote the smallest time $t<T$ such that $\xi(t)\notin B$, or $\tau=T$ if $\xi(t)\in B$ for $s\leqq t\leqq T$.

We call τ the *exit time* for $\xi(t)$ from B. Let us show that τ is a stopping time. By redefining $\xi(\cdot,\omega)$ on a set of probability 0, we may suppose that every sample path is continuous. Let $\mathcal{T}'$ be a countable dense subset of $\mathcal{T}$ and, for $m=1,2,\ldots$, K_m the closed set consisting of all x with distance $(x, \Sigma - B) \geqq m^{-1}$. Since sample paths are continuous

$$\{\tau > t\} = \bigcup_{m=1}^{\infty} \{\xi(r) \in K_m \text{ for all } r \leqq t, r \in \mathcal{T}'\},$$

for $s < t < T$; for other values of t, the condition is trivially satisfied. Since ξ is nonanticipative, $\{\xi(r) \in K_m\}$ differs from an $\mathcal{F}_r$-measurable set by a subset of some fixed Γ with $P(\Gamma)=0$. Since $\mathcal{F}_r \subset \mathcal{F}_t$, the event $\{\tau > t\}$ differs from an $\mathcal{F}_t$-measurable set by a subset of Γ. This shows that τ is a stopping time.

Brownian Motion Processes. In discrete time a process v with the simplest probabilistic structure is one for which the random vectors $v(t_1), v(t_2), \ldots, v(t_m)$ at different times $t_1, \ldots, t_m$ are independent. If one tries to consider the same kind of processes in continuous time, mathematical difficulties are encountered. These difficulties can be avoided by considering instead processes with independent increments.

Definition. A process ζ has *independent increments* if, for all $t_0 < t_1 < \cdots < t_m$ in $\mathcal{T}$, the random vectors

$$\zeta(t_0), \zeta(t_1) - \zeta(t_0), \ldots, \zeta(t_m) - \zeta(t_{m-1})$$

are independent.

The corresponding class of discrete time processes are the partial sums of independent random vectors. Many examples of processes with independent increments are known, and also general representation theorems for them. See Feller [1, Chap. VII], Gikhman-Skorokhod [1, Chap. VI].

We shall be concerned with one important class of such processes – brownian motions. For simplicity, let us begin a discussion of these processes and their properties by considering first the case of one-dimensional brownian motion.

Definition. A continuous 1-dimensional process w is a *brownian motion* on the interval $\mathcal{T} = [s, T]$ if:

1. w has independent increments;
2. The increment $w(t) - w(r)$ is gaussian with mean 0, variance $\sigma^2 |t-r|$, for any $r, t \in \mathcal{T}$; and
3. $w(s)$ is gaussian with mean 0.

In (2) σ is a positive constant. If $\sigma = 1$, we call w a *standard* brownian motion.

The fact that $w(t) - w(r)$ has a gaussian distribution in (2) actually follows from the other conditions. This can be seen from the following stronger result, proved in Gikhman-Skorokhod [2, § 1], and needed for technical reasons later. Suppose that w is a continuous process, with $w(s)=0$, and w satisfies:

4. $E[w(t) - w(r) | \mathcal{F}_r] = 0$

$$E[(w(t) - w(r))^2 | \mathcal{F}_r] = \sigma^2 (t-r), \qquad s \leqq r < t \leqq T,$$

where $\{\mathscr{F}_t\}$ is an increasing family of σ-algebras with respect to which w is nonanticipative. Then w is a brownian motion. Moreover, $w(t)-w(r)$ is independent of $\mathscr{F}_r$.

Now suppose that all conditions in the definition are satisfied, except perhaps the gaussian property of $w(t)-w(r)$. We may suppose that $w(s)=0$, by considering the process $w-w(s)$ instead of w. Take $\mathscr{F}_t=\mathscr{F}(w(v), v\leqq t)$. By (1) conditional expectations in (4) become expectations; and these are then just the conditions in (2) on the mean and variance of increments. By the result just cited, w is a brownian motion.

A useful estimate for brownian motions is the following:

$$P\left(\max_{t_0\leqq t\leqq t_1}[w(t)-w(t_0)]>a\right)=2P\left(w(t_1)-w(t_0)>a\right), \tag{2.1}$$

for any $a>0$ and $s\leqq t_0<t_1\leqq T$. See Gikhman-Skorokhod [2, §1]. Since $w(t_1)-w(t_0)$ is a gaussian random variable, with mean 0 and variance $\sigma^2(t_1-t_0)$, the right side is explicitly known.

Although brownian sample functions are continuous, their local behavior is nevertheless erratic. This is suggested by the fact that $w(r+h)-w(r)$ has variance $\sigma^2 h$. Hence this increment is typically of order $h^{1/2}$, not order h as would be true for smooth sample functions. Indeed, with probability 1 the brownian sample function $w(\cdot,\omega)$ is nowhere differentiable on $[s,T]$. For a proof of this rather remarkable fact see Breiman [1, p. 261]. It is easier to prove, using (2.1), a slightly weaker statement, namely, that with probability 1 $w(\cdot,\omega)$ is almost nowhere differentiable on $[s,T]$. See Doob [1, p. 394].

Since a function of bounded variation is differentiable almost everywhere, $w(\cdot,\omega)$ is with probability 1 of infinite variation on $[s,T]$. The latter statement is also a consequence of the convergence of sums of squares of brownian increments to a constant. See Example 3.1.

The mathematical brownian motion process was suggested by Wiener as a model for the motion of particles suspended in a fluid. The nonexistence of a derivative dw/dt implies that, in Wiener's model of physical brownian motion, particles do not have well defined velocities. This corresponds roughly to physical observation; see remarks in Nelson [1]. Nevertheless, the nonexistence of velocities was regarded for some time by physicists as a defect of the model. Another model was proposed by Ornstein and Uhlenbeck in which particles do move with continuous velocities. It is mentioned in Sect. 5.

In engineering literature, the formal time derivative of a brownian motion is called a *white noise*. If one proceeds in a purely formal way, $v=dw/dt$ is to be regarded as a stationary process in which the random variables $v(t)$ are independent for different time instants t, with $Ev(t)=0$. The covariance function $R(h)=E[v(t)v(t+h)]$ turns out to be Dirac's delta function, and its Fourier transform (the spectral density of the white noise process) is constant. {See Gikhman-Skorokhod [1, Chap. I] for terms.} Thus the average power with which various frequencies appear in the spectral resolution is constant; hence, the name "white".

By considering brownian motions instead of their (nonexistent) time derivatives, one has a way to treat white noise on a mathematically rigorous basis. In practice, white noise is an idealization of stationary noise processes of wide but finite spectral bandwidth. We shall see that Markov diffusions can be represented as the outputs of stochastic differential equations with white noise inputs. Considerable care must be used in regarding these inputs as approximations to finite bandwidth inputs. See remarks at the end of Sect. 5, and in more detail Wong-Zakai [1][2], McShane [7], Ito [1].

Brownian Motion in n-Dimensions. A process $w=(w_1, \ldots, w_n)$ is called an n-dimensional brownian motion if the components $w_1, \ldots, w_n$ are independent 1-dimensional brownian motions. The covariance matrix for the increments $w(t)-w(r)$ is diagonal, with i-th diagonal element $\sigma_i^2(t-s)$. If $\sigma_i=1$ for $i=1, \ldots, n$, then w is a *standard* brownian motion.

For a standard n-dimensional brownian motion we have the estimate

$$P\Big(\max_{t_0 \leq t \leq t_1} |w(t)-w(t_0)|>a\Big) \leq 4nP\left(w_i(t_1)-w_i(t_0)>\frac{a}{n}\right). \tag{2.2}$$

To get this, we observe that for $\eta=(\eta_1, \ldots, \eta_n)$ in E^n, $|\eta|>a$ implies either $\eta_i>n^{-1}a$ or $-\eta_i>n^{-1}a$ for some component η_i. We use this observation with $\eta=w(t)-w(t_0)$, and apply the estimate (2.1) to each of the 1-dimensional brownian motions w_i, $-w_i$ with a replaced there by $n^{-1}a$.

Wiener Measure. Let ξ be any continuous n-dimensional process on a compact interval $\mathscr{T}$; and let $\mathscr{C}^n(\mathscr{T})$ denote the space of all continuous functions from $\mathscr{T}$ into E^n with the usual sup norm. Then ξ defines a measure P^* on the space $\mathscr{C}^n(\mathscr{T})$ in the following way. By redefining $\xi(\cdot, \omega)$ for $\omega \in \Gamma$, where $P(\Gamma)=0$, we may assume that $\xi(\cdot, \omega)$ is continuous for every $\omega \in \Omega$. With ξ is associated a random vector Ξ (with values in $\mathscr{C}^n(\mathscr{T})$), defined by $\Xi(\omega)=\xi(\cdot, \omega)$. Any random vector induces a measure on the space in which it takes its values (the probability distribution of the random vector). In particular, P^* is defined by

$$P^*(D)=P(\Xi \in D) \qquad \text{for each } D \in \mathscr{B}[\mathscr{C}^n(\mathscr{T})].$$

It can be shown that $\mathscr{B}[\mathscr{C}^n(\mathscr{T})]$ is the least σ-algebra containing all sets of the form $D=\{g: g(t_1) \in B_1, \ldots, g(t_m) \in B_m\}$ where $t_1, \ldots, t_m$ is any finite subset of $\mathscr{T}$ and $B_1, \ldots, B_m$ are Borel subsets of E^n. The distribution measure P^* is determined by its values $P^*(D)=P(\xi(t_1) \in B_1, \ldots, \xi(t_1) \in B_m)$ for all such D. See Billingsley [1, p. 57].

For the case when ξ is a standard brownian motion w, P^* is called *Wiener measure.*

§ 3. Ito's Stochastic Integral

We now define the concept of stochastic integral, in the Ito sense, and give its important properties. Then in §4 we consider stochastic differential equations. Both of these sections consist of a summary of concepts, results, and examples. For proofs of these results we refer to Gikhman-Skorokhod [2, Part I, §'s 1–11]. Actually dimension $n=1$ is treated there. The extensions of the results needed to $n>1$ involve no new ideas. They appear as special cases of results in Gikhman-Skorokhod [2, Part II, Chap. 2].

Let w be a standard brownian motion on a compact interval $\mathscr{T}=[s, T]$. For the present we take w 1-dimensional. We wish to define an integral $\int_s^t e(r)\,dw(r)$ when $s<t\leqq T$, for a rather wide class of processes e (the class $\mathscr{M}_0$ below).

First of all, let us recall the Riemann-Stieltjes definition, which unfortunately does not turn out to be adequate for the purpose. Consider partitions of the interval $[s, T]$; such a partition Δ consists of points $r_0, r_1, \ldots, r_m$ with $s=r_0<r_1<\cdots<r_m=t$. Also choose arbitrarily $r_1^*, \ldots, r_m^*$ with $r_{k-1}\leqq r_k^*\leqq r_k$ for $k=1, \ldots, m$. Then

$$(RS)\int_s^t e(r)\,dw(r)=\lim_{\|\Delta\|\to 0}\sum_{k=1}^m e(r_k^*)[w(r_k)-w(r_{k-1})] \tag{3.1}$$

provided the limit exists, where

$$\|\Delta\|=\max_{1\leqq k\leqq m}(r_k-r_{k-1}).$$

Since $w(\cdot,\omega)$ is continuous, the Riemann-Stieltjes integral exists for each ω such that $e(\cdot,\omega)$ is a function of bounded variation on $[s, t]$. See Graves [1, Chap. XII, Thm. 8]. In particular this is true if $e=e(\cdot)$ is a function of bounded variation on $[s, t]$ (not a stochastic process). However, we often wish to consider integrands e which are not of bounded variation. The following example illustrates the difficulty.

Example 3.1. Let $e=w$. We show that $(RS)\int_s^t w\,dw$ does not exist. Consider the two choices $r_k^*=r_k$ and $r_k^*=r_{k-1}$. Let χ denote the difference between the corresponding Riemann-Stieltjes sums on the right side of (3.1). Then

$$\chi=\sum_{k=1}^m [w(r_k)-w(r_{k-1})]^2=\sum_{k=1}^m \eta_k^2(r_k-r_{k-1}),$$

where $\eta_1, \ldots, \eta_m$ are independent, gaussian with $E\eta_k=0$, $E\eta_k^2=1$. Then

$$E\chi=\sum_{k=1}^m (r_k-r_{r-1})=t-s,$$

$$E\chi^2=\sum_{k=1}^m (E\eta_k^4)(r_k-r_{k-1})^2+\sum_{k\neq l}(r_k-r_{k-1})(r_l-r_{l-1}),$$

$$\lim_{\|\Delta\|\to 0} E\chi^2=(t-s)^2.$$

Since $\operatorname{var}\chi=E\chi^2-(E\chi)^2$, χ tends to the nonzero constant $t-s$ in mean square as $\|\Delta\|\to 0$.

The brownian motion w is defined on some probability space $(\Omega, \mathscr{F}, P)$. Let $\{\mathscr{F}_t\}$ be an increasing family of σ-algebras, such that w is *adapted* to $\{\mathscr{F}_t\}$ in the following sense: w is nonanticipative with respect to $\{\mathscr{F}_t\}$ and, for $s\leqq r<t\leqq T$, the increment $w(t)-w(r)$ is independent of $\mathscr{F}_r$.

Let $\mathscr{M}_0$ denote the class of all real valued, measurable processes e on $[s, T]$, such that e is nonanticipative with respect to $\{\mathscr{F}_t\}$ and $\int_s^T |e(t)|^2\,dt<\infty$ with probability 1. Let $\mathscr{M}$ denote the class of $e\in\mathscr{M}_0$ such that $E\int_s^T |e(t)|^2\,dt<\infty$.

The stochastic integral $\int_s^t e\,dw$ will be defined for any $e \in \mathcal{M}_0$ and $s<t \leqq T$. First of all, a process $e \in \mathcal{M}$ is called a step process on $[s, T]$ if there is a partition Δ of $[s, T]$ such that $e(r)$ is constant for $r_{k-1} \leqq r < r_k$, $k=1, \ldots, m$. For a step process the integral equals the Riemann-Stieltjes sum with $r_k^* = r_{k-1}$:

$$\int_s^t e(r)\,dw(r) = \sum_{k=1}^m e(r_{k-1})\,\Delta_k w, \qquad \Delta_k w = w(r_k) - w(r_{k-1}). \tag{3.2}$$

By using properties of conditional expectations and independence of $\Delta_k w$ from $\mathcal{F}_{r_{k-1}}$, it is not difficult to show that for step processes

$$E \int_s^t e\,dw = 0 \tag{3.3}$$

$$E\left(\int_s^t e\,dw\right)\left(\int_s^t f\,dw\right) = E \int_s^t e f\,dr. \tag{3.4}$$

In particular, when $e=f$ we get the very useful formula

$$E\left(\int_s^t e\,dw\right)^2 = E \int_s^t e^2\,dr. \tag{3.5}$$

To define the stochastic integral for $e \in \mathcal{M}_0$, one can take a sequence of step processes e^j, $j=1, 2, \ldots$, such that $\int_s^t |e^j(r) - e(r)|^2\,dr$ tends to 0 in probability, as $j \to \infty$. The sequence of random variables $\int_s^t e^j\,dw$ converges to a limit in probability, which is defined as $\int_s^t e\,dw$. It is shown that the limit does not depend on the approximating sequence e^j, with probability 1 for each t. For $e \in \mathcal{M}$, one can find e^j such that $E \int_s^t |e^j - e|^2\,dr$ tends to 0 as $j \to \infty$. It follows that formulas (3.3)–(3.5) remain correct for any $e, f \in \mathcal{M}$. For details see Gikhman-Skorokhod [2, § 2].

The stochastic integral has been defined up to a P null set, which might depend on the upper limit t of integration. However, it can be shown Gikhman-Skorokhod [2, § 3] that the integral can be defined simultaneously for all $t \in \mathcal{T}$, in such a way that

$$\zeta(t) = \int_s^t e\,dw \tag{3.6}$$

is a continuous function of t. Moreover, the following estimates hold:

$$E \max_{s \leqq r \leqq t} |\zeta(r)|^2 \leqq 4 E \int_s^t e^2(r)\,dr, \tag{3.7a}$$

$$P\left(\max_{s \leqq r \leqq t} \left|\int_s^t e(r)\,dw(r)\right| > C\right) \leqq P\left(\int_s^t e^2(r)\,dr > N\right) + N C^{-2} \tag{3.7b}$$

for any positive C, N.

Remarks. 3.1. Gikhman-Skorokhod [2, § 2]. In the stochastic integral one can replace s by $r \in (s, t)$; moreover, (3.3)–(3.5) remain true if one takes conditional expectations with respect to $\mathscr{F}_r$ instead of expectations. In particular

$$E\left(\int_r^t e\, dw \mid \mathscr{F}_r\right) = 0$$

which implies with probability 1

$$\zeta(r) = E(\zeta(t) \mid \mathscr{F}_r), \qquad r < t,$$

since the stochastic integral on (r, t) equals $\zeta(t) - \zeta(r)$. This states that the process ζ is a *martingale*. A process η is called a *submartingale* if with probability 1

$$\eta(r) \leqq E\{\eta(t) \mid \mathscr{F}_r\}, \qquad r < t;$$

in case the inequality is reversed, η is a *supermartingale*. See Doob [1, Chap. VII] or Breiman [1, Chap. 5] for properties of martingales and submartingales.

3.2. If e is a continuous process in $\mathscr{M}_0$, then the limit of Riemann-Stieltjes sums in (3.1) exists and equals $\int_s^t e\, dw$ provided we take $r_k^* = r_{k-1}$. A different limit is usually obtained if one takes $r_k^* = \frac{1}{2}(r_k + r_{k-1})$. This limit, if it exists, is called the *Stratonovich integral*.

We come next to a formula (3.9) which is crucial to the developments to follow. In particular, it furnishes the basis for relating a stochastic differential equation and the backward partial differential operator of the corresponding Markov diffusion (Sects. 5, 6).

Suppose that β, γ, ξ are nonanticipative processes, such that $E \int_s^T |\beta|\, dt < \infty$, $\gamma \in \mathscr{M}$, ξ is continuous, and

$$\xi(t) - \xi(s) = \int_s^t \beta(r)\, dr + \int_s^t \gamma(r)\, dw(r), \qquad s \leqq t \leqq T. \tag{3.8}$$

Eq. (3.8) is a stochastic integral equation. Although the time derivative $d\xi/dt$ does not exist, it is customary to use the suggestive notation

$$d\xi = \beta(t)\, dt + \gamma(t)\, dw, \qquad s \leqq t \leqq T. \tag{3.8'}$$

The stochastic differential equation (3.8′) is to be interpreted as the integral equation (3.8), with $\xi(s)$ as data at the initial time s.

Let us say that a function $\psi(t, x)$ is of class $C^{1,2}$ if the partial derivatives $\psi_t, \psi_x, \psi_{xx}$ are continuous on $[T_0, T] \times E^n$.

Ito Stochastic Differential Rule. If ψ is of class $C^{1,2}$ and $\eta(t) = \psi[t, \xi(t)]$ where ξ satisfies (3.8), then

$$d\eta = \psi_t(t, \xi(t))\, dt + \psi_x(t, \xi(t))\, d\xi + \tfrac{1}{2} \psi_{xx}(t, \xi(t))\, \gamma^2(t)\, dt. \tag{3.9}$$

Eq. (3.9) is to be interpreted in the sense that its integrated form from s to t holds with probability 1. Note that the term $\frac{1}{2} \psi_{xx} \gamma^2\, dt$ does not appear in the corresponding formula from elementary calculus. For a proof of this result, see Gikhman-Skorokhod [1, pp. 387–9] or [2, §3, Theorem 4].

Example 3.1 (continued). Let us take $\beta=0, \gamma(t)=1, \psi(x)=\frac{1}{2}x^2, \xi=w$. Then $\frac{1}{2}d(w^2)\doteq w\,dw+\frac{1}{2}dt$. Upon integrating from s to t and rearranging terms,

$$\int_s^t w\,dw=\tfrac{1}{2}[w^2(t)-w^2(s)]-\tfrac{1}{2}(t-s).$$

By Remark 3.2 above, this also equals the limit of Riemann-Stieltjes sums with $r_k^*=r_{k-1}$. If we took instead $r_k^*=r_k$, then the fact that χ in Example 3.1 tends to $t-s$ in mean square implies that the Riemann-Stieltjes sums tend to $\frac{1}{2}[w^2(t)-w^2(s)]+\frac{1}{2}(t-s)$. If we took the Stratonovich integral, with $r_k^*=\frac{1}{2}(r_k+r_{k-1})$, then the limit is $\frac{1}{2}[w^2(t)-w^2(s)]$ as in elementary calculus. (The fact that his integral obeys the rules of elementary calculus has been cited by Stratonovich [2] as an advantage for it.)

Example 3.2. Suppose that ξ is a solution of $d\xi=a\,\xi\,dt+b\,\xi\,dw$, with fourth moment $E\,\xi^4(t)$ bounded on $[s,T]$. Let $\psi(x)=x^2$, $s=0$. Then

$$d(\xi^2)=2\,\xi\,d\xi+b^2\,\xi^2\,dt=(2a+b^2)\,\xi^2\,dt+2b\,\xi^2\,dw.$$

If we write this in integrated form and take expectations,

$$(*)\qquad E\,\xi^2(t)=E\,\xi^2(0)+(2a+b^2)\int_0^t E\,\xi^2(r)\,dr.$$

Since $E\,\xi^4(r)$ is bounded, $\xi^2\in\mathscr{M}$; hence by (3.3)

$$E\int_0^t \xi^2\,dw=0.$$

If we set $\rho(t)=E\,\xi^2(t)$, then by (*) $d\rho/dt=(2a+b^2)\,\rho$. Hence

$$\rho(t)=\rho(0)\exp[(2a+b^2)\,t].$$

In this example, we have easily calculated the second moment $\rho(t)$. This idea can be extended, to find the moments of $\xi(t)$ of every order. See Problem 2. We also note that if $2a+b^2<0$, then $E\,\xi^2(t)\to 0$ as $t\to\infty$. This is a kind of stochastic stability property (for a thorough discussion of stochastic stability using methods based on the Ito stochastic differential rule and estimates for supermartingales, see Kushner [1]).

Example 3.3. This example is of considerable interest for the Girsanov transformation formula to be discussed in § 10. Suppose $d\xi=-\frac{1}{2}\gamma^2\,dt+\gamma\,dw$, with $\xi(s)=0$, and suppose that $E\int_s^t|\gamma(r)|^4\,dr<\infty$ while

$$(3.10)\qquad E\exp 4\,\xi(t)=E\exp 4\int_s^t[\gamma\,dw-\tfrac{1}{2}\gamma^2\,dr]\leqq M$$

holds for $s\leqq t\leqq T$. We use the Ito stochastic differential rule with $\psi(x)=\exp x$ to get (in integrated form)

$$\exp\xi(t)=1+\int_s^t\gamma(r)\exp[\xi(r)]\,dw.$$

Using the bound in (3.10) and the Cauchy-Schwarz inequality,

$$E\int_s^t [\gamma(r)\exp\xi(r)]^2\,dr \leqq M\left[\int_s^t E[\gamma(r)]^4\,dr\right]^{\frac{1}{2}} < \infty.$$

Hence by (3.3)

$$E\exp\xi(t)=1,\qquad s\leqq t\leqq T.$$

An extension of this argument shows that, in fact,

$$E\left\{\exp\int_{t_1}^{t_2}[\gamma\,dw-\tfrac{1}{2}\gamma^2\,dr]\,|\,\mathscr{F}_{t_1}\right\}=1$$

for $s\leqq t_1<t_2\leqq T$, which implies that $\exp\int_s^t[\gamma\,dw-\frac{1}{2}\gamma^2\,dr]$ is a martingale with respect to $\{\mathscr{F}_t\}$.

A sufficient condition for (3.10) is that γ be bounded, $|\gamma(t)|\leqq C$ for $s\leqq t\leqq T$. In fact, the estimate

$$E\exp 4\,\xi(t)\leqq\exp[6\,C^2(t-s)] \tag{3.11}$$

then holds. To establish this, it suffices to suppose that γ is a step process with the given bound C. It is easy to show that $E\exp 8\,\xi(t)$ is bounded on $[s,T]$ when γ is a step process. We now use the stochastic differential rule with $\psi(y)=\exp 4y$ to get

$$\exp 4\,\xi(t)=1+4\int_s^t\gamma(r)\exp 4\,\xi(r)\,dw+6\int_s^t\gamma^2(r)\exp 4\,\xi(r)\,dr.$$

Using the bound on $\exp 8\,\xi(r)$, the middle term has expectation 0. Thus

$$E\exp 4\,\xi(t)=1+6\,E\int_s^t\gamma^2(r)\exp 4\,\xi(r)\,dr$$

$$\leqq 1+6\,C^2\int_s^t E\exp 4\,\xi(r)\,dr.$$

Gronwall's inequality now gives (3.11). Example 3.3 also holds in n-dimensions (Problem 4).

Vector-Valued Stochastic Integrals. Let $w=(w_1,\ldots,w_d)$ be a d-dimensional standard brownian motion, and $e=(e_{il})$, $i=1,\ldots,n$, $l=1,\ldots,d$, a matrix-valued process with each $e_{il}\in\mathscr{M}_0$. The discussion for $n=1$ extends with merely superficial changes. The stochastic integral is now the n-dimensional random vector $\zeta(t)$ with components

$$\zeta_i(t)=\sum_{l=1}^{d}\int_s^t e_{il}(r)\,dw_l(r),\qquad i=1,\ldots,n.$$

Formula (3.3) remains correct, as do the following analogues of (3.4), (3.5), provided $e_{il}, f_{il}\in\mathscr{M}$. Let

$$tr\,e'f=\sum_{i,l}e_{il}f_{il},\qquad |e|^2=tr\,e'e.$$

Then

$$E\left(\int_s^t e\,dw\right)'\left(\int_s^t f\,dw\right)=E\int_s^t tr\,e'f\,dr, \tag{3.12}$$

$$E\left|\int_s^t e\,dw\right|^2=E\int_s^t |e|^2\,dr. \tag{3.13}$$

The Ito stochastic differential rule now reads as follows. Let $C^{1,2}$ denote the class of $\psi(t,x)$, $x\in E^n$, with partial derivatives $\psi_t, \psi_{x_i}, \psi_{x_i x_j}$, $i,j=1,\dots,n$, continuous on $[T_0,T]\times E^n$. Here T_0, T are times which are fixed throughout the chapter, and $T_0\leqq s\leqq t\leqq T$. Let

$$\psi_x=(\psi_{x_1},\dots,\psi_{x_n}), \qquad a_{ij}=\sum_{l=1}^d \gamma_{il}\gamma_{jl}.$$

The gradient ψ_x is to be regarded as a row vector. If $\eta(t)=\psi[t,\xi(t)]$ where ξ satisfies the vector-matrix version of (3.8), then

$$d\eta=\psi_t(t,\xi(t))\,dt+\psi_x(t,\xi(t))\,d\xi+\tfrac{1}{2}\sum_{i,j=1}^n a_{ij}\psi_{x_i x_j}\,dt. \tag{3.14}$$

§ 4. Stochastic Differential Equations

Many natural phenomena can be modeled by a system of ordinary differential equations, containing a term v representing the effects of disturbances:

$$\frac{d\xi}{dt}=b(t,\xi(t))+v(t). \tag{4.1}$$

Here we use vector notation; thus $\xi=(\xi_1,\dots,\xi_n)$, $b=(b_1,\dots,b_n)$, etc. When such disturbances are of a very irregular or unpredictable sort, one usually takes for v some stochastic process (often called in engineering literature "random noise"). Usually one also arranges that $Ev(t)=0$, by absorbing a nonrandom term $Ev(t)$ in b.

We shall assume that v is a white noise multiplied by a coefficient σ depending perhaps on time and state: $v(t)=\sigma(t,\xi(t))\dfrac{dw}{dt}$. (This assumption will be discussed in Sect. 5.) In the stochastic differential notation (3.8′) we shall be concerned with equations of the form

$$d\xi=b(t,\xi(t))\,dt+\sigma(t,\xi(t))\,dw \tag{4.2}$$

where w is a standard brownian motion of some dimension d. When written in terms of components, (4.2) states that

$$d\xi_i=b_i(t,\xi(t))\,dt+\sum_{l=1}^d \sigma_{il}(t,\xi(t))\,dw_l, \qquad i=1,\dots,n.$$

As in (3.8) a solution of (4.2) with initial data $\xi(s)$ is to be interpreted as a solution of the stochastic integral equation

$$\xi(t)=\xi(s)+\int_s^t b(r,\xi(r))\,dr+\int_s^t \sigma(r,\xi(r))\,dw(r), \qquad s\leqq t\leqq T. \tag{4.3}$$

The purpose of this section is to state some conditions for existence and uniqueness of solutions to (4.3), and to give some estimates on moments.

Throughout the chapter we consider two fixed times T_0, T. As initial time in (4.3) we consider any s satisfying $T_0\leqq s<T$. In order to quote a theorem about existence and uniqueness, we make two assumptions (4.4), (4.5) on the coefficients b, σ. The first is a linear growth condition, and the second a local Lipschitz condition.

Let $Q^0=(T_0,T)\times E^n$. We assume that b_i, σ_{ij} are defined and Borel measurable on the closure $\bar{Q}^0$. Moreover,

(4.4) A constant C exists such that, for all $(t,x)\in\bar{Q}^0$,

$$|b(t,x)|\leqq C(1+|x|), \qquad |\sigma(t,x)|\leqq C(1+|x|).$$

(4.5) For any bounded $B\subset E^n$ and $T_0<T'<T$ there exists a constant K (perhaps depending on B and T') such that, for all $x,y\in B$ and $T_0\leqq t\leqq T'$,

$$|b(t,x)-b(t,y)|\leqq K\,|x-y|$$

$$|\sigma(t,x)-\sigma(t,y)|\leqq K\,|x-y|.$$

If b, σ satisfy (4.4) and (4.5) with constant K not depending on B or T', then the so-called *Ito conditions* hold. Sufficient conditions for the Ito conditions to hold are that b_i, σ_{il} are of class C^1 on $\bar{Q}^0$ and their first order partial derivatives in the variables $x=(x_1,\ldots,x_n)$ are bounded (using gradient notation, $|b_x|\leqq M$, $|\sigma_x|\leqq M$ for some M).

Theorem 4.1. *Let b, σ satisfy conditions* (4.4), (4.5). *Let $\xi(s)$ be independent of the brownian motion w, with $E|\xi(s)|^2<\infty$. Then a solution ξ of* (4.3) *exists. It is unique in the following sense. If ξ' is any solution of* (4.3) *with $\xi'(s)=\xi(s)$, then ξ' and ξ with probability* 1 *have the same sample functions.*

Let us indicate briefly a series of three steps by which this theorem can be proved. First, consider the case when the Ito conditions hold. In that case, rather standard techniques for ordinary differential equations can be adapted, using repeatedly property (3.13) of stochastic integrals. See Gikhman-Skorokhod [1, Chap. VIII] or [2, §6, Theorem 1]. Second, if the constant K in (4.5) does not depend on T' (and thus (4.5) holds for $T_0\leqq t\leqq T$), then one can approximate b, σ by suitable sequences b_n, σ_n satisfying the Ito conditions with $b_n=b$, $\sigma_n=\sigma$ if $|x|\leqq n$, $n=1,2,\ldots$. See Gikhman-Skorokhod [2, §6, Theorem 3]. Finally, the solution ξ is defined by that result uniquely on $[s,T']$ for any $T'<T$, hence for $s\leqq t<T$. There is a bound on $E|\xi(t)|^2$ for $s\leqq t<T$ (by the special case $k=2$ of (4.6) below). By (4.4) this implies

$$E\int_s^T |b(r,\xi(r))|\,dr<\infty, \qquad E\int_s^T |\sigma(r,\xi(r))|^2\,dr<\infty.$$

From continuity of $\int_s^t b\,dr$ and $\int_s^t \sigma\,dr$ as functions of the upper limit $t \in [s, T]$, the right side of (4.3) is continuous on $[s, T]$ with probability 1. For $\xi(T)$ we take the random vector which the right side of (4.3) defines. (The possible dependence of K on T', and hence these last remarks, will be needed for technical reasons in Chap. VI)

Remark. The linear growth condition (4.4) insures that the solution ξ of (4.3) does not explode in finite time. There is another kind of condition to exclude explosions, involving a stochastic version of the idea of Lyapunov function in ordinary differential equations. See Kushner [1, Chap. III]. Other conditions of Lyapunov type can be given to insure certain properties as $t \to \infty$ (in particular, stochastic stability or ergodicity of solutions to (4.3)). See Kushner [1, Chap. II], Gikhman-Skorokhod [2, Part II, Chap. 3], Wonham [1].

Existence and uniqueness are known under weaker assumptions than (4.5). In particular, uniqueness holds if b satisfies the Ito conditions, and if σ satisfies (4.4) with $\sigma(t, \cdot)$ a diagonal matrix satisfying a uniform Hölder condition with exponent $\geqq \frac{1}{2}$. See Yamada-Watanabe [1]. In § 10 we shall consider cases in which b may be discontinuous.

Bounds on Moments. Let us now assume that the initial data $\xi(s)$ has finite absolute moments of every order: $E|\xi(s)|^k < \infty$ for $k = 1, 2, \ldots$. For us, this will be hardly any restriction, since we usually consider constant initial data $\xi(s) = y$. Then $\xi(t)$ also has finite absolute moments of every order. In fact:

Theorem 4.2. Gikhman-Skorokhod [2, § 6, Theorem 4]. *For each* $k = 1, 2, \ldots$

$$E|\xi(t)|^k \leqq C_k(1 + E|\xi(s)|^k), \qquad s \leqq t \leqq T, \tag{4.6}$$

where the constant C_k *depends on* k, $T - s$, *and the constant* C *in* (4.4).

From (4.6) let us now derive another estimate (4.7) which will be needed later. For simplicity take $k = 2m$ even. Let $\zeta(t) = \int_s^t \sigma(r, \xi(r))\,dw$. Then $E|\zeta(t)|^{2m}$ is also bounded on $[s, T]$, for each m. [This can be seen from (4.6) together with (4.3) and the linear growth estimate on b in (4.4). It also follows directly from (4.6), the estimate on σ in (4.4), and the Ito stochastic differential rule applied to $d|\zeta(t)|^{2m}$.]

Since ζ is a martingale, $|\zeta|^{2m}$ is a submartingale which implies the estimate Doob [1, p. 296, 353]

$$P(\|\zeta\|^{2m} > c) \leqq c^{-1} E|\zeta(T)|^{2m}$$

for any $c > 0$. Here $\|g\| = \max_{s \leqq t \leqq T} |g(t)|$ is the usual sup norm. If we take $c = \rho^{2m}$, then

$$P(\|\zeta\| > \rho) \leqq M \rho^{-2m}, \qquad \rho > 0$$

where M is an upper bound for $E|\zeta(T)|^{2m}$. From (4.3) and the estimate for b in (4.4),

$$|\xi(t)| \leqq |\xi(s)| + C \int_s^t (1 + |\xi(r)|)\,dr + |\zeta(t)|.$$

By Gronwall's inequality (Appendix A)

$$\|\xi\| \leqq (|\xi(s)| + C(T - s) + \|\zeta\|)\, e^{C(T-s)}.$$

Let $C_1 = \exp[-C(T-T_0)]$ and recall that $T_0 \leqq s \leqq T$. For $r \geqq r_0 = 3CC_1^{-1}(T-T_0)$, $\|\xi\| > r$ implies either $|\xi(s)| > \frac{C_1}{3} r$ or $\|\zeta\| > \frac{C_1}{3} r$. Let

$$H(r) = P(\|\xi\| > r).$$

For $r \geqq r_0$,

$$H(r) \leqq P\left(|\xi(s)| > \frac{C_1 r}{3}\right) + P\left(\|\zeta\| > \frac{C_1 r}{3}\right) \leqq \left(\frac{C_1 r}{3}\right)^{-2m} [E|\xi(s)|^{2m} + M];$$

$$H(r) \leqq M_1 r^{-2m}, \quad r \geqq r_0, \tag{4.7}$$

for a constant M_1 depending on a bound for $E|\xi(s)|^{2m}$, $T-T_0$, m, and C. From (4.7) follows a stronger form of inequality (4.6). See Problem 5.

Random Coefficients b, σ. In the above, b and σ were given functions on $\bar{Q}^0 = [T_0, T] \times E^n$. However, most of the results can be generalized to include random, nonanticipative coefficients. Such coefficients will arise in Chap. VI when the control is some nonanticipative process (not necessarily obtained from a feedback control law).

We now suppose that $b(\cdot,\cdot,\cdot)$, $\sigma(\cdot,\cdot,\cdot)$ are functions on $\bar{Q}^0 \times \Omega$ which are $\mathscr{B}(\bar{Q}^0) \times \mathscr{F}$ measurable and satisfy:

(4.8) (i) For each $(t,x) \in \bar{Q}^0$, $b(t,x,\cdot)$, $\sigma(t,x,\cdot)$ are $\mathscr{F}_t$-measurable, where $\{\mathscr{F}_t\}$ is an increasing family of σ-algebras, $\mathscr{F}_t \subset \mathscr{F}$.

(ii) There exists a constant C such that

$$|b(t,x,\omega)| \leqq C(1+|x|), \qquad |\sigma(t,x,\omega)| \leqq C(1+|x|),$$
$$|b(t,x,\omega) - b(t,y,\omega)| \leqq C|x-y|,$$
$$|\sigma(t,x,\omega) - \sigma(t,y,\omega)| \leqq C|x-y|$$

for all $t \in [T_0, T]$, $x, y \in E^n$, $\omega \in \Omega$.

Consider initial data $\xi(s)$ which is $\mathscr{F}_s$-measurable, with $E|\xi(s)|^2 < \infty$, and a brownian motion w adapted to $\{\mathscr{F}_t\}$. We seek a solution of (4.3) which is a continuous process ξ, nonanticipative with respect to $\{\mathscr{F}_t\}$. Such a solution exists and is unique; moreover, the proof is essentially the same as for Theorem 4.1. See Gikhman-Skorokhod [2, §7, Theorem 1]. It should be noted that solutions of (4.3) need not be Markov when b or σ is random.

§ 5. Markov Diffusion Processes

Let us begin by recalling the definition of Markov process, on a time interval $\mathscr{T}$. Then we mention families of linear operators $S_{s,t}$, $\mathscr{A}(t)$ associated with a Markov process. In this book we shall consider Markov diffusion processes, for which the operator $\mathscr{A}(t)$ takes the form (5.5) of a partial differential operator.

For a stochastic process ξ, with state space some complete separable metric space Σ, to be Markov the state $\xi(r)$ at any time r must contain all probabilistic information relevant to the evolution of the process for times $t>r$. This can be expressed precisely as follows. Consider any finite set of times $t_1<t_2<\cdots<t_m<t$ in $\mathscr{T}$. Then for all $B\in\mathscr{B}(\Sigma)$,

$$P\big(\xi(t)\in B\,|\,\xi(t_1),\ldots,\xi(t_m)\big)=P\big(\xi(t)\in B\,|\,\xi(t_m)\big) \tag{5.1}$$

P-almost surely. Recall from Appendix D that these are versions of the conditional probabilities with respect to the σ-algebras $\mathscr{F}\big(\xi(t_1),\ldots,\xi(t_m)\big)$, $\mathscr{F}\big(\xi(t_m)\big)$ respectively. Let

$$\hat{P}(s,y,t,B)=P\big(\xi(t)\in B\,|\,\xi(s)=y\big).$$

The right side of (5.1) is $\hat{P}\big(t_m,\xi(t_m),t,B\big)$. We call $\hat{P}$ the *transition function.*

Definition. A stochastic process ξ on a time parameter set $\mathscr{T}$, with state space Σ, is a *Markov process* if:

(a) Eq. (5.1) holds for any $t_1<t_2<\cdots<t_m<t$ in $\mathscr{T}$ and $B\in\mathscr{B}(\Sigma)$;

(b) $\hat{P}(s,\cdot,t,B)$ is $\mathscr{B}(\Sigma)$ measurable for fixed s,t,B, and $\hat{P}(s,y,t,\cdot)$ is a probability measure on $\mathscr{B}(\Sigma)$ for fixed s,y,t;

(c) The Chapman-Kolmogorov equation

$$\hat{P}(s,y,t,B)=\int_\Sigma \hat{P}(r,x,t,B)\,\hat{P}(s,y,r,dx)$$

holds for $s<r<t$, $s,r,t\in\mathscr{T}$.

For the introductory discussion to follow the reader may find Gikhman-Skorokhod [1, Chap. 7], Feller [1, Chaps. IX, X] very useful. For treatments of the deep general theory of Markov processes, we refer to Dynkin [1, 2], Blumenthal-Getoor [1], Meyer [1].

Property (a) can be expressed in the slightly more elegant form:

$$P\big[\xi(t)\in B\,|\,\mathscr{F}\big(\xi(v),v\leq r\big)\big]=\hat{P}\big(r,\xi(r),t,B\big) \tag{5.1$'$}$$

for $r<t$, $r,t\in\mathscr{T}$.

Let us now define a family of linear operators $S_{s,t}$ associated with a Markov process. [In what follows, we shall not actually use results about these operators nor of the corresponding operators $\mathscr{A}(t)$. However, it seems conceptually helpful to think in these terms. Moreover, it shows the connection between Markov processes and semigroups.]

Let $V(\Sigma)$ denote the space of all bounded, real valued, Borel measurable functions Φ on Σ, with the norm

$$\|\Phi\|=\sup_{x\in\Sigma}|\Phi(x)|.$$

For every $\Phi\in V(\Sigma)$ and $s<t$ $(s,t\in\mathscr{T})$, let

$$S_{s,t}\Phi(y)=\int_\Sigma \Phi(x)\,\hat{P}(s,y,t,dx)=E_{sy}\Phi[\xi(t)].$$

Here the notation E_{sy} indicates the initial data $\xi(s)=y$. By the Chapman-Kolmogorov equations,

$$S_{s,r}S_{r,t}=S_{s,t}\qquad\text{for } s<r<t.$$

Define the operators $\mathscr{A}(t)$, on some subspace of $V(\Sigma)$, by

$$\mathscr{A}(t)\Phi = \lim_{h\to 0^+} h^{-1}[S_{t,t+h}\Phi - \Phi]. \tag{5.2}$$

The limit could be understood in a weak or strong sense. (Since the present discussion is purely motivational we shall not be more precise. See, however, Dynkin [2].) One expects from (5.2), under suitable restrictions on a function $\psi(t,x)$, that the following formula holds:

$$\frac{d}{dt} S_{s,t}\psi = S_{s,t}[\psi_t + \mathscr{A}(t)\psi]$$

where ψ_t is the partial derivative in the time-like variable t. If one integrates both sides over an interval $s \leqq r \leqq t$, and interchanges the time integral with expectations, one gets

$$\begin{aligned} &E_{sy}\psi[t,\xi(t)] - \psi(s,y) \\ &\quad = E_{sy}\int_s^t [\psi_t(r,\xi(r)) + \mathscr{A}(r)\psi(r,\xi(r))]\,dr \end{aligned} \tag{5.3}$$

for sufficiently well behaved ψ. This important formula coincides with one we shall get more directly from the Ito stochastic differential rule, in case ξ is a diffusion which satisfies a stochastic differential equation of the type (4.3).

A Markov process is called *autonomous* if its transition probabilities depend only on $t-s$:

$$\hat{P}(s,y,t,B) = \hat{\mathscr{P}}(t-s,y,B).$$

For the autonomous case the operators $S_t = S_{0,t}$ form a semigroup, and the corresponding $\mathscr{A}$ (not depending on t) is the generator (or in the terminology of Dynkin [2, Chap. I] the infinitesimal operator).

Diffusions. Let us suppose from now on that $\Sigma = E^n$. A Markov process on an interval $\mathscr{T}$ is called an *n-dimensional diffusion process* if:

(1) For every $\varepsilon > 0$ and $t \in \mathscr{T}$, $x \in E^n$,

$$\lim_{h\to 0^+} h^{-1} \int_{|x-z|>\varepsilon} \hat{P}(t,x,t+h,dz) = 0;$$

(2) There exist functions $a_{ij}(t,x)$, $b_i(t,x)$, $i,j = 1,\dots,n$, such that for every $\varepsilon > 0$ and $t \in \mathscr{T}$, $x \in E^n$

$$\lim_{h\to 0^+} h^{-1} \int_{|x-z|\leqq\varepsilon} (z_i - x_i)\hat{P}(t,x,t+h,dz) = b_i(t,x)$$

$$\lim_{h\to 0^+} h^{-1} \int_{|x-z|\leqq\varepsilon} (z_i - x_i)(z_j - x_j)\hat{P}(t,x,t+h,dz) = a_{ij}(t,x).$$

The vector function $b = (b_1,\dots,b_n)$ is called the *local drift coefficient* and the matrix-valued function $a = (a_{ij})$ the *local covariance matrix*. The justification for these names is as follows. Suppose that instead of (1) the slightly stronger condition

$$\lim_{h\to 0^+} h^{-1} \int_{|x-z|>\varepsilon} |z-x|^2 \hat{P}(t,x,t+h,dz) = 0 \tag{1'}$$

holds. Then from (1′) and (2), $b(t, \xi(t))h$ and $a(t, \xi(t))h$ are good approximations to the mean and covariance matrix of the increment $\xi(t+h)-\xi(t)$ conditioned on $\xi(t)$.

In what follows, we shall consider a fixed time interval $\mathscr{T}_0=[T_0, T]$, and shall be interested in various initial data $s, y=\xi(s)$ with $T_0 \leqq s < T$, $y \in E^n$. Thus, given the coefficients a, b we consider not just one diffusion but rather the family of diffusions corresponding to all such initial data (s, y). [Many authors, including Dynkin [2], use the term Markov process for this family, or more generally for the family of processes associated with a given transition function $\hat{P}(s, y, t, B)$.]

Diffusions Represented as Solutions of a Stochastic Differential Equation. Suppose now that we are given a vector-valued function b and matrix valued σ, which are continuous on the closed strip $\bar{Q}^0=\mathscr{T}_0 \times E^n$ and satisfy the conditions (4.4), (4.5). It can be shown Gikhman-Skorokhod [2, §10, Theorem 2] that, for each (s, y), the solution of (4.3) with $\xi(s)=y$ is a diffusion. The local drift coefficient is b, and the local covariance matrix a satisfies

$$a_{ij}=\sum_{l=1}^{d} \sigma_{il}\sigma_{jl}, \qquad i, j=1, \ldots, n. \tag{5.4}$$

In other words, $\sigma\sigma'=a$. The matrices $a(t, x)$ are symmetric nonnegative definite. For a diffusion process, $\mathscr{A}(t)$ takes the form of a second order partial differential operator, in the sense that the domain of $\mathscr{A}(t)$ contains all Φ of class C^2 and

$$\mathscr{A}(t)\,\Phi=\tfrac{1}{2}\sum_{i,j=1}^{n} a_{ij}(t, x)\frac{\partial^2 \Phi}{\partial x_i\,\partial x_j}+\sum_{i=1}^{n} b_i(t, x)\frac{\partial \Phi}{\partial x_i}. \tag{5.5}$$

Given such operators $\mathscr{A}(t)$, to find corresponding diffusions by the method of stochastic differential equations we need a square root $\sigma(t, x)$ of the nonnegative definite symmetric matrix $a(t, x)$, as in (5.4). Moreover, we would like b, σ to satisfy the conditions (4.4), (4.5). If a_{ij} is of class C^2 and a_{ij} together with its first and second order partial derivatives is bounded for $i, j=1, \ldots, n$, then such a $\sigma(t, x)$ exists which is Lipschitz in (t, x). See Freidlin [2, Theorem 1] or Phillips-Sarason [1, Lemma 1.1]. In that case σ satisfies the Ito conditions, and therefore (4.4), (4.5).

Remark. A different question is whether a given diffusion process is a solution of (4.3) for suitable brownian motion w. For $n=1$, this is proved in Doob [1, p. 287] assuming that (1′) (2) above hold, that $a>0$, and that $b, \sigma=a^{1/2}$ satisfy the growth conditions (4.4). For an extension of this result, valid for any finite dimension n, see Wong [2, Theorem 4.3]. (We do not use these results here.)

For a solution of (4.3), Eq. (5.3) can now be derived formally, as follows. We apply the Ito stochastic differential rule, and rewrite the right side of (3.14) using $d\xi=b\,dt+\sigma\,dw$. In integrated form, we get

$$\psi(t, \xi(t))-\psi(s, \xi(s))=\int_s^t [\psi_t+\mathscr{A}(r)\,\psi]\,dr+\int_s^t \psi_x\,\sigma\,dw. \tag{*}$$

Upon taking expectations and using (3.3) with $e=\psi_x\sigma$ we get (5.3), in case of fixed initial data $\xi(s)=y$. This argument is correct provided ψ is $C^{1,2}$, all expectations and integrals in (5.3) exist, and $E\int_s^t |\psi_x \sigma|^2\, dr < \infty$. (The last condition implies (3.3)). However, we shall now prove a result which implies (5.3) under less restrictive assumptions. Moreover, we replace the fixed upper limit of integration t by a stopping time τ, namely, the exit time from some open set Q.

If τ is a stopping time, $s \leqq \tau \leqq T$, and $e \in \mathscr{M}_0$, then $\int_s^\tau e\, dw$ is defined to be $\zeta(\tau)$, with ζ as in (3.6). Another expression for it is

$$\int_s^\tau e\, dw = \int_s^T \chi e\, dw,$$

where $\chi(t)=1$ if $t<\tau$, $\chi(t)=0$ if $t \geqq \tau$. See Gikhman-Skorokhod [2, §4]. From (3.3) we have for $e \in \mathscr{M}$

$$(**) \qquad E\int_s^\tau e\, dw = E\int_s^T \chi e\, dw = 0.$$

The *exit time* τ from an open set Q is defined as the first time t with $(t, \xi(t)) \notin Q$. By the reasoning in Example 2.2 the exit time is a stopping time.

Lemma 5.1. *Let ψ be of class $C^{1,2}$, Q open and bounded, $(s, \xi(s)) \in Q$, and τ the exit time from Q. Then*

$$(5.6) \qquad E\psi(\tau, \xi(\tau)) - E\psi(s, \xi(s)) = E\int_s^\tau \left[\psi_t(r, \xi(r)) + \mathscr{A}(r)\psi(r, \xi(r))\right] dr.$$

Proof. Formula (*) holds except for a fixed set of probability 0 not depending on t. Let us replace t by $\tau(=\tau(\omega))$ in (*), take expectations, and use (**) with $e = \chi \psi_x \sigma$. All expectations and integrals involved exist, since all terms are bounded. □

Let us denote by $C^{1,2}(Q)$ the space of functions $\psi(t,x)$ continuous on Q together with the partial derivatives $\psi_t, \psi_{x_i}, \psi_{x_i x_j}$, $i,j=1,\dots,n$. We say that ψ satisfies a *polynomial growth condition* on Q if, for some constants D, k, $|\psi(t,x)| \leqq D(1+|x|^k)$ when $(t,x) \in Q$. {This condition is trivially satisfied if Q is bounded and ψ is continuous on $\bar{Q}$.} Let $C_p^{1,2}(Q)$ denote the class of ψ in $C^{1,2}(Q)$ which satisfy a polynomial growth condition on Q.

In the next theorem Q is open, $Q \subset Q^0$; and ξ denotes a solution of (4.3) with initial data $\xi(s)=y$, $(s,y) \in Q$. The exit time from Q is τ.

Theorem 5.1. *Assume that:*

(a) *b, σ satisfy* (4.4);

(b) *ψ is in $C_p^{1,2}(Q)$, ψ is continuous on $\bar{Q}$; and*

(c) *$\psi_t + \mathscr{A}(t)\psi + M(t,x) \geqq 0$ for all $(t,x) \in Q$, where $E_{sy}\int_s^\tau |M(t, \xi(t))|\, dt < \infty$ for each $(s,y) \in Q$. Then*

$$(5.7) \qquad \psi(s,y) \leqq E_{sy}\left\{\int_s^\tau M(t, \xi(t))\, dt + \psi(\tau, \xi(\tau))\right\}.$$

Proof. Take a sequence of bounded open sets $Q_1, Q_2, \dots$ with $\bar{Q}_j \subset Q_{j+1} \subset Q$, $j=1,2,\dots$, and Q the union of the Q_j. For $(s,y) \in Q_j$, let τ_j be the exit time of

$(t, \xi(t))$ from Q_j. Then τ_j is an increasing sequence, which tends with probability 1 to τ as $j \to \infty$ since ξ is a continuous process. Let α_j be a C^2 function such that $\alpha_j = 1$ on $\bar{Q}_j$ and $\alpha_j = 0$ outside a compact subset of Q. The function $\psi_j = \alpha_j \psi$ is of class $C^{1,2}$, and $\psi = \psi_j$ on $\bar{Q}_j$. By Lemma 5.1, with τ replaced by τ_j:

$$\psi(s, y) = -E \int_s^{\tau_j} [\psi_t + \mathscr{A}(t)\psi]\, dt + E\psi(\tau_j, \xi(\tau_j)),$$

$$(\#) \qquad \psi(s, y) \leqq E \int_s^{\tau_j} M(t, \xi(t))\, dt + E\psi(\tau_j, \xi(\tau_j)).$$

Here, $E = E_{sy}$.

The first term on the right side of $(\#)$ tends to $E \int_s^\tau M\, dt$ as $j \to \infty$. Let us show that

$$(\#\#) \qquad \lim_{j \to \infty} E\psi(\tau_j, \xi(\tau_j)) = E\psi(\tau, \xi(\tau)).$$

Now $\psi(\tau_j, \xi(\tau_j)) \to \psi(\tau, \xi(\tau))$ with probability 1 as $j \to \infty$, since $\tau_j \to \tau$ and ψ is continuous on $\bar{Q}$. For $R > 0$, let

$$\chi_R = 1 \text{ if } \|\xi\| \leqq R, \qquad \chi_R = 0 \text{ if } \|\xi\| > R.$$

By the dominated convergence theorem,

$$\lim_{j \to \infty} E[\psi(\tau_j, \xi(\tau_j))\chi_R] = E[\psi(\tau, \xi(\tau))\chi_R]$$

for each $R > 0$. Since ψ has polynomial growth, $|\psi(t, x)| \leqq D(1 + |x|^k)$ for some D, k. Let $H(r) = P(\|\xi\| > r)$, and take $2m > k$ in (4.7). Then $\int^\infty r^{k-1} H(r)\, dr < \infty$, which implies (upon integrating by parts on (R, ∞))

$$\lim_{R \to \infty} \int_R^\infty (1 + r^k)\, dH(r) = 0.$$

Then

$$|\psi(\tau_j, \xi(\tau_j))| \leqq D(1 + |\xi(\tau_j)|^k) \leqq D(1 + \|\xi\|^k),$$

$$E[|\psi(\tau_j, \xi(\tau_j))|(1 - \chi_R)] \leqq -D \int_R^\infty (1 + r^k)\, dH(r),$$

and the right side tends to 0 as $R \to \infty$. This proves $(\#\#)$, and hence the theorem. □

In particular, if $M = -(\psi_t + \mathscr{A}(t)\psi)$ satisfies the integrability condition (c), we have by applying Theorem 5.1 to $\pm\psi$, $\mp M$:

Theorem 5.2. *Assume that* (a), (b) *of Theorem* 5.1 *hold, and that*

$$E_{sy} \int_s^\tau |\psi_t + \mathscr{A}(t)\psi|\, dt < \infty$$

for each $(s, y) \in Q$. *Then*

$$(5.8) \qquad \psi(s, y) = -E_{sy} \int_s^\tau [\psi_t + \mathscr{A}(t)\psi]\, dt + E_{sy}\psi(\tau, \xi(\tau)).$$

Remark 5.1. From the proof of the two theorems, one can see that they remain true for random coefficients b, σ satisfying (4.8). Condition (4.8)(ii) can even be replaced in these theorems by any weaker assumption which guarantees that

$$E_{sy} \int_s^T |b|^{2m}\,dt < \infty, \qquad E_{sy} \int_s^T |\sigma|^{2m}\,dt < \infty \qquad \text{for each } m = 1, 2, \ldots .$$

Diffusions as Approximations. In many continuous time models, an approximation is made in which certain non-Markovian processes are replaced by diffusions. Other models deal with Markov processes in which the time parameter set is a discrete but rather finely spaced subset of E^1. As an idealization these Markov processes are often replaced by diffusions. We now discuss some cases where such approximations arise.

(a) *Wideband Noise.* Suppose that some physical process, if unaffected by random disturbances, can be described by a (vector) ordinary differential equation $d\xi = b(t, \xi(t))\,dt$. If, however, such disturbances enter the system in an additive way, then one might take as a model

$$(5.9) \qquad d\xi = b(t, \xi(t))\,dt + v(t)\,dt,$$

where v is some stationary process with mean 0 and known autocovariance matrix $R(r)$:

$$R_{ij}(r) = E\{v_i(t)\, v_j(t+r)\}, \qquad i, j = 1, \ldots, n.$$

If $R(r)$ is nearly 0 except in a small interval near $r=0$, then v is called wide band noise. White noise corresponds to the ideal case when R_{ij} is a constant a_{ij} times a Dirac delta function. Then $v(t)\,dt$ is replaced by $\sigma\,dw$, where σ is a constant matrix such that $\sigma\sigma^T = a$, $a = (a_{ij})$. The corresponding diffusion, satisfying (4.3), is then an approximation to the solution to (5.9). To indicate how this is proved, suppose that $v = v^c$ where c is a small positive parameter; and let $w^c(t) = w(s) + \int_s^t v^c(r)\,dr$. Suppose that $\|w^c - w\|$ tends to 0 in probability as $c \to 0$ ($\|\ \|$ is the sup norm on $[s, T]$). Let ξ^c be the solution of (5.9) with $v = v^c$ and $\xi^c(s) = \xi^0(s)$ where ξ^0 is a solution of (4.3) with the given (constant) σ. If we assume that $|b_x| \leq K$, then $\|\xi^c - \xi^0\|$ tends to 0 in probability as $c \to 0$. To show this, write the equations for ξ^0 and ξ^c in integrated form and subtract. We get

$$|\xi^c(t) - \xi^0(t)| \leq K \int_s^t |\xi^c(r) - \xi^0(r)|\,dr + w^c(t) - w(t).$$

The desired conclusion follows from Gronwall's inequality.

Example 5.1. Let $n=1$. For $c>0$ define v^c by

$$c\,dv^c + v^c\,dt = dw.$$

The process w^c furnishes a model, due to Ornstein and Uhlenbeck, for physical brownian motion. The velocity $v^c = \dot{w}^c$ is a continuous process. The process w^c is not Markov, but (w^c, v^c) is a 2-dimensional Markov process. As $c \to 0$, $\|w^c - w\| \to 0$ in probability, no matter what (fixed) initial data $v^c(s)$ are given (Problem 8).

Warning. If the coefficient σ is allowed to depend on system states, then the above approximation is incorrect. Suppose that

$$d\xi^c = b(t, \xi^c(t))\, dt + \sigma(t, \xi^c(t))\, dw^c.$$

As $c \to 0$, ξ^c tends to a solution ξ^0 not of (4.3), but rather of the stochastic differential equation

$$d\xi^0 = (b + \tfrac{1}{2}\sigma_x \sigma)\, dt + \sigma\, dw.$$

Here the i-th component of the vector $\sigma_x \sigma$ is $\sum_{j,k} (\partial \sigma_{ij}/\partial x_k) \sigma_{kj}$, and b, σ, σ_x are evaluated at $(t, \xi^0(t))$. For a discussion of this matter see Wong-Zakai [1][2], McShane [2][7], Ito [1].

(b) *Discrete Parameter Models.* For $\delta > 0$ consider the discrete time parameter set

$$\mathscr{T}^\delta = \{t = k\delta, k = 0, 1, \ldots, N\}, \qquad N\delta = T,$$

and ξ^δ a Markov process on $\mathscr{T}^\delta$. The process ξ^δ may have continuous or discrete state space, contained in E^n. Suppose that the conditional distribution of the increment $\xi^\delta(t+\delta) - \xi^\delta(t)$ given $\xi^\delta(t)$ has mean $\delta b(t, \xi^\delta(t)) + o(\delta)$ and covariance matrix $\delta a(t, \xi^\delta(t)) + o(\delta)$, where $\delta^{-1} o(\delta)$ tends to 0 as $\delta \to 0$. One may ask whether, under suitable conditions, ξ^δ tends to a limit ξ^0 which is a diffusion with local drift b and local covariance a. The assumptions about means and covariances do not exclude the possibility that ξ^δ tends to some ξ^0 which is not a diffusion. For instance, if the increments $\xi^\delta(t+\delta) - \xi^\delta(t)$ are independent, then ξ^0 could be some process with independent increments other than a brownian motion. Roughly speaking, for ξ^0 to be a diffusion some condition is needed which insures that $\xi^0(t+h) - \xi^0(t)$ is approximately gaussian.

Perhaps the earliest application of the technique of diffusion approximation was to the case when the process ξ^δ consists of sums of independent random variables with mean 0, suitably scaled according to space and time. In that case, ξ^0 is a brownian motion. The approximation of ξ^δ by ξ^0 yields so-called invariance principles for various functionals of ξ^δ. See for instance Donsker [1].

The diffusion approximation method has proved useful in a wide variety of applications, including sequential decision problems in statistics, highway traffic problems, queues, and population genetics theory. We refer to Chernoff [1], Khas'minskii and Nevel'son [1], Iglehart [1][2], Newell [1][2], Schach [1], Crow-Kimura [1]. For further convergence results see Gikhman-Skorokhod [1], Kushner [7].

§ 6. Backward Equations

In the next three sections we discuss some partial differential equations associated with diffusion processes. The local drift b and local covariance a determine two partial differential operators of second order. They are called the backward and forward operators; and they are (at least formally) adjoint to each other. It is useful to distinguish notationally between variables (s, y) which appear in the backward operator, and variables (t, x) in the forward operator.

The *backward operator* is

$$(6.1) \qquad \frac{\partial}{\partial s}+\mathscr{A}(s)=\frac{\partial}{\partial s}+\frac{1}{2}\sum_{i,j=1}^{n} a_{ij}(s,y)\frac{\partial^2}{\partial y_i \partial y_j}+\sum_{i=1}^{n} b_i(s,y)\frac{\partial}{\partial y_i}.$$

Note that $\mathscr{A}(s)$ is the same operator as in (5.5) provided $a=\sigma\sigma'$. If the matrices $a(s,y)=(a_{ij}(s,y))$ are positive definite, the backward operator is *parabolic*. The backward operator is *uniformly parabolic* if the characteristic values of $a(s,y)$ are bounded below by some $c>0$. When the $a(s,y)$ are merely nonnegative definite, it is called a *degenerate* parabolic operator.

Given a real-valued function Λ, the partial differential equation

$$(6.2) \qquad \frac{\partial \psi}{\partial s}+\mathscr{A}(s)\psi+\Lambda(s,y)=0$$

is called a *backward equation*. If $\Lambda=0$, it is the *homogeneous* backward equation. (Forward equations are discussed in Sect. 8.)

We are concerned in Sects. 6 and 7 with solutions of (6.2) in some open set Q, with boundary data for ψ on the appropriate part of the boundary ∂Q. From probabilistic formulas for the solution $\psi(s,y)$ it will become apparent that boundary data should be assigned only on those parts of Q through which a diffusion starting at some $(s,y)\in Q$ exits with positive probability.

The Cauchy Problem. In this section we consider the case when Q is the strip $Q^0=(T_0,T)\times E^n$ in $n+1$ dimensions, and boundary data are imposed at the final time T:

$$(6.3) \qquad \psi(T,y)=\Psi(y), \qquad y\in E^n.$$

Such data at a fixed time T are called Cauchy data. Suppose that b, σ satisfy (4.4), and that ψ is a solution of (6.2)–(6.3) with $\psi\in C_p^{1,2}(Q^0)$ and ψ continuous in $\bar{Q}^0$. By Theorem 5.2

$$(6.4) \qquad \psi(s,y)=E_{sy}\left\{\int_s^T \Lambda[t,\xi(t)]\,dt+\Psi[\xi(T)]\right\},$$

provided $E_{sy}\int_s^T|\Lambda|\,dt<\infty$. This is a probabilistic formula for $\psi(s,y)$ in terms of the diffusion which has initial data (s,y).

Example 6.1. Let $b=0$, $a=$identity matrix. The backward operator is

$$\frac{\partial \psi}{\partial s}+\frac{1}{2}\Delta_y\psi$$

where $\Delta_y=\sum_{i=1}^n \partial^2/\partial y_i^2$ is the Laplace operator. This is the simplest example of a uniformly parabolic operator. To get a diffusion process ξ corresponding to these coefficients a, b and initial data $\xi(s)=y$, we simply translate an n-dimensional brownian motion w: $\xi(t)=y+w(t)-w(s)$.

Example 6.2. Consider a second order linear differential equation with white noise forcing term

$$\ddot{\eta}+\mu\dot{\eta}+\nu\eta=\dot{w}$$

where $\cdot = d/dt$. Let $\xi_1 = \eta$, $\xi_2 = \dot{\eta}$, and rewrite this equation as

$$d\xi_1 = \xi_2\, dt, \qquad d\xi_2 = -(\mu\xi_2 + \nu\xi_1)\, dt + dw.$$

The backward operator is

$$\frac{\partial}{\partial s} + \frac{1}{2}\frac{\partial^2}{\partial y_2^2} + y_2 \frac{\partial}{\partial y_1} - (\mu y_2 + \nu y_1)\frac{\partial}{\partial y_2}.$$

It is degenerate parabolic since the only second order derivative appearing is $\partial^2/\partial y_2^2$. This example will be discussed further in Sect. 9.

Existence of Solutions to the Cauchy Problem (6.2)–(6.3). If the backward operator is uniformly parabolic, then an existence theorem from the theory of parabolic partial differential equations can be quoted. See Friedman [1, p. 25], Ladyzenskaya-Solonnikov-Ural'seva [1, Chap. IV], also Appendix E. In that case only rather weak smoothness assumptions on the coefficients a_{ij}, b_i in (6.1), and on Λ, Ψ are needed. If the backward operator is degenerate parabolic, then existence and uniqueness theorems for the Cauchy problem have been obtained both by methods of partial differential equations (see for instance Kohn-Nirenberg [1], Oleinik [1], Phillips-Sarason [1]) and by a probabilistic method. In the latter method, one defines ψ by (6.4) and then shows that ψ satisfies (6.2) by using the smooth dependence of the solution ξ to (4.3) on the initial state y, Gikhman-Skorokhod [2, §'s 8, 11]. Without uniform parabolicity, more smoothness on a_{ij}, b_i, Λ, Ψ must be assumed.

We quote the following result when $\Lambda = 0$, which will be needed later. See Gikhman-Skorokhod [2, §11, Theorem 1].

Theorem 6.1. *Assume that:*

(a) *b, σ are continuous, and satisfy* (4.4);

(b) *$b(t,\cdot)$, $\sigma(t,\cdot)$ are C^2 for each $t \in [T_0, T]$; moreover, b_x, σ_x are bounded on Q^0 and the second order partial derivatives b_{x_i}, $b_{x_i x_j}$, σ_{x_i}, $\sigma_{x_i x_j}$, $i, j = 1, \ldots, n$ satisfy a polynomial growth condition on Q^0;*

(c) *Ψ is C^2, and satisfies together with its partial derivatives Ψ_{x_i}, $\Psi_{x_i x_j}$, $i, j = 1, \ldots, n$, a polynomial growth condition.*

Then $\psi(s, y) = E_{sy}\Psi[\xi(T)]$ is a solution in $C_p^{1,2}(\bar{Q}^0)$ of the homogeneous backward equation $\frac{\partial \psi}{\partial s} + \mathscr{A}(s)\psi = 0$, with the Cauchy data (6.3).

§ 7. Boundary Value Problems

Let us now consider a backward Eq. (6.2) in some open set $Q \subset Q^0$. We again get a probabilistic formula for a solution ψ, with boundary data on an appropriate part $\partial^* Q$ of the boundary ∂Q. For $\partial^* Q$ we take a closed subset of ∂Q, such that $(\tau, \xi(\tau)) \in \partial^* Q$ for every choice of initial point $(s, y) \in Q$. Again, τ denotes the exit time from Q. As boundary data, we specify

(7.1) $$\psi(s, y) = \Psi(s, y), \qquad (s, y) \in \partial^* Q.$$

Under the conditions of Theorem 5.2, we get from (5.8) the following probabilistic formula for the solution to (6.2)–(7.1):

$$\psi(s,y)=E_{sy}\left\{\int_s^\tau \Lambda(t,\xi(t))\,dt+\Psi(\tau,\xi(\tau))\right\}. \tag{7.2}$$

For $Q=Q^0$, $\partial^* Q$ is the hyperplane $\{T\}\times E^n$; the data (7.1) are then just the Cauchy data (6.3). Another example is $Q=(T_0,T)\times G$, where $G\subset E^n$ is open. We call such a Q a cylindrical domain. We may then take

$$\partial^* Q=([T_0,T]\times\partial G)\cup(\{T\}\times G).$$

If the backward operator is uniformly parabolic, this turns out to be smallest possible choice for $\partial^* Q$.

Boundary Problems—Autonomous Case. Let us now suppose that (4.3) has the autonomous form

$$d\xi=b[\xi(t)]\,dt+\sigma[\xi(t)]\,dw, \qquad t\geqq 0; \tag{4.3a}$$

and consider the operator

$$\mathscr{A}=\tfrac{1}{2}\sum_{i,j=1}^n a_{ij}(y)\frac{\partial^2}{\partial y_i\,\partial y_j}+\sum_{i=1}^n b_i(y)\frac{\partial}{\partial y_i}.$$

If the characteristic values of the matrices $a(y)=(a_{ij}(y))$ are bounded below by $c>0$, then $\mathscr{A}$ is *uniformly elliptic.*

For each $y\in E^n$ the solution of (4.3a) with $\xi(0)=y$ is a Markov process. The transition densities are autonomous. For $\Phi\in C^2$ with compact support, $\mathscr{A}\Phi$ agrees with the infinitesimal operator of the associated semigroup Dynkin [2, Chap. V.5].

For $G\subset E^n$, G open, we now consider the equation

$$\mathscr{A}\Phi+\Lambda(y)=0, \qquad y\in G \tag{7.3}$$

with boundary data

$$\Phi(y)=\Psi(y), \qquad y\in\partial^* G. \tag{7.4}$$

For $y\in G$, let τ denote the exit time from G of $\xi(t)$; if $\xi(t)\in G$ for all $t>0$, let $\tau=+\infty$. (In the autonomous case we impose no upper bound T on time t.) Here $\partial^* G$ denotes a closed subset of ∂G such that $P_y(\xi(\tau)\notin\partial^* G,\ \tau<\infty)=0$ for each $y\in G$.

In the following theorem we assume that $b(x)$, $\sigma(x)$ satisfy the linear growth condition (4.4); and that ξ satisfies (4.3a) with initial data $\xi(0)=y$ in G. By Φ in $C^2(G)$ we mean Φ has continuous second order partial derivatives in G.

Theorem 7.1. *Let Φ be in $C^2(G)$, with Φ continuous and bounded in $\bar{G}$. Suppose that $P_y(\tau<\infty)=1$ for each $y\in G$, where τ is the exit time from G.*

(a) *If $\mathscr{A}\Phi+M(x)\geqq 0$ for all $x\in G$, where $E_y\int_0^\tau |M[\xi(t)]|\,dt<\infty$, then*

$$\Phi(y)\leqq E_y\left\{\int_0^\tau M[\xi(t)]\,dt+\Phi[\xi(\tau)]\right\}, \qquad y\in G. \tag{7.5}$$

(b) *If Φ is a solution of (7.3)–(7.4) and $E_y \int_0^\tau |\Lambda[\xi(t)]| dt < \infty$, then*

$$\Phi(y) = E_y \left\{ \int_0^\tau \Lambda[\xi(t)] dt + \Psi[\xi(\tau)] \right\}. \tag{7.6}$$

Proof of (a). For $T > 0$, let $\tau \wedge T = \min(\tau, T)$ and $Q = (0, T) \times G$. Apply Theorem 5.1 in Q, with $\psi = \Phi$:

$$\Phi(y) \leqq E_y \left\{ \int_0^{\tau \wedge T} M[\xi(t)] dt + \Phi[\xi(\tau \wedge T)] \right\}.$$

Let $T \to \infty$. Using the boundedness of Φ and assumption about M, the right side tends to the right side of (7.5).

Part (b) is proved similarly, using Theorem (5.2). □

A unique solution to (7.3)–(7.4) with the properties required in Theorem 7.1(b) exists in case $\mathscr{A}$ is uniformly elliptic, G is bounded, and some additional technical conditions are satisfied. This is a result in the theory of elliptic partial differential equations; see Ladyzhenskaya-Ural'seva [1, Chap. 3]. The probabilistic method used to prove Theorem 6.1 does not seem to work for the boundary value problem (7.3)–(7.4). The difficulty is the nondifferentiability of the exit time τ with respect to the initial state y.

Some results about existence of solutions to boundary problems for degenerate elliptic (or parabolic) equations are known Kohn-Nirenberg [1], Oleinik [1]. However, the assumptions are more restrictive than for the uniformly elliptic case. For a probabilistic method see Freidlin [1].

Example. Let $\Lambda = 1$, $\Psi = 0$. Then $\Phi(y) = E_y \tau$ is the expected time to reach ∂G. This is always finite if G is bounded and $\mathscr{A}$ uniformly elliptic. Otherwise, $E_y \tau$ might be infinite. A condition for positive recurrence of the process ξ is $E_y \tau < \infty$ whenever G is the exterior of an n-dimensional sphere and $y \in G$. Positive recurrence implies that an equilibrium distribution exists. See Wonham [1].

Remark 7.1. For unbounded solutions Φ of (7.3)–(7.4), Theorem 7.1 remains true if one assumes

$$\lim_{R \to \infty} E_y \{\Phi[\xi(\tau)]\, \theta_R\} = 0, \qquad y \in G, \tag{7.7}$$

where $\theta_R = 1$ if $\max_{0 \leqq t \leqq \tau} |\xi(t)| > R$, $\theta_R = 0$ otherwise.

Remark 7.2. Theorem 7.1 extends to random nonanticipative coefficients b, σ in the same way noted in Remark 5.1.

§ 8. Forward Equations

Let us now consider the situation when the solution of (4.3) with initial data $\xi(s) = y$ has a probability density $p(s, y, t, x)$:

$$P_{sy}(\xi(t) \in B) = \hat{P}(s, y, t, B) = \int_B p(s, y, t, x)\, dx, \qquad \text{all } B \in \mathscr{B}(E^n).$$

Here $T_0 \leqq s < t \leqq T$. We first show that, under suitable assumptions, p satisfies in the variables (t, x) a partial differential Eq. (8.2) adjoint to the homogeneous backward equation ((6.2) with $\Lambda = 0$). Then we show, under different assumptions, that p satisfies the homogeneous backward equation. Finally, we discuss the idea of fundamental solution for the backward equation.

For sufficiently smooth coefficients $a_{ij}(t, x)$, $b_i(t, x)$, and function $q(t, x)$, let

$$\mathscr{A}^*(t)\, q = \tfrac{1}{2} \sum_{i,j=1}^{n} \frac{\partial^2}{\partial x_i\, \partial x_j} (a_{ij}\, q) - \sum_{i=1}^{n} \frac{\partial}{\partial x_i} (b_i\, q). \tag{8.1}$$

The operator $-\partial/\partial t + \mathscr{A}^*(t)$ is called the *forward operator*. Let $Q_{s,t} = (s, t) \times E^n$.

Theorem 8.1. *Assume that b, σ satisfy (4.4), (4.5) and for $i, j = 1, \ldots, n$, $b_i(t, \cdot)$ is C^1 and $\sigma_{ij}(t, \cdot)$ is C^2 for $T_0 \leqq t \leqq T$. If $p(s, y, \cdot, \cdot)$ is in $C^{1,2}(Q_{s,T})$, then*

$$-\frac{\partial p}{\partial t} + \mathscr{A}^*(t)\, p = 0. \tag{8.2}$$

Proof. Consider any $\psi \in C^{1,2}$ such that ψ has compact support in $Q_{s,T}$. By (5.8), with $\tau = T$,

$$\begin{aligned} 0 &= E_{sy} \int_s^T [\psi_t + \mathscr{A}(t)\, \psi]\, dt \\ &= \int_{Q_{s,T}} [\psi_t(t, x) + \mathscr{A}(t)\, \psi(t, x)]\, p(s, y, t, x)\, dx\, dt. \end{aligned}$$

On the right side, integrate $\psi_t\, p$ by parts in t, and $(\mathscr{A}(t)\, \psi)\, p$ by parts in the variables $x_1, \ldots, x_n$ to obtain

$$0 = \int_{Q_{s,T}} \psi(-p_t + \mathscr{A}^*(t)\, p)\, dt\, dx.$$

Since this is true for all such ψ, we get Theorem 8.1. □

Theorem 8.2. *Suppose that:*

(1) *b, σ satisfy assumptions* (a), (b) *of Theorem* 6.1;

(2) *For each $(t, x) \in Q^0$, $p(\cdot, \cdot, t, x)$ is in $C^{1,2}(Q_{T_0 t})$;*

(3) *For $T_0 < s < t$, the partial derivatives $p_s(s, y, t, \cdot)$, $p_{y_i}(s, y, t, \cdot)$, $p_{y_i y_j}(s, y, t, \cdot)$ are integrable over any bounded subset of E^n.*

Then $p(\cdot, \cdot, t, x)$ satisfies the homogeneous backward equation, (6.2) with $\Lambda = 0$.

Proof. Consider any Ψ of class C^2 with compact support. For $T_0 < s < t$, let

$$\psi(s, y) = \int_{E^n} p(s, y, t, x)\, \Psi(x)\, dx = E_{sy}\, \Psi[\xi(t)]. \tag{8.3}$$

By Theorem 6.1, with T replaced by t,

$$0 = \psi_s + \mathscr{A}(s)\, \psi, \qquad T_0 < s < t.$$

On the other hand, by assumption (3) we may differentiate under the integral sign in (8.3) Fleming [1, p. 197] to get

$$\psi_s + \mathscr{A}(s)\,\psi = \int_{E^n} (p_s + \mathscr{A}(s)\,p)\,\Psi(x)\,dx.$$

Since this is 0 for all such Ψ, $p_s + \mathscr{A}(s)\,p = 0$ as required. □

Fundamental Solutions. The transition density p in Theorem 8.2 is, in terms of the theory of linear partial differential equations, just the fundamental solution. To show this, we note that for any continuous bounded function Ψ the function $\psi(s, y)$ defined by (8.3) satisfies the homogeneous backward equation in $Q_{T_0,t}$. Moreover, ψ is continuous in $\bar{Q}_{T_0,t}$ and $\Psi(y) = \psi(t, y)$. This means that p is a fundamental solution, according to the definition Friedman [1, p. 3]. (There the sense of time is opposite to ours, with Cauchy data imposed at an initial time rather than a final time.) On the other hand, suppose that there is a fundamental solution $p(s, y, t, x)$ satisfying (2), (3) of Theorem 8.2. The left hand equality in (8.3) defines a solution ψ of $0 = \partial\psi/\partial s + \mathscr{A}(s)\,\psi$ of class $C^{1,2}(Q_{T_0,t})$ continuous and bounded in $\bar{Q}_{T_0,t}$, with $\Psi(y) = \psi(t, y)$. By Theorem 5.2, with τ replaced by t, the right hand inequality in (8.3) holds.

Therefore, the fundamental solution p is the transition density.

If the backward equation is uniformly parabolic, with bounded Hölder continuous coefficients a, b, then a fundamental solution p exists. See Friedman [1, Chap. 1, §'s 4, 6] or Ladyzenskaya-Solonnikov-Ural'seva [1, Chap. IV, § 11, 13]. For C^∞ coefficients a, b, the existence of a C^∞ fundamental solution implies that the backward operator is hypoelliptic. See Hormander [1].

In the next section we consider linear stochastic differential equations. For such equations, all the conditions in Theorem 8.1, 8.2 will be satisfied if the system is completely controllable.

§ 9. Linear System Equations; the Kalman-Bucy Filter

Let us now suppose that the stochastic differential equation (4.3) has the special linear form:

$$d\xi = A(t)\,\xi(t)\,dt + \sigma(t)\,dw, \qquad s \leqq t \leqq T. \tag{9.1}$$

We can then find a system of ordinary differential equations for the means and covariances, as follows. Let

$$\mu_i(t) = E\,\xi_i(t), \qquad \rho_{ij}(t) = E\big[(\xi_i(t) - \mu_i(t))(\xi_j(t) - \mu_j(t))\big].$$

They are the components of the mean vector $\mu(t)$ and the covariance matrix $R(t)$. For $t = s$, $\mu(s)$ and $R(s)$ give the means and covariances for the initial data $\xi(s)$.

To derive ordinary differential equations which they satisfy, we take special choices for ψ in formula (*), Sect. 5. First, we take $\psi(x) = x_i$ and take expected values. Then

$$\mu_i(t) = E\,\xi_i(t) = E\,\xi_i(s) + E\int_s^t b_i(r, \xi(r))\,dr, \qquad i = 1, \ldots, n.$$

Since $b(t,x)=A(t)\,x$, this system of equations becomes

$$\mu(t)=\mu(s)+E\int_s^t A(r)\,\xi(r)\,dr=\mu(s)+\int_s^t A(r)\,\mu(r)\,dr.$$

Thus, in differential notation

$$d\mu=A\,\mu\,dt. \tag{9.2}$$

We next take $\psi(t,x)=(x_i-\mu_i(t))(x_j-\mu_j(t))$. Then from (5.8) one finds that

$$\rho_{ij}(t)=\rho_{ij}(s)-\int_s^t E[\dot{\mu}_i(\xi_j-\mu_j)+\dot{\mu}_j(\xi_i-\mu_i)]\,dr$$

$$+\int_s^t E[b_i(\xi_j-\mu_j)+b_j(\xi_i-\mu_i)]\,dr+\int_s^t a_{ij}\,dr,$$

where $\cdot$ denotes derivative. The terms involving $\dot{\mu}_i$, $\dot{\mu}_j$ are 0 since $\mu_i=E\,\xi_i$. Now b_i is the i-th component $(A\,\xi)_i$ of the vector

$$b=A\,\xi=A(\xi-\mu)+A\,\mu.$$

We have

$$E(A\,\mu)_i(\xi_j-\mu_j)=(A\,\mu)_i\,E(\xi_j-\mu_j)=0,$$

$$E[A(\xi-\mu)]_i(\xi_j-\mu_j)=\sum_{k=1}^n A_{ik}\,\rho_{kj}.$$

Thus $E[b_i(\xi_j-\mu_j)]$ is the (i,j)-th element of the matrix AR. Similarly, $E[b_j(\xi_i-\mu_i)]$ is the (i,j)-th element of RA'. In differential notation we have

$$dR=(AR+RA'+a)\,dt. \tag{9.3}$$

The ordinary differential equations (9.2), (9.3) together with the initial data $\mu(s)$, $R(s)$ determine the means and covariances.

Now suppose that the initial data $\xi(s)$ is gaussian. We allow $\xi(s)$ to be degenerate (see Appendix D, Example 3 for the definition). Then $\xi(t)$ is also gaussian (again possibly degenerate). This can be proved using the representation (Hermes-LaSalle [1, p. 38])

$$\xi(t)=W(t,s)\,\xi(s)+\int_s^t W(t,r)\,\sigma(r)\,dw, \tag{9.4}$$

where W is a fundamental matrix ($dW=A(t)\,W\,dt$, $W(s,s)=\text{identity}$). The means and covariances determine the distribution of a gaussian random vector. Thus, *for gaussian initial data and linear equation* (9.1), *the distribution of* $\xi(t)$ can be found by solving the ordinary differential equations (9.2), (9.3). In this case the problem of solving the forward partial differential equation (8.2) reduces to a much easier one.

Let us now give conditions under which the distribution of $\xi(t)$ is nondegenerate for $t>s$. Nondegeneracy means that $R(t)$ is positive definite, and thus $\xi(t)$ has a gaussian density of the form in Appendix D, Example 3. We call the pair (A,σ) *completely controllable* at s if, for any $t>s$ and $x,y\in E^n$, there exists a function

$v \in L^2[s, t]$ and corresponding solution X of

$$dX = (AX + \sigma v)\, dt$$

such that $X(s) = y$, $X(t) = x$. In Hermes-LaSalle [1, p. 91] the term fully controllable is used for the same concept.

Theorem 9.1. *The following statements are equivalent:*

(1) *(A, σ) is completely controllable at s;*

(2) *A gaussian transition density $p(s, y, t, x)$ exists,*

(3) *For arbitrary gaussian initial data $\xi(s)$, the random vector $\xi(t)$ is nondegenerate gaussian for $s < t$.*

Proof. Let

$$R^0(t) = \int_s^t W(r, s)\,\sigma(r)\,[W(r, s)\,\sigma(r)]'\, dr.$$

Then complete controllability of (A, σ) at s is equivalent to positive definiteness of $R^0(t)$ for all $t > s$. See Hermes-LaSalle [1, p. 93]. The two terms on the right side of (9.4) are independent. Using (3.13), we get

$$R(t) = W(t, s)\, R(s)\, W(t, s)' + R^0(t).$$

For initial data $\xi(s) = y$, $R(s) = 0$, and $R(t) = R^0(t)$. Positive definiteness of $R(t)$ for $t > 0$ is equivalent to existence of gaussian density $p(s, y, t, x)$. Thus (1) and (2) are equivalent. Positive definiteness of $R^0(t)$ implies that of $R(t)$. Hence (2) implies (3). Clearly (3) implies (2). □

For constant matrices A, σ, complete controllability is equivalent to the condition

$$\operatorname{rank}[\sigma, A\sigma, \ldots, A^{n-1}\sigma] = n. \tag{9.5}$$

See Hermes-LaSalle [1, p. 74].

Example. Consider again Example 6.2. Condition (9.5) is satisfied, and thus a gaussian transition density exists. This example can be generalized to any l-th order, constant coefficient differential equation

$$\eta^{(l)} + \mu_{l-1}\,\eta^{(l-1)} + \cdots + \mu_1\,\dot{\eta} + \mu_0\,\eta = \dot{w}.$$

The Kalman-Bucy Filter Model. In many applications it is important to estimate the state of a system from measurements of quantities related to it. For stationary linear systems this problem was treated by Wiener and Kolmogorov. For non-stationary systems of linear differential equations with linear measurements Kalman and Bucy showed that the estimate of the state satisfies a linear differential equation. This equation is called the Kalman-Bucy filter. To describe this problem let $\xi(t)$ be a solution of linear stochastic differential equation (9.1) and suppose that $\xi(t)$ cannot be observed. Instead, one can observe $\eta(t)$ which satisfies

$$d\eta = H(t)\,\xi(t)\, dt + \sigma_1(t)\, dw_1, \qquad s \leqq t \leqq T, \tag{9.6}$$

with $\eta(s)=0$. The observation process $\eta=(\eta_1,\ldots,\eta_k)$ is k dimensional. The matrices $A(t)$, $H(t)$, $\sigma(t)$, $\sigma_1(t)$ are respectively $n\times n$, $k\times n$, $n\times d$, $k\times k$ dimensional; moreover, they are C^1 on $[T_0, T]$. The initial state $\xi(s)$ in (9.1) is a gaussian random vector; w, w_1 are standard brownian motions of dimensions d, k and $\xi(s)$, w, w_1 are mutually independent. The solutions ξ, η are gaussian processes defined on the same probability space $(\Omega, \mathscr{F}, P)$, and nonanticipative with respect to a family $\{\mathscr{F}_t\}$ of σ-algebras to which w and w_1 are adapted, with $\xi(s)$ a $\mathscr{F}_s$ measurable random vector. We use the notations

$$a=\sigma\,\sigma', \qquad a_1=\sigma_1\,\sigma_1'$$

for the noise covariance coefficients in (9.1), (9.6) respectively. It is assumed that $\sigma_1(t)$ is a nonsingular matrix.

This is the Kalman-Bucy filter model. Let

$$\mathscr{G}_t=\mathscr{F}(\eta(r),\ r\leq t)$$

be the σ-algebra corresponding to the data available at time t. Then $\{\mathscr{G}_t\}$ is an increasing family of σ-algebras, and $\mathscr{G}_t\subset\mathscr{F}_t$. The filtering problem is to find an estimate $\gamma(t)$ for $\xi(t)$, such that $\gamma(t)$ is $\mathscr{G}_t$-measurable, $E|\gamma(t)|^2<\infty$, and the mean square error $E\{b'[\xi(t)-\gamma(t)]\}^2$ is minimum for each n dimensional vector b. Let $\hat{\xi}(t)=(\hat{\xi}_1(t),\ldots,\hat{\xi}_n(t))$ be the conditional expectation vector:

$$\hat{\xi}_i(t)=E\{\xi_i(t)\,|\,\mathscr{G}_t\}, \qquad i=1,\ldots,n.$$

Let us show that the optimal estimate is $\gamma(t)=\hat{\xi}(t)$. To see this, given b and t let $\xi_b=b'\xi(t)$ with γ_b, $\hat{\xi}_b$ defined similarly. Write $\xi_b-\gamma_b=\xi_b-\hat{\xi}_b+\hat{\xi}_b-\gamma_b$. Then

$$E(\xi_b-\gamma_b)^2=E(\xi_b-\hat{\xi}_b)^2+2E(\xi_b-\hat{\xi}_b)(\hat{\xi}_b-\gamma_b)+E(\hat{\xi}_b-\gamma_b)^2.$$

Since

$$E(\xi_b-\hat{\xi}_b)(\hat{\xi}_b-\gamma_b)=E\{[E(\xi_b-\hat{\xi}_b)\,|\,\mathscr{G}_t](\hat{\xi}_b-\gamma_b)\}=0,$$
$$E(\xi_b-\gamma_b)^2\geq E(\xi_b-\hat{\xi}_b)^2.$$

Thus, $\hat{\xi}_b$ gives minimum mean square error, as asserted.

Let us take, in particular, for b the standard basis vectors $(1,0,\ldots,0),\ldots,$ $(0,\ldots,0,1)$ and sum from 1 to n. We see that $\hat{\xi}(t)$ also minimizes $E|\xi(t)-\gamma(t)|^2$.

The basic results about the Kalman-Bucy filter model are as follows. Let $\tilde{\xi}=\xi-\hat{\xi}$ denote the error process, and

$$R(t)=E\{\tilde{\xi}(t)\,\tilde{\xi}'(t)\}$$

the covariance matrix of the error $\tilde{\xi}(t)$ at time t.

Theorem 9.2. (a) *The conditional expectation vector $\hat{\xi}(t)$ satisfies the stochastic differential equation*

$$d\hat{\xi}=A(t)\,\hat{\xi}(t)\,dt+F(t)[d\eta-H(t)\,\hat{\xi}(t)\,dt] \tag{9.7}$$

with initial data $\hat{\xi}(s)=E\,\xi(s)$, where

$$F(t)=R(t)\,H'(t)[a_1(t)]^{-1}. \tag{9.8}$$

(b) *The error* $\tilde{\xi}(t)$ *is independent of* $\mathscr{G}_t$. *Moreover, the error covariance* $R(t)$ *satisfies the matrix differential equation*

$$dR = (AR + RA' - RH' a_1^{-1} HR + a)\,dt, \tag{9.9}$$

with $R(s)$ *the covariance matrix of* $\xi(s)$; *and*

(c)
$$\eta(t) - \int_s^t H(r)\,\hat{\xi}(r)\,dr = \int_s^t \sigma_1(r)\,d\hat{w}(r),$$

where $\hat{w}$ *is a standard k-dimensional brownian motion and* $\hat{w}(t)$ *is* $\mathscr{G}_t$*-measurable.*

For the original treatment of this result see Kalman [1] and Kalman-Bucy [1]. In those papers a duality between the filtering problem and the linear regulator control problem was first pointed out. This idea will be used in proving Theorem 9.2. The particular proof which we give was pointed out to us by *A*. Lindquist, and follows a proof of L.E. Zachrisson. Before embarking on it, let us make the following observations.

Remark 9.1. By Theorem 9.2(c) one can rewrite (9.7) as

$$d\hat{\xi} = A\,\hat{\xi}\,dt + F\,\sigma_1\,d\hat{w}, \tag{9.7'}$$

which is a linear stochastic differential equation of the same type as (9.1). The process $\int_s^t \sigma_1\,d\hat{w}$ is called an *innovation*. See Kailath [1], Kailath-Frost [1].

Remark 9.2. Since $\xi, \hat{\xi}$ satisfy the linear equations (9.1), (9.7′) and $\xi(s)$ is gaussian, $\xi(t)$, $\hat{\xi}(t)$ are gaussian random vectors for each t. Therefore, the error $\tilde{\xi}(t) = \xi(t) - \hat{\xi}(t)$ is also gaussian; $\tilde{\xi}(t)$ has mean 0 and covariance $R(t)$. The distribution of $\xi(t)$, conditioned on the data $\mathscr{G}_t$, is gaussian with mean $\hat{\xi}(t)$ and covariance $R(t)$.

Remark 9.3. If there were no data ($\mathscr{G}_t$ the trivial σ-algebra), then one would have $\hat{\xi}(t) = \mu(t)$ and $R(t)$ the covariance of $\xi(t)$ as in (9.2) and (9.3).

Proof of Theorem 9.2. First of all, we make some preliminary simplifications. Without loss of generality we may take $s=0$. Let $\mu(t) = E\,\xi(t)$, and

$$\xi^* = \xi - \mu, \qquad \eta^*(t) = \eta(t) - \int_0^t H(r)\,\mu(r)\,dr.$$

Then ξ^* is also a solution of (9.1), with mean 0; and

$$d\eta^* = H\,\xi^*\,dt + \sigma_1\,dw_1, \qquad \eta^*(0) = 0.$$

Thus we may suppose that $\mu(t) = 0$.

To prove (a) and (b) of the theorem we first consider the estimates

$$\gamma(t) = \int_0^t v(t,\tau)\,d\eta(\tau) \tag{9.10}$$

with $v(t,\tau)$ a continuous $n \times k$ matrix valued function on $[0,T] \times [0,T]$ and determine the optimal estimate in this class. Then it is shown that this optimal estimate is the conditional expectation $\hat{\xi}(t)$ which minimizes

$$E\{b'[\xi(t)-\gamma(t)]\}^2$$

in the wider class of estimates stated in the theorem.

Let us begin by determining the optimal estimate in the class of the form (9.10). For fixed $t>0$, define $X(t,\tau)$ as the $n\times n$ matrix solution of the (deterministic) matrix differential equation

$$dX=(-XA+vH)\,d\tau, \qquad 0\leqq\tau\leqq t, \tag{9.11}$$

with terminal condition $X(t,t)=I=$identity matrix. Let us multiply on the right by $\xi(\tau)$ and integrate by parts; this is justified (Problem 6) since X is of class C^1 in τ.) We get by (9.1), (9.11)

$$\int_0^t(-XA\xi+vH\xi)\,d\tau=X\xi|_0^t-\int_0^t X(A\xi\,d\tau+\sigma\,dw),$$

$$\xi(t)=X(t,0)\,\xi(0)+\int_0^t vH\xi\,d\tau+\int_0^t X\sigma\,dw.$$

Then from (9.6)

$$\xi(t)-\gamma(t)=X(t,0)\,\xi(0)+\int_0^t X(t,\tau)\,\sigma(\tau)\,dw-\int_0^t v(t,\tau)\,\sigma_1(\tau)\,dw_1. \tag{9.12}$$

The three terms on the right side are independent, since $\xi(0)$, w, w_1 are independent.

Given a vector b, let us multiply by b' in (9.12) and take $E\{\ \}^2$. We get

$$\begin{aligned}E\{b'[\xi(t)-\gamma(t)]\}^2&=b'X(t,0)\,R(0)\,X(t,0)'\,b+\int_0^t b'XaX'b\,d\tau\\&\quad+\int_0^t b'va_1v'b\,d\tau.\end{aligned} \tag{9.13}$$

For fixed t, the problem of choosing $v(t,\cdot)$ to minimize (9.13) becomes of the linear regulator type, Example II.2.3, after the following substitutions. Let

$$r=t-\tau, \qquad x(r)=X(t,t-r)'\,b, \qquad u(r)=v(t,t-r)'\,b.$$

Then by (9.11)

$$dx=[A'(t-r)\,x(r)-H'(t-r)\,u(r)]\,dr, \qquad 0\leqq r\leqq t,$$

with initial data $x(0)=b$, while (9.13) becomes

$$\begin{aligned}E\{b'[\xi(t)-\gamma(t)]\}^2&=x(t)'\,R(0)\,x(t)\\&\quad+\int_0^t[x(r)'\,a(t-r)\,x(r)+u(r)'\,a_1(t-r)\,u(r)]\,dr.\end{aligned} \tag{9.13'}$$

This is of the form of the linear regulator, if we replace in II.2 and IV.5: $A(r)$ by $A'(t-r)$, $B(r)$ by $-H'(t-r)$, D by $R(0)$, $M(r)$ by $a(t-r)$, and $N(r)$ by $a_1(t-r)$. By IV (5.5), (9.13') is minimized by taking

$$u(r)=a_1(t-r)^{-1}\,H(t-r)\,K(r)\,x(r), \tag{9.14}$$

where $K(r)$ satisfies the matrix Riccati equation IV(5.4) with $K(t)=R(0)$. Let us set $R(\tau)=K(t-\tau)$. Then R satisfies the matrix Riccati equation (9.9). Moreover, by Theorem IV(5.1) and IV(5.6), the minimum value of (9.13) is

$$E\{b'[\xi(t)-\gamma(t)]\}^2 = x'(0)\,K(0)\,x(0) = b'\,R(t)\,b.$$

From (9.8) and (9.14) it is attained by taking $v(t,\tau)=X(t,\tau)\,F(\tau)$ in (9.10). Since

$$E\{b'[\xi(t)-\gamma(t)]\}^2 = b'\,E\{[\xi(t)-\gamma(t)]\,[\xi(t)-\gamma(t)]'\}\,b,$$

$R(t)$ is the covariance matrix of the random vector $\xi(t)-\gamma(t)$.

Let us now consider (for t still fixed) a slightly larger class of estimates than in (9.10). Let $\mathscr{H}$ be the space of all

$$\gamma = \int_0^t v(\tau)\,d\eta(\tau),$$

where now $v(\tau)$ is any square integrable, $n\times k$ matrix-valued function on $[0,t]$. We give $\mathscr{H}$ the usual L_2 norm. From (9.6)

$$\gamma = \int_0^t v(\tau)\,H(\tau)\,\xi(\tau)\,d\tau + \int_0^t v(\tau)\,\sigma_1(\tau)\,dw_1(\tau).$$

Since the processes ξ, w_1 are independent, we have

$$E|\gamma|^2 = E\left|\int_0^t v(\tau)\,H(\tau)\,\xi(\tau)\,d\tau\right|^2 + E\left|\int_0^t v(\tau)\,\sigma_1(\tau)\,dw_1(\tau)\right|^2.$$

On the right side we apply Cauchy-Schwarz to the first term and interchange expectation with time integral; the second term is evaluated using (3.13):

$$\text{(9.15)}\qquad \begin{aligned} E|\gamma|^2 &\leqq \left(\int_0^t |v(\tau)\,H(\tau)|^2\,d\tau\right)\left(\int_0^t E|\xi(\tau)|^2\,d\tau\right) + \int_0^t |v(\tau)\,\sigma_1(\tau)|^2\,d\tau,\\ E|\gamma|^2 &\leqq C\int_0^t |v(\tau)|^2\,d\tau, \end{aligned}$$

where C is a constant depending on t and on bounds for H, σ_1 and a bound (4.6) for $E|\xi(\tau)|^2$.

From (9.15) it follows that $\mathscr{H}$ is a Hilbert space. Moreover, estimates with v continuous as in (9.10) are dense in $\mathscr{H}$. Thus the optimal estimate found above by taking $v=XF$ also minimizes (9.13) among all estimates in $\mathscr{H}$.

For the optimal $\gamma=\gamma(t)$, corresponding to $v=XF$, and for any $\lambda\in\mathscr{H}$, real ε

$$E\{b'[\xi(t)-\gamma(t)]\}^2 \leqq E\{b'[\xi(t)-\gamma(t)-\varepsilon\lambda]\}^2.$$

In particular, we may take $\lambda=V\eta(\tau_1)$, where V is a fixed $n\times k$ matrix and $0\leqq\tau_1\leqq t$. This corresponds to $v=V\chi$, with χ the indicator function of $[0,\tau_1]$. The above inequality implies that the right side of the inequality considered as a function of ε attains its minimum at $\varepsilon=0$. Thus the derivative with respect to ε of the right side is zero at $\varepsilon=0$. Setting this derivative equal to zero gives

$$0 = b'\,E\{[\xi(t)-\gamma(t)]\,[V\eta(\tau_1)]'\}\,b.$$

This is true for each b and V. This implies that each component of $\xi(t)-\gamma(t)$ is orthogonal to each component of $\eta(\tau_1)$, for $0 \leqq \tau_1 \leqq t$.

The process (ξ, η) satisfies the linear stochastic differential Eqs. (9.1), (9.6). Therefore, the joint finite dimensional distributions are gaussian. It follows that the random vectors $\xi(t)-\gamma(t)$, $\eta(\tau_1), \ldots, \eta(\tau_m)$ have a joint gaussian distribution for $0 \leqq \tau_1 < \cdots < \tau_m \leqq t$. From orthogonality it then follows that $\xi(t)-\gamma(t)$ is independent of $\mathscr{G}_t$. (See Gikhman-Skorokhod [1, pp. 228–9] for further discussion of this point.)

We now write $\xi(t)=\xi(t)-\gamma(t)+\gamma(t)$ and take conditional expectations:

$$\hat{\xi}(t)=E\{[\xi(t)-\gamma(t)]|\mathscr{G}_t\}+E\{\gamma(t)|\mathscr{G}_t\}=E\{\xi(t)-\gamma(t)\}+\gamma(t).$$

Here we use $\mathscr{G}_t$-measurability of $\gamma(t)$, and independence of $\mathscr{G}_t$ from $\xi(t)-\gamma(t)$. But $\xi(t)$, $\gamma(t)$ have expectation 0. Thus,

$$\hat{\xi}(t)=\gamma(t), \qquad \tilde{\xi}(t)=\xi(t)-\gamma(t).$$

To complete the proof of parts (a), (b) of Theorem 9.2 it remains only to show that $\hat{\xi}$ obeys (9.7). We now consider $X(t,\tau)$ as a function of t. By (9.11), $v=XF$, and $X(t,t)=I$, X is a fundamental matrix solution of

$$\frac{\partial X}{\partial \tau}=X(-A+FH).$$

Therefore, it satisfies in t the adjoint equation, which we write in integrated form:

$$X(t,\tau)=\int_\tau^t [A(\theta)-F(\theta)H(\theta)]X(\theta,\tau)\,d\theta+I.$$

Let us multiply each term by $F(\tau)\,d\eta(\tau)$ and integrate from 0 to t:

$$\hat{\xi}(t)=\int_0^t X(t,\tau)F(\tau)\,d\eta(\tau),$$

$$\begin{aligned}\int_0^t\int_\tau^t (A-FH)X\,d\theta\,F(\tau)\,d\eta(\tau)&=\int_0^t (A-FH)\int_0^\theta XF(\tau)\,d\eta(\tau)\,d\theta\\&=\int_0^t [A(\theta)-F(\theta)H(\theta)]\hat{\xi}(\theta)\,d\theta.\end{aligned}$$

The interchange of order of integration is justified since the integrand is C^1 in (θ,τ); see Problem 7. Thus

$$\hat{\xi}(t)=\int_0^t [A(\theta)-F(\theta)H(\theta)]\hat{\xi}(\theta)\,d\theta+\int_0^t F(\tau)\,d\eta(\tau),$$

which is just the integrated form of (9.7).

It remains to prove Theorem 9.2(c). We define the process $\hat{w}$ by

(9.16) $$\hat{w}(t)=\int_0^t \sigma_1(\tau)^{-1}[d\eta(\tau)-H(\tau)\hat{\xi}(\tau)\,d\tau].$$

Since $\eta(\tau)$ and $\tilde{\xi}(\tau)$ are $\mathcal{G}_t$-measurable for $0 \leqq \tau \leqq t$, $\hat{w}(t)$ is $\mathcal{G}_t$-measurable. Also

$$d\eta - H\hat{\xi}\,dt = H\tilde{\xi}\,dt + \sigma_1\,dw_1.$$

Thus, for $0 \leqq r \leqq t$,

$$\hat{w}(t) - \hat{w}(r) = \int_r^t \sigma_1(\tau)^{-1} H(\tau)\tilde{\xi}(\tau)\,d\tau + w_1(t) - w_1(r).$$

By (b), $\tilde{\xi}(\tau)$ is independent of $\mathcal{G}_\tau$, hence also of $\mathcal{G}_r$. Moreover, $w_1(t) - w_1(r)$ is independent of $\mathcal{G}_r$. Since increments of $\hat{w}$ for times $\leqq r$ are $\mathcal{G}_r$-measurable and $\hat{w}(t) - \hat{w}(r)$ is independent of $\mathcal{G}_r$, $\hat{w}$ is a process with independent increments. Moreover, the increments of $\hat{w}$ are gaussian and have mean 0. Given r and $B \in E^k$, let $\zeta_B(t) = B'[\hat{w}(t) - \hat{w}(r)]$. Then

$$d\zeta_B = B'\sigma_1^{-1} H\tilde{\xi}\,dt + B'\,dw_1.$$

Given vectors $B, C \in E^k$, the Ito stochastic rule implies

$$d(\zeta_B\zeta_C) = \zeta_C\,d\zeta_B + \zeta_B\,d\zeta_C + B'C\,dt.$$

By (3.3) and the independence of $\tilde{\xi}(\tau)$ from $\zeta_B(\tau)$ and $\zeta_C(\tau)$, we have

$$E\int_r^t \zeta_C(\tau)\,d\zeta_B(\tau) = E\int_r^t \zeta_B(\tau)\,d\zeta_C(\tau) = 0.$$

Therefore,

$$B'E\{[\hat{w}(t) - \hat{w}(r)][\hat{w}(t) - \hat{w}(r)]'\}C = E\zeta_B(t)\zeta_C(t) = B'C(t-r).$$

Since this is true for all B, C, the covariance matrix of $\hat{w}(t) - \hat{w}(r)$ is $I(t-r)$. Therefore, $\hat{w}$ is a standard k-dimensional brownian motion. □

§ 10. Absolutely Continuous Substitution of Probability Measures

In this section we state a transformation formula of Girsanov and show that it can often be used to show that the stochastic differential Eq. (4.3) has a solution when the local drift coefficient is discontinuous.

In Girsanov's result, Ω, $\mathcal{F}$, and an increasing family $\{\mathcal{F}_t\}$ of σ-algebras are given. However, different probability measures on $\mathcal{F}$ are considered. Let $\tilde{\xi}$ be a solution of

$$\tilde{\xi}(t) - \tilde{\xi}(s) = \int_s^t \tilde{\beta}(r)\,dr + \int_s^t \gamma(r)\,d\tilde{w}(r), \tag{10.1}$$

where $\tilde{w} = (\tilde{w}_1, \ldots, \tilde{w}_n)$ is a standard n-dimensional brownian motion with respect to some probability measure $\tilde{P}$. We are interested in finding a solution of the corresponding equation (10.4) when the coefficient $\tilde{\beta}$ is replaced by β and $\tilde{w}$ by some other brownian motion w.

The coefficients $\tilde{\beta}$, γ are nonanticipative processes with $\tilde{\beta}_i$, γ_{ij}, $i, j = 1, \ldots, n$, in $\mathcal{M}$. Let $\theta = (\theta_1, \ldots, \theta_n)$ be a nonanticipative process which is bounded ($|\theta| \leqq M$ for some M); and let

$$\zeta_s^t(\theta) = \int_s^t \theta'(r)\,d\tilde{w}(r) - \tfrac{1}{2}\int_s^t |\theta(r)|^2\,dr. \tag{10.2}$$

We write $P(d\omega)=G(\omega)\tilde{P}(d\omega)$ to indicate that P is absolutely continuous with respect to $\tilde{P}$, and G is the Radon-Nikodym derivative.

Theorem 10.1. *Let P be absolutely continuous with respect to $\tilde{P}$, with*

$$P(d\omega)=\exp\zeta_s^T(\theta)\tilde{P}(d\omega). \tag{10.3}$$

Then:

(a) $P(\Omega)=1$ *(hence $(\Omega,\mathscr{F},P)$ is a probability space.)*

(b) *Let $w(t)=\tilde{w}(t)-\int_s^t\theta(r)\,dr$. Then w is a standard n-dimensional brownian motion with respect to P.*

(c) *Let $\beta=\tilde{\beta}+\gamma\theta$. Then*

$$\xi(t)-\xi(s)=\int_s^t\beta(r)\,dr+\int_s^t\gamma(r)\,dw(r). \tag{10.4}$$

[Here ξ and $\tilde{\xi}$ have the same sample functions, but are considered as stochastic processes with respect to different probability measures P and $\tilde{P}$.]

For a proof we refer to Girsanov [1, pp. 287–296]. Actually what is proved there is that (b), (c) follow from (a); see Girsanov [1, Fundamental Theorem 1]. In the proof it is shown that for bounded θ, $\tilde{E}\exp c\,\tilde{\zeta}_s^t(\theta)$ is bounded on $[s,T]$ for any $c>1$ (Girsanov [1, Lemma 1]). Here $\tilde{E}$ denotes expectation with respect to the probability measure $\tilde{P}$. From the Ito stochastic differential rule

$$\exp\tilde{\zeta}_s^T(\theta)=1+\int_s^T\exp\tilde{\zeta}_s^r(\theta)\theta'(r)\,dw(r).$$

(Compare with Example 3.3). Let $c=2$. By (3.13) and boundedness of θ,

$$\tilde{E}\exp\tilde{\zeta}_s^T(\theta)=1 \tag{$*$}$$

which by (10.3) is condition (a) of Theorem 10.1. By replacing T by $T_1<T$ in (10.3) and then letting $T_1\to s^+$, one sees that $\xi(s)$ has the same distribution with respect to P and $\tilde{P}$.

For other treatments of Girsanov's theorem see Gikhman-Skorokhod [2, §12], Benes [1, §4], Liptser-Shiryaev [1, Chap. 6], Friedman [2]. Actually, condition (*) holds under the weaker assumption that

$$\tilde{E}\exp\{(1+\delta)\}\int_s^T|\theta(r)|^2\,dr<\infty \qquad \text{for some } \delta>0.$$

See Gikhman-Skorokhod [2, p. 82].

Remark. It is easy to verify that $\tilde{P}$ is absolutely continuous with respect to P, with

$$\tilde{P}(d\omega)=\exp\zeta_s^T(-\theta)P(d\omega). \tag{$\widetilde{10.3}$}$$

In the definition of ζ_s^T, $\tilde{w}$ is replaced by w in (10.2).

We turn to an important special case. Let $\tilde{\xi}$ be a solution of

$$d\tilde{\xi}=\tilde{b}(t,\tilde{\xi}(t))\,dt+\sigma(t,\tilde{\xi}(t))\,d\tilde{w}, \tag{$\widetilde{4.3}$}$$

and suppose that b is related to $\tilde{b}$ by

(10.5) $$b(t,x)=\tilde{b}(t,x)+\sigma(t,x)\,\Theta(t,x),$$

for all $(t,x)\in\bar{Q}^0$. We assume

(10.6) (i) $\tilde{b}$, σ satisfy the Ito conditions and σ is bounded.
(ii) Θ is bounded and Borel measurable.

Theorem 10.2. *Assume* (10.5), (10.6), *and* $E|\xi(s)|^2<\infty$. *Then a solution* ξ *of* (4.3) *exists.*

Proof. In Theorem 10.1 take

$$\tilde{\beta}(r)=\tilde{b}(r,\tilde{\xi}(r)),\qquad \gamma(r)=\sigma(r,\tilde{\xi}(r)),$$
$$\theta(r)=\Theta(r,\tilde{\xi}(r)),\quad \beta(r)=b(r,\xi(r)),$$

and recall that ξ, $\tilde{\xi}$ are the same functions on $[s,T]\times\Omega$. Then (10.4) becomes (4.3). □

Note that the brownian motion w is not given in advance (in contrast to Sect. 4). It arises in the course of changing the probability measure.

Let π_t denote the distribution of the n-dimensional random vector $\xi(t)$, $\tilde{\pi}_t$ that of $\tilde{\xi}(t)$, and

$$\rho(t,x)=E[\exp\tilde{\zeta}_s^t(\theta)\,|\,\tilde{\xi}(t)=x],\qquad x\in E^n.$$

For any $B\in\mathscr{B}(E^n)$,

$$P[\xi(t)\in B]=\int_{\xi(t)\in B}\exp\tilde{\zeta}_s^t(\theta)\,\tilde{P}(dw)=\int_B\rho(t,x)\,\tilde{\pi}_t(dx).$$

The left side is $\pi_t(B)$. Therefore, π_t is absolutely continuous with respect to $\tilde{\pi}_t$, and

(10.7) $$\pi_t(dx)=\rho(t,x)\,\tilde{\pi}_t(dx).$$

If $\tilde{\pi}_t$ has a density $\tilde{q}(t,x)$, then π_t has the density

(10.8) $$q(t,x)=\rho(t,x)\,\tilde{q}(t,x).$$

In particular, consider initial data $\tilde{\xi}(s)=\xi(s)=y$, in which q, $\tilde{q}$ become the transition densities. If there is a transition density $\tilde{p}$ corresponding to $\tilde{b}$, σ, then there is also a transition density p corresponding to b, σ satisfying

(10.9) $$p(s,y,t,x)=Z(s,y,t,x)\,\tilde{p}(s,y,t,x),$$

where Z is ρ for initial data $\tilde{\xi}(s)=y$.

Remarks. Since b may be discontinuous, condition (2) of the definition of diffusion process in Sect. 5 need not hold everywhere. In the terminology of Krylov [1], ξ is a quasidiffusion.

We need in Sect. 11 the following result about the densities in (10.8). We write $L^\mu(K)$ for the space of μ-th power integrable functions on $K\subset E^{n+1}$.

Lemma 10.1. *Suppose that* $\tilde{\xi}(t)$ *has a density* $\tilde{q}(t,x)$ *in* $L^{\tilde{\mu}}(Q_{s'T})$ *for any* $s'>s$. *Then, if* $1<\mu<\tilde{\mu}$, $q(t,x)$ *is in* $L^\mu(Q_{s'T})$ *for any such* s'.

Proof. For $c>1$ let

$$N_c=\int_\Omega \exp[c\tilde{\zeta}_s^t(\theta)]\tilde{P}(d\omega).$$

By (a) of Theorem 10.1, with $c\theta$ instead of θ,

$$1=\int_\Omega \exp\tilde{\zeta}_s^t(c\theta)\tilde{P}(d\omega).$$

Since $|\theta|\leqq M$ and

$$\tilde{\zeta}_s^t(c\theta)-c\tilde{\zeta}_s^t(\theta)=-\frac{c^2-c}{2}\int_s^t|\theta(r)|^2\,dr, \qquad N_c\leqq\exp\tfrac{1}{2}M^2(t-s)(c^2-c).$$

Since conditional expectation does not increase L^c norm with respect to $\tilde{P}$,

$$\int_{E^n}\rho(t,x)^c\tilde{q}(t,x)\,dx=\tilde{E}(|\tilde{E}\{\exp\tilde{\zeta}_s^t(\theta)|\tilde{\xi}(t)\}|^c)$$
$$\leqq\tilde{E}(\tilde{E}\{\exp c\tilde{\zeta}_s^t(\theta)|\tilde{\xi}(t)\}).$$

The right side is N_c, and thus

$$\int_{E^n}\rho(t,x)^c\tilde{q}(t,x)\,dx\leqq N_c.$$

From (10.8)

$$(*)\qquad \int_{Q_{s'T}} q(t,x)^c\,\tilde{q}(t,x)^{1-c}\,dt\,dx\leqq(T-s)\exp\tfrac{1}{2}M^2(T-s)(c^2-c).$$

Choose h, h', c, and v such that

$$h>(\tilde{\mu}-1)(\tilde{\mu}-\mu)^{-1}, \qquad h^{-1}+(h')^{-1}=1, \qquad c=h\mu, \qquad v=h^{-1}(c-1).$$

An elementary argument shows that $1<v h'<\tilde{\mu}$. Write $q^\mu=(q^\mu\tilde{q}^{-v})\tilde{q}^v$, and use Holder's inequality with parameters h, h'. The finiteness of $\int_{Q_{s'T}} q^\mu\,dt\,dx$ follows from (*) and finiteness of $\int_{Q_{s'T}}\tilde{q}^{\tilde{\mu}}\,dt\,dx$. □

Let us next mention two interesting particular cases.

(1) σ is nonsingular with σ^{-1} bounded. Given $b(t,x)$ bounded, Borel measurable, let $\tilde{b}=0$, $\Theta=\sigma^{-1}b$. The backward operator is uniformly elliptic. This case was treated extensively in Stroock-Varadhan [1].

(2) Suppose that the stochastic differential equations (4.3) split into v equations with no noise term, and $n-v$ with noise terms $(0\leqq v<n)$

$$\begin{aligned} d\xi_i&=\tilde{b}_i(t,\xi(t))\,dt, \qquad i=1,\dots,v,\\ d\xi_i&=\tilde{b}_i(t,\xi(t))\,dt+g_i(t,\xi(t))\,dt\\ &\quad+\sum_{j=v+1}^{n}\sigma_{ij}(t,\xi(t))\,dw_j, \qquad i=v+1,\dots,n, \end{aligned} \tag{10.10}$$

where $(w_{v+1},\dots,w_n)$ is an $(n-v)$-dimensional standard brownian motion. Consider also the corresponding equations without the g_i terms:

$$\begin{aligned} d\tilde{\xi}_i&=\tilde{b}_i(t,\tilde{\xi}(t))\,dt, \qquad i=1,\dots,v,\\ d\tilde{\xi}_i&=\tilde{b}_i(t,\tilde{\xi}(t))\,dt+\sum_{j=v+1}^{n}\sigma_{ij}(t,\tilde{\xi}(t))\,d\tilde{w}_j, \qquad i=v+1,\dots,n. \end{aligned} \tag{$\widetilde{10.10}$}$$

Assume that the $(n-v)\times(n-v)$ matrices $\hat{\sigma}=(\sigma_{ij})$ are non-singular with $\hat{\sigma}^{-1}$ bounded. Let

$$\Theta_i=0, \qquad i=1,\dots,v,$$

$$\Theta_i=\sum_{j=v+1}^{n}(\hat{\sigma}^{-1})_{ij}\,g_j, \qquad i=v+1,\dots,n.$$

Let us arbitrarily introduce additional brownian components $\tilde{w}_1,\dots,\tilde{w}_v$ so that $(\tilde{w}_1,\dots,\tilde{w}_n)$ is an n-dimensional standard brownian motion. If $\tilde{b}$, $\hat{\sigma}$ satisfy (10.6)(i), and g is bounded and Borel measurable, then Theorem 10.2 implies that (10.10) has a solution ξ with given initial data.

We shall be interested in the case where $\widetilde{(10.10)}$ has a transition density $\tilde{p}$ with the integrability property in Lemma 10.1. In particular, this will be true if $\widetilde{(10.10)}$ is a linear system satisfying the controllability condition (Theorem 9.1).

Uniqueness in Probability Law. In Sect. 4 we stated uniqueness results for solutions of (4.3). In those results the random variable $\xi(s)$ and the brownian motion w are given. Let us now consider a different kind of uniqueness, namely, that the coefficients b, σ in (4.3) together with the distribution π of $\xi(s)$ uniquely determine the distribution of the diffusion ξ.

We say that two continuous processes ξ, ξ' on $[s, T]$ are *identical in law* if, when considered as random vectors they have the same distribution P^* on $\mathscr{C}^n[s, T]$. (For the definition of P^*, see end of Sect. 2.) Note that ξ and ξ' need not be defined on the same probability space. It can be shown that sample path uniqueness of the kind in Theorem 4.1 implies uniqueness in probability law. See Yamada-Watanabe [1].

Suppose now that w, w' are two standard brownian motions. Consider the corresponding solutions ξ, ξ' of (4.3) for various initial data. In particular, by taking initial state $\xi(s)=\xi'(s)=y$ we get the transition distributions:

$$\hat{P}(s, y, t, B)=P_{sy}(\xi(t)\in B)$$

$$\hat{P}'(s, y, t, B)=P'_{sy}(\xi'(t)\in B).$$

As before, the subscript on the right side indicates that the initial data are (s, y).

We first note uniqueness in probability law under the Ito conditions on b, σ. Then it is proved under much weaker assumptions using the Girsanov transformation. In the latter argument we follow essentially an argument of Nisio and Watanabe.

Lemma 10.2. *Let b, σ satisfy the Ito conditions. Let ξ, ξ' be solutions of* (4.3), *such that $\xi(s)$, $\xi'(s)$ have the same distribution π with $\int_{E^n}|y|^2\,\pi(dy)<\infty$. Then ξ and ξ' are identical in law.*

Lemma 10.2 follows easily from the standard method of successive approximations, Gikhman-Skorokhod [2, § 6]. It is also a special case of sample path uniqueness and the result of Yamada-Watanabe [1] already mentioned.

Theorem 10.3. *Let $b=\tilde{b}+\sigma\,\Theta$ where $\tilde{b}$, σ, Θ satisfy* (10.6). *Let ξ, ξ' be solutions of* (4.3), *such that $\xi(s)$, $\xi'(s)$ have the same distribution π with $\int_{E^n}|y|^2\,\pi(dy)<\infty$. Then ξ and ξ' are identical in law.*

Proof. Let us adjoin to (4.3) the n equations $dw_i=dw_i$, $i=1,\ldots,n$, obtaining $2n$ stochastic differential equations satisfied by the process (ξ, w). The solutions ξ, ξ' are defined with respect to brownian motions w, w' on probability spaces $(\Omega,\mathscr{F},P)$, $(\Omega',\mathscr{F}',P')$, and on the time interval $\mathscr{T}=[s,T]$. We shall show that the $2n$-dimensional processes (ξ,w), (ξ',w') are identical in law, which implies ξ, ξ' are identical in law. We do this first in a special case, to which we afterward show that the general case can be reduced.

Special Case. $\Omega=\mathscr{C}^{2n}(\mathscr{T})$, $\mathscr{F}=\mathscr{B}[\mathscr{C}^{2n}(\mathscr{T})]$. Let us write elements of Ω as $\omega=(g,h)$, with $g,h\in\mathscr{C}^n(\mathscr{T})$. Suppose that

$$\xi(t,g,h)=g(t), \qquad w(t,g,h)=h(t).$$

The random vector defined on Ω by the process (ξ,w) is just the identity map. Therefore, in this special case, $P=P^*$ where P^* denotes the distribution measure of (ξ,w). Let $\mathscr{F}_t$ denote the least σ-algebra containing all events $g(r)\in B_1$, $h(r)\in B_2$ for $r\leqq t$, B_1, B_2 Borel subsets of E^n.

Now suppose that w is a brownian motion, and ξ a solution of (4.3), with respect to $\Omega,\mathscr{F},P^*$. Let us apply Theorem 10.1, with the roles of ξ, $\tilde{\xi}$ reversed, to get a solution $\tilde{\xi}$ of

$$d\tilde{\xi}=\tilde{b}(t,\tilde{\xi}(t))\,dt+\sigma(t,\tilde{\xi}(t))\,d\tilde{w} \tag{4.3}$$

with $\xi(s)=\tilde{\xi}(s)$. According to $\widetilde{(10.3)}$ the distribution measure $\tilde{P}^*$ is absolutely continuous with respect to P^* and satisfies

$$\tilde{P}^*(d\omega)=\exp\int_s^T\left[-\Theta'(r,\xi(r))\,dw-\tfrac{1}{2}\left|\Theta(r,\xi(r))\right|^2 dr\right]P^*(d\omega); \tag{10.11}$$

while according to (10.3), P^* is absolutely continuous with respect to $\tilde{P}^*$ and satisfies

$$P^*(d\omega)=\exp\int_s^T\left[\Theta'(r,\tilde{\xi}(r))\,d\tilde{w}-\tfrac{1}{2}\left|\Theta(r,\tilde{\xi}(r))\right|^2 dr\right]\tilde{P}^*(d\omega). \tag*{$\widetilde{(10.11)}$}$$

If (ξ',w') is another solution of (4.3), with respect to $\Omega,\mathscr{F},(P^*)'$, we define in the same way a solution $(\tilde{\xi}',\tilde{w}')$ of $\widetilde{(4.3)}$ with corresponding distribution measure $(\tilde{P}^*)'$. By Lemma 10.2, $(\tilde{P}^*)'=\tilde{P}^*$. Since all sample paths are the same (only the probability measures change) we now write $\tilde{w}'=\tilde{w}$, $\tilde{\xi}'=\tilde{\xi}$. But $(P^*)'$ also satisfies $\widetilde{(10.11)}$, and hence $(P^*)'=P^*$. This proves the theorem in the special case.

General Case. Consider any solution ξ of (4.3), and denote by H the following mapping from Ω into $\Omega^*=\mathscr{C}^{2n}(\mathscr{T})$: $H(\omega)=(\xi(\cdot,\omega),w(\cdot,\omega))$. Let $\mathscr{F}^*=\mathscr{B}(\Omega^*)$ and $\mathscr{F}_t^*$ be σ-algebras defined as in the special case above. The mapping H is measurable with respect to $\mathscr{F},\mathscr{F}^*$ and also for each $t\in\mathscr{T}$ measurable with respect to $\mathscr{F}_t$, $\mathscr{F}_t^*$ since w, ξ are nonanticipative processes. Here $\{\mathscr{F}_t\}$ is a family of σ-algebras to which the brownian motion w is adapted (see Sect. 3). The distribution measure P^* of the process (ξ, w) is just the measure on Ω^* induced from P by H. Define the processes ξ^*, w^* as components of the identity map as above; then $\xi^*(H)=\xi$, $w^*(H)=w$. Moreover, w^* is a brownian motion (with respect to P^*) adapted to $\{\mathscr{F}_t^*\}$; and

$$\int_s^t b(r, \xi(r))\, dr = \int_s^t b(r, \xi^*(r))\, dr.$$

Here the left side is evaluated at ω, the right side at $H(\omega)$. Also, with probability 1,

$$\int_s^t \sigma(r, \xi(r))\, dw(r) = \int_s^t \sigma(r, \xi^*(r))\, dw^*(r);$$

here we can use Remark 3.2, that each stochastic integral is the limit of Riemann-Stieltjes sums and note that corresponding Riemann-Stieltjes sums

$$\sum_{k=1}^m \sigma(r_{k-1}, \xi(r_{k-1}))\, \Delta_k w \quad \text{and} \quad \sum_{k=1}^m \sigma(r_{k-1}, \xi^*(r_{k-1}))\, \Delta_k w^*$$

are the same. Thus, ξ^* is a solution of

$$d\xi^* = b(t, \xi^*(t)) + \sigma(t, \xi^*(t))\, dw^*,$$

and (ξ^*, w^*) is identical in probability law with (ξ, w). Taken together with the special case, this proves Theorem 10.3. □

§ 11. An Extension of Theorems 5.1, 5.2

In Theorem 10.2 we stated a condition under which (4.3) has a solution even when the local drift coefficient b is discontinuous. For such b, we cannot expect solutions ψ of the corresponding backward equation (6.2) to be of class $C^{1,2}(Q)$. Thus, Theorems 5.1 and 5.2 need not apply. However, we now show that if the processes ξ defined in Theorem 10.2 have a suitable transition density, then these theorems remain true when the partial derivatives of ψ are locally integrable to a sufficiently high power λ.

We suppose that the system equations have the special form (10.10). For $1<\lambda<\infty$ let, $0 \leqq \nu < n$, $\mathscr{E}^{\lambda\nu}(Q)$ denote the space of functions $\psi(t, x)$ such that:

(i) ψ is continuous on $\bar{Q}$ and satisfies a polynomial growth condition;

(ii) The following partial derivatives are in $L^\lambda(K)$ for any bounded $K \subset Q$: ψ_t, ψ_{x_i} for $i=1, \dots, n$, $\psi_{x_i x_j}$ for $i, j = \nu+1, \dots, n$.

Here the partial derivatives are generalized derivatives (see Appendix E). If α is C^2 with compact support, then $\alpha\psi \in \mathscr{E}^{\lambda\nu}(Q)$ provided $\psi \in \mathscr{E}^{\lambda\nu}(Q)$. The following lemma is proved by a standard approximation technique (Appendix E).

Lemma 11.1. *Suppose that $\psi \in \mathscr{E}^{\lambda\nu}(Q)$ and ψ has support in a compact set $K \subset Q$. Then there is a sequence ψ^l, $l=1, 2, \dots$, of C^∞ functions with support in a compact set $K' \subset Q$, such that ψ^l tends to ψ uniformly on Q while each partial derivative*

$\psi_t^l, \psi_{x_i}^l, \psi_{x_i}^l, \psi_{x_i x_j}^l$ *appearing in* (ii) *tends in* L^λ *norm on* Q *to the corresponding* $\psi_t, \psi_{x_i}, \psi_{x_i x_j}$ *as* $l \to \infty$.

Let us suppose that the transition distributions for the Markov processes satisfying $\widetilde{(10.10)}$ have a transition density $\tilde{p}$ satisfying, for some $\tilde{\mu} > 1$,

$$\int_{Q_{s'T}} |\tilde{p}(s, y, t, x|^{\tilde{\mu}}\, dt\, dx < \infty, \quad \text{if } s' > s. \tag{11.1}$$

This assumption holds for instance for a linear, completely controllable system $d\xi = A\,\tilde{\xi}\, dt + \sigma\, dw$ by Theorem 9.1.

For solutions of (10.10) the transition density satisfies (10.9). By Lemma 10.1, for $1 < \mu < \tilde{\mu}$, we have

$$\int_{Q_{s'T}} |p(s, y, t, x)|^{\mu}\, dt\, dx < \infty, \quad \text{if } s' > s. \tag{11.2}$$

Let us extend Lemma 5.1. The extension of Theorems 5.1, 5.2 will then follow.

Lemma 11.2. *Let* ψ *be of class* $\mathscr{E}^{\lambda\nu}(Q^0)$ *with* $\lambda^{-1} + \mu^{-1} = 1$, *and suppose that* ψ *has support in a compact subset of* Q^0. *If* Q *is an open subset of* Q^0 *and* $s' > s$, *then*

$$\begin{aligned} &E_{sy}\, \psi(\tau, \xi(\tau)) - E_{sy}\, \psi(s' \wedge \tau, \xi(s' \wedge \tau)) \\ &\qquad = E_{sy} \int_{s' \wedge \tau}^{\tau} [\psi_t(r, \xi(r))\, dr + \mathscr{A}(r)\, \psi(r, \xi(r))]\, dr, \end{aligned} \tag{5.6}$$

where $(s, y) \in Q$ *and* τ *is the exit time from* Q, $s' \wedge \tau = \min(s', \tau)$.

Proof. Let ψ^l, $l = 1, 2, \ldots$, be as in Lemma 11.1, and

$$\Lambda^l = -\left[\psi_t + \sum_{i,j=\nu+1}^{n} a_{ij}\, \psi_{x_i x_j}^l + \sum_{i=1}^{n} b_i\, \psi_{x_i}^l\right] = -[\psi_t^l + \mathscr{A}(t)\, \psi^l].$$

Then Λ^l tends in L^λ norm to $\Lambda = -[\psi_t + \mathscr{A}(t)\, \psi]$ as $l \to \infty$. Then

$$\begin{aligned} E \int_{s' \wedge \tau}^{\tau} |\Lambda^l - \Lambda|\, dt &\leq E \int_{s'}^{T} |\Lambda^l - \Lambda|\, dt \\ &= \int_{s'}^{T} \int_{E^n} |\Lambda^l - \Lambda|\, p(s, y, t, x)\, dt\, dx. \end{aligned} \tag{*}$$

Here $E = E_{sy}$. By (11.2), and Hölder's inequality, the right side tends to 0 as $l \to \infty$. By Lemma 5.1,

$$E\, \psi^l(s' \wedge \tau, \xi(s' \wedge \tau)) = E\left\{ \int_{s' \wedge \tau}^{\tau} \Lambda^l(t, \xi(t))\, dt + \psi^l(\tau, \xi(\tau)) \right\}.$$

From (*), uniform boundedness of ψ^l, and uniform convergence of ψ^l to ψ, we get

$$E\, \psi(s' \wedge \tau, \xi(s' \wedge \tau)) = E\left\{ \int_{s' \wedge \tau}^{\tau} \Lambda(t, \xi(t))\, dt + \psi(\tau, \xi(\tau)) \right\}. \quad \square$$

Theorem 11.1. *Let the system equations have the form* (10.10). *Assume* (10.5), (10.6) *and* (11.1). *The conclusions of Theorems* 5.1, 5.2 *remain true if, instead of* $\psi \in C_p^{1,2}(Q)$, *we assume that* $\psi \in \mathscr{E}^{\lambda v}(Q)$ *with* $\lambda^{-1}+\mu^{-1}=1$, $1<\mu<\tilde{\mu}$.

To prove Theorem 11.1 only minor changes in the proofs of Theorems 5.1, 5.2 are needed, using Lemma 11.2 instead of Lemma 5.1. In the proof of Theorem 5.1, we first get the formula corresponding to (#) in which s is replaced by $s' \wedge \tau_j$ and $\psi(s, y)$ by $E\psi(s' \wedge \tau_j, \xi(s' \wedge \tau_j))$. Then let $s' \to s$. Since ξ is a continuous process, $\xi(s)=y$, ψ is bounded and continuous on $\bar{Q}_j$, and $E\int_s^{\tau_j}|M|\,dt<\infty$, we have

$$\lim_{s' \to s} E\psi(s' \wedge \tau_j, \xi(s' \wedge \tau_j))=\psi(s, y)$$

$$\lim_{s' \to s} E \int_{s' \wedge \tau_j}^{\tau_j} M(t, \xi(t))\,dt=E\int_s^{\tau_j} M(t, \xi(t))\,dt.$$

This gives (#). The proof then continues as before.

Existence of Solutions $\psi \in \mathscr{E}^{\lambda v}(Q)$. When the backward operator (6.1) is uniformly parabolic ($v=0$) and Q is a bounded cylinder, the existence of solutions to (6.2) with boundary data is mentioned in Appendix E. For $v>0$ we do not know a corresponding general existence theorem. However, in Chap. VI, Sect. 8, we shall apply Theorem 11.1 to a certain function V which is in $\mathscr{E}^{\lambda v}(Q)$ and satisfies the dynamic programming equation.

Problems—Chapter V

(1) Let $n=1$, $d\xi=k\,\xi\,dt+\sigma\,\xi\,dw$ for $t \geqq 0$, with $\xi(0)=x$. Here k and $\sigma>0$ are constants. Show that $\log[x^{-1}\xi(t)]$ is gaussian with mean $(k-\frac{1}{2}\sigma^2)\,t$ and variance $\sigma^2 t$. *Hint.* Apply the Ito stochastic differential rule to $d[\log \xi(t)]$.

Note. This is a simple population growth model with randomly fluctuating growth constant.

(2) In Example 3.2 find $E\,\xi(t)^k$ for any $k=1, 2, \ldots$.

(3) (A randomly perturbed harmonic oscillator). Let $n=2$, $d\xi_1=\xi_2\,dt$, $d\xi_2=-\xi_1\,dt+c\,\xi_1\,dw$, $c>0$. Use the method of Example 3.2 find a system of three linear differential equations for the moments $m_1(t)=E\,\xi_1(t)^2$, $m_2(t)=E\,\xi_1(t)\,\xi_2(t)$, $m_3(t)=E\,\xi_2(t)^2$. Show that the "total energy" $E\,|\xi(t)|^2=m_1(t)+m_3(t)$ is for large t approximately $K\,e^{\lambda_1 t}$ where λ_1 is the real root of $\lambda^3+4\lambda-2c^2=0$ and $\lambda_1>0$.

(4) (A vector extension of Example 3.3). Let $\gamma=(\gamma_1, \ldots, \gamma_n)$ be a nonanticipative process with $|\gamma(t)| \leqq C$, and $w=(w_1, \ldots, w_n)$ a standard n-dimensional brownian motion. Show that $\exp\int_s^t[\gamma'\,dw-\frac{1}{2}|\gamma|^2\,dr]$ is a martingale.

(5) Let ξ be a solution of (4.3) with $E\,|\xi(s)|^m<\infty$ for each $m>0$. Show using (4.7) that $E\,\|\xi\|^m<\infty$ for each $m>0$.

(6) Let $g(t)=(g_1(t), \ldots, g_k(t))$ be a vector valued C^1 function of t (not a stochastic process). Let $\eta=(\eta_1, \ldots, \eta_k)$ satisfy a vector-matrix version of (3.8). Show using

(3.14) that $d(g'\eta)=\dot{g}'\eta\,dt+g'\,d\eta$, and hence

$$\int_0^t g(\tau)'\,d\eta(\tau)=g(t)'\,\eta(t)-g(0)'\,\eta(0)-\int_0^t \dot{g}(\tau)'\,\eta(\tau)\,d\tau.$$

(Here $'$ denotes vector transpose and $\dot{}=d/dt$.)

(7) Let $G(\tau,\theta)$ be matrix valued and C^1, and η as in Problem 6. Show that

$$\int_0^t \int_\tau^t G(\tau,\theta)\,d\theta\,d\eta(\tau)=\int_0^t \int_0^\theta G(\tau,\theta)\,d\eta(\tau)\,d\theta.$$

Hint. Integrate by parts (Problem 6) and use Fubini's theorem.

(8) Let w^c be as in Example 5.1. Show that the sup norm $\|w^c-w\|$ tends to 0 in probability as $c\to 0$. *Hint.* Integrate the differential equation $c\,dv^c+v^c\,dt=dw$ once and solve the resulting linear differential equation for w^c in terms of w.

Chapter VI. Optimal Control of Markov Diffusion Processes

§ 1. Introduction

Many control systems occurring in practice are subject to imperfectly known disturbances, which may be taken as random. Such disturbances have been ignored in the study of deterministic control models in Chaps. II–IV. Optimal stochastic control theory attempts to deal with models in which random system disturbances are allowed.

In describing a stochastic control model, the kind of information available to the controller at each instant of time plays an important role. Several situations are possible:

(1) The controller has no information during the system operation. In this case he chooses as control a function of time as in Chaps. II, III. Such controls are often called "open loop".

(2) The controller knows the state of the system at each instant of time t. We call this the case of *complete observations.*

(3) The controller has partial knowledge of the system states. This is called the case of *partial observations.*

For deterministic optimal control problems, we saw in Chap. IV that controls could be taken either in open loop or feedback form. Feedback controls will not give a smaller minimum in such problems. The reason is essentially the following. For a deterministic control model, the state of the system at any time t can be deduced from the initial data and the control used up to time t, by solving the differential Eq. II(3.1). Thus, observing the current state of the system at each time t does not really give more information than knowing the initial data.

For stochastic control systems, there are many paths which the system states may follow given the control and initial data. In the stochastic case, the best system performance depends on the information available to the controller at each time t.

In this chapter we consider mainly case (2)—complete observations of system states. It is then natural to formulate controls in feedback form. (For technical reasons, it will be convenient to consider later another class of controls, which are nonanticipative stochastic processes.)

In case (2) the method of dynamic programming can be used. It is that approach which will be developed here. Other approaches are mentioned in §'s 2 and 10, with references. To illustrate the dynamic programming approach, we begin in §2 with a formal derivation for the dynamic programming Eq. (2.5) associated with a general kind of optimal control problem for continuous time Markov processes. The remainder of the chapter is devoted to the study of controlled diffusion processes, for which a rather satisfactory theory has been obtained on a mathematically rigorous basis. For such processes the dynamic programming equation becomes a nonlinear partial differential equation of second order. In §4 we give sufficient conditions for an optimum, supposing that this partial differential equation with the appropriate boundary conditions has a "well-behaved" solution. This turns the stochastic control problem into a problem about a nonlinear partial differential equation. The sufficient condition is applied to several examples, including the important problem of the linear regulator (§5). In §6 we shall see that, under stronger assumptions, the dynamic programming equation indeed has a solution with the desired behavior; and that an optimal feedback control law exists. Crucial among these assumptions is the uniform parabolicity of the dynamic programming equation. In §8 we replace that assumption by a weaker one encountered in many cases of interest. The notion of solution of the dynamic programming equation then must be taken in some generalized sense.

In §9 we consider the case when the stochastic control problem is nearly deterministic, since the coefficient of the noise entering the system equations is small. In that case the second order terms in the dynamic programming equation are multiplied by a small positive parameter ε. For the deterministic problem ($\varepsilon=0$) it becomes the first order partial differential equation already considered in Chap. IV.

In §10 we discuss briefly open loop problems, and problems with partial observations.

One of the most useful results in stochastic control theory is the separation principle (§11). When it holds, the stochastic control problem can be split into two others. The first is a mean square optimal estimate for the system state, using a Kalman-Bucy filter. The second is a stochastic control problem with complete observations, whose solution for the important case of the linear regulator appears in §5.

§ 2. The Dynamic Programming Equation for Controlled Markov Processes

In this section we give a formal derivation of the dynamic programming Eq. (2.5), for a rather general class of controlled Markov processes. Neither a precise statement of conditions under which the dynamic programming equation holds nor any proofs are given here. A mathematically rigorous treatment for the case of controlled diffusions is given in the rest of the chapter, beginning with §3. For a corresponding discrete time model, a rigorous treatment is given in Kushner [2, Chap. 3].

The problem is optimal control of a class of Markov processes ξ. Let Σ denote the state space of the processes being controlled. As in Chaps. II–IV let U denote a set from which the control applied at any time t is chosen. A *feedback control law* is a function $\mathbf{u}=\mathbf{u}(\cdot,\cdot)$ from $\mathscr{T}_0\times\Sigma$ into U, where $\mathscr{T}_0=[T_0, T]$ is an interval containing all times t being considered. In the rigorous treatment to follow, we admit only feedback controls satisfying some further conditions; see (3.5) below.

We suppose that to each feedback control $\mathbf{u}$ (satisfying suitable analytic assumptions) corresponds a transition function $\hat{P}^{\mathbf{u}}(s, y, t, B)$ associated with a family of Markov processes as in Chap. V.5. Given initial data (s, y), with $s\in\mathscr{T}_0$, $y\in\Sigma$ we suppose that there is a corresponding Markov process ξ which depends on the choice of $\mathbf{u}$ as well as the initial data.

Associated with the transition function $\hat{P}^{\mathbf{u}}$ are operators $S^{\mathbf{u}}_{s,t}$, and $\mathscr{A}^{\mathbf{u}}(t)$ defined by V(5.2). In our formal treatment we suppose that

$$[\mathscr{A}^{\mathbf{u}}(s)\Phi](y)=[\mathscr{A}^{v}(s)\Phi](y), \quad \text{if } v=\mathbf{u}(s,y). \tag{2.1}$$

Here $\mathscr{A}^{v}(s)$ is obtained using the constant feedback control law v.

As criterion to be minimized, we take the expected value of a criterion of Bolza type II(4.3).

$$J(s,y,\mathbf{u})=E_{sy}\left\{\int_s^T L(t,\xi(t),u(t))\,dt+\Psi[\xi(T)]\right\}, \tag{2.2}$$

where the control applied at time t using feedback control $\mathbf{u}$ is

$$u(t)=\mathbf{u}(t,\xi(t)). \tag{2.3}$$

In Chaps. II and IV we took the Mayer form $(L=0)$. However, certain results below are more conveniently stated for the Bolza form.

In dynamic programming, the optimal expected system performance is considered as a function of the initial data:

$$W^0(s,y)=\inf_{\mathbf{u}} J(s,y,\mathbf{u}). \tag{2.4}$$

An *optimal* feedback control law $\mathbf{u}^*$ has the property that $W^0(s,y)=J(s,y,\mathbf{u}^*)$ for all (s,y). To derive formally an equation for W^0, given $\mathbf{u}$ we have by V(5.3) with $\psi=W^0$, $\mathscr{A}=\mathscr{A}^{\mathbf{u}}$

$$W^0(s,y)=-E_{sy}\int_s^t [W^0_s+\mathscr{A}^{\mathbf{u}}(r)\,W^0]\,dr+E_{sy}\,W^0(t,\xi(t)), \tag{*}$$

for $s<t\leqq T$. The integrand is evaluated at $(r,\xi(r))$. Here we denote the time partial derivative of W^0 by W^0_s rather than W^0_t, to indicate the "backward" role of s and y. See V.6.

On the other hand, suppose that the controller uses $\mathbf{u}$ for times $s\leqq r\leqq t$ and uses an optimal control $\mathbf{u}^*$ after time t. His expected performance can be no less than $W^0(s,y)$. Thus let

$$\mathbf{u}_1(r,x)=\begin{cases}\mathbf{u}(r,x) & \text{for } r\leqq t\\ \mathbf{u}^*(r,x) & \text{for } r>t.\end{cases}$$

By properties of conditional expectations,

$$J(s,y,\mathbf{u}_1)=E_{sy}\int_s^t L(r,\xi(r),u(r))\,dr+E_{sy}J(t,\xi(t),\mathbf{u}^*).$$

Then

$$W^0(s, y) \leqq J(s, y, \mathbf{u}_1), \qquad W^0(t, \xi(t)) = J(t, \xi(t), \mathbf{u}^*),$$

$$(**) \qquad W^0(s, y) \leqq E_{sy} \int_s^t L(r, \xi(r), u(r))\,dr + E_{sy} W^0(t, \xi(t)).$$

Equality holds if an optimal control law $\mathbf{u} = \mathbf{u}^*$ is used during $[s, t]$. Let us subtract $(*)$ from $(**)$ and divide by $t - s$. Since $y = \xi(s)$, we get (formally) as $t \to s^+$, using (2.1), (2.3) with $v = \mathbf{u}(s, y)$:

$$0 \leqq W_s^0 + \mathscr{A}^v(s)\, W^0 + L(s, y, v).$$

Equality holds if $v = \mathbf{u}^*(s, y)$, where $\mathbf{u}^*$ is an optimal feedback control law. Thus, for W^0 we have derived formally the *continuous-time dynamic programming equation of optimal stochastic control theory:*

$$(2.5) \qquad 0 = W_s^0 + \min_{v \in U} [\mathscr{A}^v(s)\, W^0 + L(s, y, v)].$$

The boundary condition

$$(2.6) \qquad W^0(T, y) = \Psi(y)$$

follows from the definition of W^0.

When $L = 0$ there is a stochastic version of the monotonicity property of the value function given in Theorem IV.3.1. As in $(**)$,

$$W^0(t_1, \xi(t_1)) \leqq E_{sy}\{W^0(t, \xi(t)) \,|\, \xi(t_1)\}, \qquad s \leqq t_1 \leqq t,$$

which implies that $W^0(t, \xi(t))$ is a submartingale. For an optimal control policy $\mathbf{u}^*$, and corresponding state process ξ^*, $W^0(t, \xi^*(t))$ is a martingale.

Often one does not wish to control the Markov process on a fixed time interval $[s, T]$. Instead, the control occurs up to the time τ when $(t, \xi(t))$ exits from some given open set Q. In that case, instead of the criterion (2.2) we shall take

$$(2.7) \qquad J(s, y, \mathbf{u}) = E_{sy}\left\{\int_s^\tau L(t, \xi(t), u(t))\,dt + \Psi(\tau, \xi(\tau))\right\},$$

where Ψ is given on a portion Z of the complement of Q in which the exit place must occur. Thus, $(\tau, \xi(\tau)) \in Z$. For diffusions, Z is a subset of ∂Q. In later sections, we write $Z = \partial^* Q$ as in V.7.

Under suitable assumptions, t can be replaced by the random time $\min(t, \tau)$ in $(*)$ and $(**)$; moreover, $P_{sy}(s < \tau < t)$ is small if $t - s$ is small. The dynamic programming equation is still (2.5); however, instead of (2.6) one now imposes the condition $W^0(s, y) = \Psi(s, y)$ for $(s, y) \in Z$.

Substantial difficulties are encountered if one seeks to put all of the above formal discussion on a precise basis. It has been implicitly assumed that $W^0(t, \cdot)$ is in the domain of $\mathscr{A}^{\mathbf{u}}(t)$, which is not always the case. Moreover, the argument preceding (2.5) involving passage to the limit as $t \to s^+$ has not been justified.

Up to now, mathematically correct treatments of dynamic programming have been given for various classes of controlled continuous-time Markov processes. In the sections to follow we treat controlled diffusions, using results from the theory of parabolic partial differential equations. For alternate, more general but more complex approaches see Rishel [2], Davis-Varaiya [2]. These approaches also give dynamic programming type necessary and sufficient conditions for problems with partial observations. Optimal control of jump Markov processes is treated in Boel [1], Rishel [4–6], Varaiya [1], Stone [1], Sworder [1].

Still another approach is to view the problem as the optimal control of the coefficients of the linear parabolic operators $\mathscr{A}^{\mathbf{u}}(t)$. This approach was taken in Fleming [6] and also in Bensoussan-Lions [2] using a variational inequality technique.

§ 3. Controlled Diffusion Processes

In the remainder of this chapter we consider a system model in which the state process ξ is a finite dimensional diffusion which evolves according to a system of stochastic differential equations. We begin the present section by formulating precisely these equations and the class of feedback control laws to be admitted. Later in the section we formulate the minimum problem which will be studied. In the notation of § 2, we now take $\Sigma = E^n$. When written in vector matrix notation, the equations describing the state process ξ are assumed to have the form

$$d\xi = f(t, \xi(t), u(t))dt + \sigma(t, \xi(t), u(t))dw, \quad t \geqq s. \tag{3.1}$$

Here $\xi(t)$ is the state of the system at time t, $u(t)$ the control applied at time t, and w a d-dimensional standard brownian motion.

The following notation, already introduced in Chap. V, will be used throughout this chapter:

T_0 and T are fixed with $T_0 < T$; $\mathscr{T}_0 = [T_0, T]$, $Q^0 = (T_0, T) \times E^n$; $C^j(D)$ is the class of functions with continuous partial derivatives of all orders $\leqq j$ on D (on an open set containing D if D is not open); for $Q \subset E^{n+1}$, $C^{1,2}(Q)$ is the set of $\psi(t, x)$ with ψ_t, ψ_{x_i}, $\psi_{x_i x_j}$, $i, j = 1, \ldots, n$, continuous on Q; $C_p^{1,2}(Q)$ is the class of ψ in $C^{1,2}(Q)$ which satisfy a polynomial growth condition on Q.

We make the following assumptions about the functions $f(t, x, u)$, $\sigma(t, x, u)$. These functions are of class $C^1(\bar{Q}^0 \times U)$. Moreover, for some constant C

$$\begin{aligned} &|f(t,0,0)| \leqq C, \quad |\sigma(t,0,0)| \leqq C \\ &|f_x| + |f_u| \leqq C, \quad |\sigma_x| + |\sigma_u| \leqq C. \end{aligned} \tag{3.2}$$

Example 1. The linear case. Let $f(t, x, u) = A(t)x + B(t)u$, $\sigma = \sigma(t)$ with A, B, σ of class $C^1(\mathscr{T}_0)$. Then the above assumptions hold.

For technical reasons, further restrictions will be imposed on f and σ in later sections as appropriate.

As already mentioned in §2, observations of the states $\xi(t)$ are complete. In particular, given an initial time s the initial state $\xi(s)$ is known. Thus, we may as well consider (3.1) with initial data

$$\xi(s) = y, \tag{3.3}$$

where $y \in E^n$ is the initial state vector. At each time t, the control $u(t)$ is applied via a feedback control law $\mathbf{u}$:

(3.4) $$u(t) = \mathbf{u}(t, \xi(t)).$$

Given $\mathbf{u}$ and a function $g(t, x, u)$, we use the notation

$$g^{\mathbf{u}}(t, x) = g(t, x, \mathbf{u}(t, x)).$$

Then (3.1) can be rewritten as

(3.1′) $$d\xi = f^{\mathbf{u}}(t, \xi(t))\,dt + \sigma^{\mathbf{u}}(t, \xi(t))\,dw, \qquad t \geqq s.$$

This is of the form V.(4.3); the coefficients $f^{\mathbf{u}}$, $\sigma^{\mathbf{u}}$ now may depend on the control law $\mathbf{u}$.

We admit only feedback controls with the following properties.

Definition. A feedback control law $\mathbf{u}$ is *admissible* if $\mathbf{u}$ is a Borel measurable function from $\bar{Q}_0$ into U, such that:

(3.5) (a) For each (s, y), $T_0 \leqq s < T$, there exists a brownian motion w such that (3.1′) with (3.3) have a solution ξ, unique in probability law; and

(b) For each $k > 0$, $E_{sy}|\xi(t)|^k$ is bounded for $s \leqq t \leqq T$ and

$$E_{sy} \int_s^T |u(t)|^k\,dt < \infty.$$

Here $u(t) = \mathbf{u}(t, \xi(t))$. The bound in (b) may depend on (s, y).

We denote by $\mathscr{V}$ the class of all admissible feedback control laws.

Any of the following three conditions (i)–(iii) are sufficient for admissibility of $\mathbf{u}$:

(i) For some constant M_1, $|\mathbf{u}(t, x)| \leqq M_1(1 + |x|)$ for all $(t, x) \in \bar{Q}^0$. Moreover, for any bounded $B \subset E^n$ and $T_0 < T' < T$ there exists a constant K_1 such that, for all $x, y \in B$ and $T_0 \leqq t \leqq T'$,

$$|\mathbf{u}(t, x) - \mathbf{u}(t, y)| \leqq K_1 |x - y|.$$

(The constant K_1 may depend on B, T'; and both M_1, K_1 may depend on $\mathbf{u}$.)

(ii) $\mathbf{u}$ satisfies a Lipschitz condition on $\bar{Q}^0$.

(iii) Suppose that $f(t, x, u) = \tilde{b}(t, x) + \sigma(t, x)\,\theta(t, x, u)$, where $\tilde{b}$, σ satisfy V(10.6), θ is of class $C^1(\bar{Q}^0 \times U)$, and θ is bounded on $\bar{Q}^0 \times U_1$ for any bounded $U_1 \subset U$. Then any bounded, Borel measurable $\mathbf{u}$ is admissible.

If (i) holds, then $f^{\mathbf{u}}$, $\sigma^{\mathbf{u}}$ satisfy hypotheses (4.4), (4.5) of the existence and uniqueness Theorem V.4.1. In that case the brownian motion w can be specified in advance. By Yamada-Watanabe [1, Proposition 1] the kind of sample path uniqueness in that theorem implies uniqueness in probability law. Condition (ii) implies that there is a K_1 in (i) not depending on B or T'. Since (ii) is a Lipschitz condition on $\mathbf{u}$ in both variables (t, x), it also implies continuity of $\mathbf{u}$. Thus $\mathbf{u}(t, 0)$ is bounded on $\mathscr{T}_0$; and this together with the uniform Lipschitz constant K_1 for $\mathbf{u}(t, \cdot)$ implies the growth condition in (i) for suitable M_1. Thus (ii) is a special case of (i). Condition (3.5)(b) follows from V(4.6).

If (iii) holds, then $\Theta^{\mathbf{u}}(t,x)=\theta(t,x,\mathbf{u}(t,x))$ is bounded and Borel measurable. Admissibility of $\mathbf{u}$ follows from Theorems V.10.2 and V.10.3 together with V(4.6).

Example 1. (Continued). Let $\tilde{b}(t,x)=A(t)x$ and suppose that the matrix equation $B(t)=\sigma(t)F(t)$ has a C^1 solution F. We may take $\theta(t,u)=F(t)u$. The conditions (iii) are satisfied.

Remark. Uniqueness in probability law implies that the expected system performance J in (2.2) or (2.7) indeed depends only on $\mathbf{u}$ and on the initial data (s,y). If one wished to avoid the technical difficulty associated with proving uniqueness in probability law, one could instead allow the possibility that J might differ for different solutions of (3.1′)–(3.3). This point will arise again in the definition of admissible nonanticipative control (§4), where a general uniqueness theorem does not seem to be available. We shall see that, when a verification theorem holds (Theorem 4.1 or 8.1), no solution of the system equations corresponding to a feedback or nonanticipative control can give less than the optimal system performance $W(s,y)$.

Let L, Ψ be continuous functions which satisfy the polynomial growth conditions:

(3.6)
$$\text{(a)}\quad |L(t,x,u)|\leqq C(1+|x|+|u|)^k,$$
$$\text{(b)}\quad |\Psi(x)|\leqq C(1+|x|)^k$$

for suitable constants C, k.

We now formulate the problem to be considered throughout the rest of the chapter:

Minimum Problem. Find a feedback control $\mathbf{u}^*$ which minimizes

$$J(s,y,\mathbf{u})=E_{sy}\left\{\int_s^\tau L(t,\xi(t),u(t))\,dt+\Psi(\tau,\xi(\tau))\right\},$$

among all feedback controls $\mathbf{u}\in\mathscr{V}$.

This is the same criterion as (2.7). As before τ is the exit time from a given open set $Q\subset Q^0$ and the initial data (s,y) are in Q. If $Q=Q^0$, then $\tau=T$ and (2.7) has the form (2.2).

Conditions (3.5b) and (3.6) insure that $J(s,y,\mathbf{u})$ is finite for each $\mathbf{u}\in\mathscr{V}$. (Further smoothness assumptions about L, Ψ will be made later as needed.)

Let us mention a few examples. In the linear regulator problem to be considered in §5, f is linear in x,u, $\sigma=\sigma(t)$, and L, Ψ are quadratic in x,u. In the linear regulator, there are no control constraints ($U=E^m$ for some m). Another interesting, but more difficult, example is obtained by imposing in the linear regulator problem saturation control constraints. For instance, one might take U an m-dimensional cube.

Example 2. Take $Q=(T_0,T)\times G$ a cylinder, where $G\subset E^n$ is open. Let $L=0$,

$$\Psi(t,x)=1 \quad \text{for } t<T,\quad x\in\partial G$$
$$\Psi(t,x)=0 \quad \text{for } t=T,\quad x\in G.$$

Then $J = P_{sy}(\tau < T)$, and $1 - J = P_{sy}(\tau = T)$ is the *inclusion probability*. When J is minimum the inclusion probability is maximum. This function Ψ is discontinuous at points (T, x), $x \in \partial G$, which necessitates minor changes in the theory. This example is mentioned again in §4. It has been studied by Dorato and Van Maelert [1].

Autonomous Version. Suppose that f, σ do not explicitly depend on t; then (3.1) takes the form

$$d\xi = f(\xi(t), u(t))\, dt + \sigma(\xi(t), u(t))\, dw, \qquad t \geqq 0. \tag{3.1^a}$$

The initial data are $\xi(0) = y$.

The control $u(t)$ is now applied through an autonomous feedback control $\mathbf{u}$:

$$u(t) = \mathbf{u}[\xi(t)]. \tag{3.4^a}$$

By admissible autonomous feedback control we mean a Borel measurable function $\mathbf{u}$ from E^n into U, such that for each y there exists a brownian motion w such that (3.1) with $\xi(0) = y$ has a solution ξ, unique in probability law on $[0, T]$ and satisfying (3.5b) with $s = 0$ and any $T > 0$.

Let G be a given open set. Starting from $y \in G$, let τ denote the exit time from G. (We shall always impose conditions guaranteeing that $\tau < \infty$ with probability 1.) As expected performance criterion we take

$$J = E_y \left\{ \int_0^\tau L[\xi(t), u(t)]\, dt + \Psi[\xi(\tau)] \right\}. \tag{2.7^a}$$

As an example, let $L = 1$, $\Psi = 0$. Then J is the mean exit time from G.

Other Problems. In using the criterion (2.7^a), we are essentially stopping the process ξ when ∂G is reached. Thus, in our problem, points of ∂G are absorbing. Various other autonomous problems have been treated using dynamic programming. For 1-dimensional diffusions, autonomous optimal control has been treated for other kinds of boundary conditions by Mandl [1–4] and by Morton [1], and for multidimensional diffusions and reflecting boundary conditions by Puterman [1].

It is also interesting to consider the ξ process in $G = E^n$, with either a discounted criterion (see Kushner [4])

$$J = E_y \int_0^\infty e^{-\rho t} L[\xi(t), u(t)]\, dt, \qquad \rho > 0,$$

or a steady state cost criterion (see Wonham [2])

$$\int_{E^n} L(y, \mathbf{u}(x))\, d\mu^{\mathbf{u}}(x),$$

where $\mu^{\mathbf{u}}$ is the equilibrium measure. In the latter case, only feedback controls $\mathbf{u}$ are admitted for which the process ξ starting from any y is ergodic.

§ 4. The Dynamic Programming Equation for Controlled Diffusions; a Verification Theorem

For diffusions, the operators $\mathscr{A}^v(s)$ in § 2 take the form (see V(5.5))

$$\mathscr{A}^v(s)=\tfrac{1}{2}\sum_{i,j=1}^{n} a_{ij}(s,y,v)\frac{\partial^2}{\partial y_i\,\partial y_j}+\sum_{i=1}^{n} f_i(s,y,v)\frac{\partial}{\partial y_i},$$

where the symmetric matrices $a=(a_{ij})$ are defined by $a=\sigma\,\sigma'$ as in V(5.4). Here we use the notation of "backward" variables (s, y) rather than "forward" variables (t, x); see V.6, V.8. The reason is that these operators will be applied to functions of the initial data (s, y) for a controlled diffusion.

Given a feedback control law $\mathbf{u}$, the corresponding operators $\mathscr{A}^{\mathbf{u}}(s)$ satisfy according to (3.1′)

$$\mathscr{A}^{\mathbf{u}}(s)=\tfrac{1}{2}\sum_{i,j=1}^{n} a_{ij}^{\mathbf{u}}(s,y)\frac{\partial^2}{\partial y_i\,\partial y_j}+\sum_{i=1}^{n} f_i^{\mathbf{u}}(s,y)\frac{\partial}{\partial y_i}.$$

Therefore, condition (2.1) holds.

The main result of this section is a sufficient condition for a minimum (Theorem 4.1, or 4.2 for the autonomous case). The sufficient condition requires a suitably behaved solution W of the dynamic programming equation (4.1), and a control law $\mathbf{u}^*$ satisfying (4.3). We call such a result a *verification theorem*. In § 6 we shall see that the sufficient condition is also necessary, under sufficiently strong assumptions.

We again let Q be an open set in which the initial data (s, y) lie, with $Q\subset Q^0$.

Let $\partial^* Q$ be a closed subset of ∂Q such that $(\tau, \xi(\tau))\in\partial^* Q$ with probability 1, for every choice of initial data $(s, y)\in Q$ and every admissible $\mathbf{u}$. Examples of $\partial^* Q$ were given in V.7 for $Q=Q^0$ and Q a cylindrical domain. We recall from V.5 the notation $C_p^{1,2}(Q)$.

Theorem 4.1. *Let $W(s, y)$ be a solution of the dynamic programming equation*

$$0=W_s+\min_{v\in U}\,[\mathscr{A}^v(s)\,W+L(s,x,v)],\qquad (s,y)\in Q, \tag{4.1}$$

with the boundary data

$$W(s,y)=\Psi(s,y),\qquad (s,y)\in\partial^* Q, \tag{4.2}$$

such that W is in $C_p^{1,2}(Q)$ and continuous on the closure $\bar{Q}$. Then:

(a) *$W(s, y)\leqq J(s, y, \mathbf{u})$ for any admissible feedback control $\mathbf{u}$ and any initial data $(s, y)\in Q$.*

(b) *If $\mathbf{u}^*$ is an admissible feedback control such that*

$$\mathscr{A}^{\mathbf{u}^*}(s)\,W+L^{\mathbf{u}^*}(s,y)=\min_{v\in U}\,[\mathscr{A}^v(s)\,W+L(s,y,v)] \tag{4.3}$$

for all $(s, y)\in Q$, then $W(s, y)=J(s, y, \mathbf{u}^*)$ for all $(s, y)\in Q$. Thus $\mathbf{u}^*$ is optimal.

Proof of (a). For each $v\in U$, $(s, y)\in Q$,

$$0\leqq W_s+\mathscr{A}^v(s)\,W+L(s,y,v).$$

Let us now replace s, y, v by $t, \xi(t), u(t)=\mathbf{u}(t, \xi(t)), s \leqq t \leqq \tau$. We get

$$(*) \qquad 0 \leqq W_s + \mathscr{A}^{\mathbf{u}}(t)\, W + L^{\mathbf{u}}.$$

We apply Theorem V.5.1 with $M = L^{\mathbf{u}}$, $\psi = W$. Conditions (3.5b) and (3.6a) imply that

$$E_{sy} \int_s^\tau |M(t, \xi(t))|\, dt \leqq E_{sy} \int_s^T |M(t, \xi(t))|\, dt < \infty.$$

We get

$$W(s, y) \leqq E_{sy} \left\{ \int_s^\tau L^{\mathbf{u}}(t, \xi(t))\, dt + \Psi(\tau, \xi(\tau)) \right\}.$$

Since the right side is $J(s, y; \mathbf{u})$, this proves (a).

To prove (b), we have equality in (*) for $\mathbf{u} = \mathbf{u}^*$. Therefore, $W(s, y) = J(s, y, \mathbf{u}^*)$ using Theorem V.5.2. □

The minimum expected performance $W^0(s, y)$ in (2.4) equals $W(s, y)$ in Theorem 4.1, when the conditions of this theorem hold. The Verification Theorem 4.1 reduces the solution of the optimal stochastic control problem to two other problems. The first is to solve the nonlinear second order partial differential equation (4.1) with the data (4.2). The second is to find $\mathbf{u}^*$ by performing the minimization indicated in (4.3) for each (s, y). In many instances, the second problem is easy once the gradient W_y is known, since $\mathbf{u}^*$ turns out to be an explicitly known function of W_y.

In § 5 we apply Theorem 4.1 to solve the important linear regulator problem. In the meantime let us consider some other examples.

Example 1. An interesting example in one dimension ($n = 1$) arises in Merton's model of optimal portfolio selection. In a simplified version Merton [1, pp. 388–390], the stock portfolio consists of two assets, one "risk free" and the other "risky". The price $p_1(t)$ per share for the risk free asset changes according to $dp_1 = p_1 r\, dt$, while the price $p_2(t)$ per share for the risky asset changes according to $dp_2 = p_2(\alpha\, dt + \sigma\, dw)$. Here r, α, σ are constants, with $r < \alpha$, $\sigma > 0$, and w is a 1-dimensional standard brownian motion. The wealth $\xi(t)$ at time t then changes according to the stochastic differential equation $d\xi = (1 - u_1)\xi r\, dt + u_1 \xi(\alpha\, dt + \sigma\, dw) - u_2\, dt$, where $u_1(t)$ is the fraction of the wealth invested in the risky asset at time t and $u_2(t)$ the consumption rate. Let $Q = \{(t, y)\colon 0 < t < T,\ y > 0\}$ and τ be the first time $(t, \xi(t))$ leaves Q. In economic terms τ is either the first time that the wealth goes to zero, or the final time T if the wealth remains positive on $[0, T]$. The control is the 2-dimensional vector $u(t) = (u_1(t), u_2(t))$. It is natural to impose the control constraints $0 \leqq u_1(t) \leqq 1$, $u_2(t) \geqq 0$. {See Merton [1, § 3] for a detailed derivation of a more general model involving several kinds of assets.}

The problem is to maximize expected discounted total utility

$$J = E_{sy} \int_s^T e^{-\rho t} F[u_2(t)]\, dt,$$

given a utility function F and discount rate $\rho > 0$. Here $y = \xi(s)$ is the wealth at initial time s, and no value is assigned to wealth $\xi(T)$ remaining at the final time T.

Let us for the moment ignore the constraints on the controls u_1, u_2. The dynamic programming equation (4.1) then has the form

$$(\#) \qquad 0 = W_s + \max_{v_1, v_2} \left[\frac{(v_1 y \sigma)^2}{2} W_{yy} + \{(1-v_1) y r + v_1 y \alpha - v_2\} W_y + e^{-\rho s} F(v_2) \right], \quad y > 0,$$

with boundary conditions $W(t,0) = 0$ and $W(T, y) = 0$. The optimal $\mathbf{u}^* = (\mathbf{u}_1^*, \mathbf{u}_2^*)$ is found by maximizing the expression in brackets as a function of $v = (v_1, v_2)$. By elementary calculus

$$\mathbf{u}_1^* = -\frac{(\alpha - r) W_y}{\sigma^2 y W_{yy}}, \qquad F'(\mathbf{u}_2^*) = e^{\rho s} W_y.$$

Let us suppose that $F' > 0$, $F'(0) = +\infty$, $F'' < 0$; and seek a solution W of (#) with $W_y > 0$, $W_{yy} < 0$ for $y > 0$. Then $\mathbf{u}_1^* > 0$, $\mathbf{u}_2^* > 0$; Theorem 4.1 will then give a solution to the problem with control constraints $u_1 \geqq 0$, $u_2 \geqq 0$. If it happens that $\mathbf{u}_1^* \leqq 1$, then this is also a solution of the problem with constraints $0 \leqq u_1 \leqq 1$, $u_2 \geqq 0$.

To get a solution we choose the particular utility function $F(v_2) = v_2^\gamma$ with $0 < \gamma < 1$. Let us try

$$W(s, y) = g(s)\, y^\gamma, \qquad y \geqq 0,$$

$$\mathbf{u}_1^* = \frac{\alpha - r}{\sigma^2 (1-\gamma)}, \qquad \mathbf{u}_2^*(s, y) = [e^{\rho s} g(s)]^{\frac{1}{\gamma - 1}} y.$$

If we put these expressions for W, $\mathbf{u}_1^*$, $\mathbf{u}_2^*$ in (#), we get after some calculations

$$0 = \left[\frac{dg}{ds} + \nu \gamma g + (1-\gamma) g (e^{\rho s} g)^{\frac{1}{\gamma - 1}} \right] y^\gamma,$$

$$\nu = \frac{(\alpha - r)^2}{2 \sigma^2 (1-\gamma)} + r.$$

Since this holds for all $y > 0$, the expression in brackets must be 0. Since $W(T, y) = 0$, $g(T) = 0$. The substitution $h = (e^{\rho s} g)^{\frac{1}{1-\gamma}}$ leads to a linear differential equation for h, which is solved to get

$$g(s) = e^{-\rho s} \left[\frac{1-\gamma}{\rho - \nu \gamma} \left(1 - e^{-\frac{(\rho - \nu\gamma)(T-s)}{1-\gamma}}\right) \right]^{1-\gamma}$$

In this example the optimal $\mathbf{u}_1^*$ is constant, and $\mathbf{u}_2^*$ is a linear function of y. If $\alpha - r \leqq \sigma^2 (1-\gamma)$, then $\mathbf{u}_1^* \leqq 1$ and we have a solution to the constrained control problem. A similar calculation can be made for a class of utility functions F of the hyperbolic absolute risk aversion type. For such F, $\mathbf{u}_1^* y$ and $\mathbf{u}_2^*$ turn out to be linear functions of y. See Merton [1, p. 390].

Example 2. Maximum inclusion probability (continued from § 3). Now the dynamic programming equation is $0 = W_s + \min_{v \in U} \mathscr{A}^v W$ with boundary data $W(s, y) = 1$ for $T_0 \leqq s < T$, $y \in \partial G$, and $W(T, y) = 0$ for $y \in G$. In this example, W must be discontinuous at points (T, y) with $y \in \partial G$. By inspection of the proof of Theorem 4.1, one can see that its conclusions hold here also provided W is in

$C^{1,2}(Q)$, W is continuous on $\bar{Q}-\{T\}\times\partial G$, and $P_{sy}(\xi(T)\in\partial G)=0$. [It can be shown by methods of parabolic partial differential equations, that such a W exists if (6.1), (6.2), (6.3)(a), (b) below are satisfied.]

In Sects. 6, 8 we shall use a slightly stronger form of Theorem 4.1 in which (4.3) is merely required to hold almost everywhere in Q (with respect to $(n+1)$-dimensional Lebesgue measure). Let ξ^* be the solution of (3.1′)–(3.3) corresponding to $\mathbf{u}^*$, and τ^* the corresponding exit time from Q.

Corollary 4.1. *Let W be as in Theorem 4.1 and let $\mathbf{u}^*$ be an admissible feedback control such that (4.3) holds for almost all $(s, y)\in Q$. Moreover, suppose that the transition function $\hat{P}^{\mathbf{u}^*}$ has a density $p^*(s, y, t, x)$. Then $W(s, y)=J(s, y, \mathbf{u}^*)$ for all $(s, y)\in Q$; and thus $\mathbf{u}^*$ is optimal.*

Proof. There exists a set N of $(n+1)$-dimensional measure 0 such that (4.3) holds for all points of $Q-N$. If $(t, \xi^*(t))\in Q-N$, then

$$\mathscr{A}^{\mathbf{u}^*}(t)\,W+L^{\mathbf{u}^*}(t, \xi^*(t))=\min_{v\in U}\left[\mathscr{A}^{v}(t)\,W+L(t, \xi^*(t)), v\right].$$

By the proof of Theorem 4.1, it suffices to show that, with probability 1, $(t, \xi^*(t))\in N$ for almost no $t\in[s, T]$. Let $\chi(t)=1$ if $(t, \xi^*(t))\in N$, and $\chi(t)=0$ otherwise. Then

$$E_{sy}\int_s^T \chi(t)\,dt=\int_s^T P_{sy}\{(t, \xi^*(t))\in N\}\,dt=\iint_N p^*(s, y, t, x)\,dx\,dt=0,$$

from which the conclusion follows.

We recall the discussion of transition densities in Chap. V.10, especially formula V(10.9) and Lemma V.10.1. The hypotheses on f and σ to be made in either §6 or §8 will insure the existence of a transition density, and hence the applicability of Corollary 4.1.

Admissible Nonanticipative Controls. A control policy uses the data available to the controller at time t only through the current state $\xi(t)$ of the process being controlled. Let us show that, under the conditions of the Verification Theorem 4.1, the controller could not get better expected performance using past data in some more complicated way.

By *admissible nonanticipative control* for initial time s let us mean a stochastic process u, defined on some probability space $(\Omega, \mathscr{F}, P)$, with the following properties:

(1) $u(t)\in U$ for $s\leqq t\leqq T$;

(2) $E\int_s^T |u(t)|^k\,dt<\infty$ for every $k>0$;

(3) u is nonanticipative with respect to an increasing family $\{\mathscr{F}_t\}$ of σ-algebras, $\mathscr{F}_t\subset\mathscr{F}$ for $s\leqq t\leqq T$; and

(4) Given $y\in E^n$, there exists a brownian motion w adapted to $\{\mathscr{F}_t\}$, such that (3.1) with (3.3) have a solution ξ and $E_{sy}|\xi(t)|^k$ is bounded for any $k>0$, $s\leqq t\leqq T$.

We denote by $\mathscr{U}(s)$ the class of all admissible nonanticipative controls, for initial time s. Some examples of $u\in\mathscr{U}(s)$ are:

(a) Suppose that u is bounded and satisfies (1), (3), with w a brownian motion adapted to $\{\mathscr{F}_t\}$. Let us take at the end of Sect. V.4 the random coefficients

$$b(t,x,\omega)=f(t,x,u(t,\omega)), \qquad \sigma(t,x,\omega)=\sigma(t,x,u(t,\omega)).$$

Here we indicate explicitly dependence on $\omega\in\Omega$ for clarity. The conditions (3.2) imply V(4.8). Hence $u\in\mathscr{U}(s)$.

(b) Let $\mathbf{u}$ be an admissible feedback control law; and define the process u by (3.4), where ξ is the solution of (3.1) with (3.3). Then $u\in\mathscr{U}(s)$.

We must emphasize that this nonanticipative control u depends not only on $\mathbf{u}$ but on the initial data $(s,\xi(s))$. In particular, for initial data $\xi(s)=y$, different choices of y lead to different nonanticipative controls.

There are, of course, many processes u in (a) which do not arise via a feedback control law as in (b). For instance, suppose that $\Omega=\mathscr{C}^d(\mathscr{T})$ where $\mathscr{T}=[s,T]$, $w(t,\omega)=\omega(t)$, $\mathscr{F}=\mathscr{B}[\mathscr{C}^d(\mathscr{T})]$, P is Wiener measure and

$$u(t,\omega)=\gamma(t,\omega(t_1),\ldots,\omega(t_m))$$

with $s\leqq t_1<t_2<\cdots<t_m\leqq t$. Then u is admissible if γ is bounded and Borel measurable.

Let us write the expected performance (2.7) as $J(s,y,u)$, for $u\in\mathscr{U}(s)$ and $\xi(s)=y$. In the definition of admissibility we have not included uniqueness of ξ. Thus J actually might depend on the particular solution of (3.1)–(3.3) selected, i.e., on the particular brownian motion w in (4) of the definition above of nonanticipative control.

Let us repeat the proof of Theorem 4.1(a), now taking $v=u(t)$ and recalling the remark in V.5 that Theorem V.5.1 remains true for random nonanticipative coefficients b, σ. We get

$$W(s,y)\leqq J(s,y,u)$$

for any $u\in\mathscr{U}(s)$. If $\mathbf{u}^*$ is an optimal feedback control and ξ^* the corresponding solution of (3.1′)–(3.3), then

$$u^*(t)=\mathbf{u}^*(t,\xi^*(t))$$

is admissible, nonanticipative, and

$$J(s,y,u^*)=J(s,y,\mathbf{u}^*)=W(s,y).$$

Thus $W(s,y)$ is the minimum expected performance not only in the class $\mathscr{V}$ of admissible feedback controls, but also among nonanticipative controls.

Corollary 4.2. *Under the conditions of Theorem* 4.1 *or Corollary* 4.1,

$$W(s,y)=W^0(s,y)=\min_{\mathscr{V}} J(s,y,\mathbf{u})=\min_{\mathscr{U}(s)} J(s,y,u). \tag{4.4}$$

In the definition of nonanticipative control, we have not fixed the probability space, the family $\{\mathscr{F}_t\}$, or the brownian motion w. If one does regard them as given, the infimum of $J(s,y,u)$ turns out to be the same under hypotheses which we impose later in §'s 6.7. See Problem 8.

Autonomous Problems. Let us next give a corresponding verification theorem for the autonomous version of the problem, formulated at the end of § 3 with J

given by (2.7^a). In this case

$$\mathscr{A}^v = \tfrac{1}{2} \sum_{i,j=1}^{n} a_{ij}(y,v) \frac{\partial^2}{\partial y_i \partial y_j} + \sum_{i=1}^{n} f_i(y,v) \frac{\partial}{\partial y_i}.$$

The dynamic programming equation is the autonomous form of (4.1):

$$(4.1^a) \qquad 0 = \min_{v \in U} [\mathscr{A}^v W + L(y,v)], \qquad y \in G,$$

with the boundary data

$$(4.2^a) \qquad W(y) = \Psi(y), \qquad y \in \partial^* G.$$

Here $\partial^* G$ is a portion of ∂G through which $\xi(t)$ must exit (often $\partial^* G = \partial G$). The fact that the exit time τ need not be bounded creates an additional difficulty. To avoid it, we assume that

$$(4.5) \qquad L(y,v) \geqq c > 0$$

for some constant c; moreover, we consider only bounded solutions of (4.1^a).

Theorem 4.2. *Let W be a solution of* (4.1^a)–(4.2^a) *such that W is in $C^2(G)$, and W is bounded and continuous in the closure $\bar{G}$. Then:*

(a) *$W(y) \leqq J(y, \mathbf{u})$ for any admissible autonomous feedback control law $\mathbf{u}$ and any $y \in G$.*

(b) *If $\mathbf{u}^*$ is admissible, $J(y, \mathbf{u}^*) < \infty$ and*

$$(4.3^a) \qquad \mathscr{A}^{\mathbf{u}^*} W + L^{\mathbf{u}^*}(y) = \min_{v \in U} [\mathscr{A}^v W + L(y,v)]$$

for all $y \in G$, then $W(y) = J(y, \mathbf{u}^*)$. Thus $\mathbf{u}^*$ is optimal among all admissible autonomous feedback control laws, for all choices of initial data $y \in G$.

Proof. We need consider only those $\mathbf{u}$ for which $J(y, \mathbf{u}) < \infty$. Then by (4.5) and boundedness of Ψ on $\partial^* G$, $P_y(\tau < \infty) = 1$. We now repeat the proof of Theorem 4.1, using Theorem V.7.1 instead of Theorem V.5.1. Note that $M = L^{\mathbf{u}} \geqq 0$, and that finiteness of $J(y, \mathbf{u})$ implies $E_y \int_0^\tau M\, dt < \infty$. □

Remark 1. If G is bounded, then boundedness of W follows from its continuity in $\bar{G}$. Since $L \geqq 0$ and Ψ is bounded on $\partial^* G$ finiteness of $J(y, \mathbf{u}^*)$ follows if it is known that $J(y, \mathbf{u}^0) < \infty$ for some particular admissible $\mathbf{u}^0$. The latter condition can often be verified in advance, by judicious choice of $\mathbf{u}^0$. Under the assumptions in Theorem 6.1^a below, we always have $J(y, \mathbf{u}^*) < \infty$.

Remark 2. The same discussion as for the nonautonomous case shows that $W(y)$ is also the minimum expected performance among all admissible non-anticipative controls.

Example 3. Let $f(x,u) = \alpha(x) + \beta(x)\, u$, $\sigma =$ identity, $L(x) = 1$, $\Psi(x) = 0$. Then $W(y)$ is the minimum expected time to reach ∂G starting at y. The dynamic programming equation is

$$\tfrac{1}{2} \Delta W + W_y\, \alpha(y) + \min_{v \in U} W_y\, \beta(y)\, v + 1 = 0$$

with $W=0$ on ∂G. If for instance U is the 1-dimensional interval $[-1,1]$, then

$$\begin{aligned}\mathbf{u}^*(y)&=\ \ 1 \quad \text{if } W_y\,\beta(y)<0\\ &=-1 \quad \text{if } W_y\,\beta(y)>0\\ &=\text{arbitrary if } W_y\,\beta(y)=0.\end{aligned}$$

Any part of the set $W_y\,\beta=0$ which turns out to be a smooth $n-1$-dimensional manifold is then called a *switching surface* for the problem.

§ 5. The Linear Regulator Problem (Complete Observations of System States)

Generally the dynamic programming partial differential equation (4.1) is difficult to solve. A case in which a solution of (4.1) is readily obtained is the linear regulator problem. This problem also has important applications in many practical situations. Because of the ease in determining the solution of (4.1) for the linear regulator problem, many control problems not of this form are approximately modeled as a linear regulator problem and the optimal control for this problem is used as an approximation of the optimal control for the original situation. In the present section we suppose that observations are complete. A solution for the partially observable linear regulator is given in § 11.

For the linear regulator problem the system equations have the form

$$d\xi=[A(t)\,\xi(t)+B(t)\,u(t)]\,dt+\sigma(t)\,dw, \tag{5.1}$$

and the expected system performance is

$$J(s,y,u)=E_{sy}\left\{\int_s^T[\xi(t)'\,M(t)\,\xi(t)+u(t)'\,N(t)\,u(t)]\,dt+\xi(T)'\,D\,\xi(T)\right\}. \tag{5.2}$$

The $n\times n$ matrices $M(t)$ and D are symmetric, nonnegative definite, and the $m\times m$ matrices $N(t)$ are symmetric, positive definite. There are no control constraints ($U=E^m$).

To apply the Verification Theorem 4.1, let us seek a solution W of the dynamic programming equation (4.1), which has the special form

$$W(s,y)=y'\,K(s)\,y+q(s), \qquad T_0\leqq s\leqq T, \tag{5.3}$$

with $K(s)$ symmetric and nonnegative definite. Here T_0 and T are arbitrary. When $s=T$, we must have $W(T,y)=y'\,D\,y$. A computation shows that

$$\begin{aligned}W_s+\mathscr{A}^v(s)\,W+L(s,y,v)=y'\,\dot K(s)\,y+\dot q(s)&\\ +y'\,K(s)[A(s)\,y+B(s)\,v]&\\ +[y'\,A(s)'+v'\,B(s)']\,K(s)\,y&\\ +\operatorname{tr} a(s)\,K(s)+y'\,M(s)\,y+v'\,N(s)\,v,&\end{aligned} \tag{5.4}$$

$$\operatorname{tr} a\,K=\sum_{i,j=1}^n a_{ij}\,K_{ij}, \qquad a=\sigma\,\sigma'.$$

Here $\cdot = d/ds$ and $'$ denotes as usual vector or matrix transpose. The right side of (5.4) is minimum with respect to v when the gradient in v is 0. The minimum occurs for $v = \mathbf{u}^*$, with

$$\mathbf{u}^*(s, y) = -N(s)^{-1} B'(s) K(s) y. \tag{5.5}$$

If W is a solution of (4.1), the minimum of (5.4) must be 0. From (5.5) we then get

$$0 = y'[\dot{K} + KA + A'K - KBN^{-1} B'K + M] y + \dot{q} + \operatorname{tr} aK,$$

for all $y = (y_1, \ldots, y_n)$ and $T_0 \leqq s \leqq T$. Thus a sufficient condition that (5.3) give a solution to (4.1) is that K satisfy the matrix Riccati equation (same as IV(5.4))

$$\dot{K}(s) = -K(s) A(s) - A'(s) K(s) + K(s) B(s) N^{-1}(s) B(s)' K(s) - M(s), \tag{5.6}$$

with the boundary condition $K(T) = D$. By Theorem IV.5.2, the solution $K(s)$ exists, and $\mathbf{u}^*$ is the same as in IV(5.5). The function $q(s)$ is defined by

$$q(s) = \int_s^T \operatorname{tr} a(s) K(s)\, ds = \sum_{i,j=1}^n \int_s^T a_{ij}(s) K_{ij}(s)\, ds, \tag{5.7}$$

with $a = \sigma \sigma'$.

According to Theorem 4.1 we then have:

Theorem 5.1. *Suppose that W is defined by* (5.3), (5.6), (5.7) *with $K(T) = D$, and $\mathbf{u}^*$ is defined by* (5.5). *Then $W(s, y)$ is the minimum expected system performance for the linear regulator, and $\mathbf{u}^*$ is the optimal feedback control law.*

Note that the noise coefficient σ enters Eq. (5.7) for q, but not Eq. (5.6) for K. Since K is not affected by σ neither is $\mathbf{u}^*$. The optimal feedback control law is the same for the stochastic linear regulator problem as for the deterministic linear regulator (with $\sigma = 0$) already treated in Chap. IV.5. Moreover, $\mathbf{u}^*(s, y)$ is a linear function of the state y. The minimum expected performance $W(s, y)$ differs from the minimum performance in the deterministic model by the non-negative term $q(s)$ in (5.3).

§ 6. Existence Theorems

In this section we shall see that a function W with the properties needed in the Verification Theorem 4.1 (or Theorem 4.2) indeed exists, under suitable assumptions. Among these assumptions is the existence of a bounded inverse σ^{-1}, which implies that the dynamic programming equation is uniformly parabolic. In §8 we shall weaken this assumption, and obtain a different existence theorem.

Once the existence of W is known, the existence of a Borel measurable optimal $\mathbf{u}^*$ follows from a selection theorem (see Theorem 6.3). The corresponding solution ξ^* of the system equations (3.1′) with (3.3) is then defined with the aid of the Girsanov formula (Theorem V.10.2). Under certain linearity and convexity assumptions, it is shown that $\mathbf{u}^*(t, \cdot)$ satisfies the kind of local Lipschitz condition needed to define ξ^* in the sense of the more traditional existence result Theorem V.4.1 (see Theorem 6.4).

To get existence of W, we quote existence theorems from the theory of parabolic second order partial differential equations. In the first result (Theorem 6.1), Q is a bounded cylinder; then we consider in Theorem 6.2 the fixed stopping time case ($\tau=T$, $Q=Q^0$).

For Theorem 6.1 we make the following assumptions:

$$Q=(T_0,T)\times G, \qquad (6.1)$$
$$\partial^* Q=([T_0,T]\times\partial G)\cup(\{T\}\times G)$$

where G is bounded with ∂G a manifold of class C^2.

$$U \text{ is compact.} \qquad (6.2)$$

The noise coefficient $\sigma=\sigma(t,x)$ is assumed not to depend on the control variable u, and $\sigma(t,x)$ is an $n\times n$ matrix. Thus w is an n-dimensional standard brownian motion. Moreover, we assume:

(6.3) (a) σ is of class $C^{1,2}(\bar{Q}^0)$; moreover $\sigma(t,x)$ has an inverse $\sigma^{-1}(t,x)$ for each $(t,x)\in\bar{Q}^0$;

(b) f, L are of class $C^1(\bar{Q}^0\times U)$;

(c) $\Psi(T,x)$ is of class $C^2(\bar{G})$; and $\Psi(t,x)=\tilde{\Psi}(t,x)$ for $T_0\leqq t\leqq T$, $x\in\partial G$, where $\tilde{\Psi}$ is of class $C^{1,2}(\bar{Q})$.

Let us define $a=\sigma\sigma'$ as in V(5.4). Since $\bar{Q}$ is compact, the inverses $\sigma^{-1}(t,x)$ are bounded on $\bar{Q}$. Thus the symmetric matrices $a(t,x)$ have characteristic values bounded below by $c>0$. This implies that Eq. (6.6) below is uniformly parabolic on Q (recall the definition in V.6).

Note. The values of σ, f, L, Ψ for $(t,x)\notin\bar{Q}$ are irrelevant, for purposes of Theorem 6.1. By suitably defining these functions for $(t,x)\in\bar{Q}^0-\bar{Q}$, we can arrange that the conditions (iii) §3 hold with $b=0$, $\theta=\sigma^{-1}f$. Then any Borel measurable feedback control $\mathbf{u}$ is admissible. Since U is compact and feedback controls $\mathbf{u}$ take values in U, any such $\mathbf{u}$ is bounded. By V(10.9) and remarks at the end of V.8, there is a transition density $p^{\mathbf{u}}(s,x,t,y)$.

Let us rewrite Eq. (4.1). For $(s,y)\in\bar{Q}$, $p\in E^n$, let

$$H(s,y,p)=\min_{v\in U}[p'f(s,y,v)+L(s,y,v)]. \qquad (6.4)$$

The function H satisfies a local Lipschitz condition, and the linear growth condition

$$|H(s,y,p)|\leqq C(1+|p|), \quad (s,y)\in\bar{Q}. \qquad (6.5)$$

See Problem 11; (6.5) uses the boundedness of Q and U assumed in (6.1), (6.2). The dynamic programming equation (4.1) now takes the form

$$0=W_s+\tfrac{1}{2}\sum_{i,j=1}^{n}a_{ij}(s,y)W_{y_iy_j}+H(s,y,W_y). \qquad (6.6)$$

An equation of this form is often called semilinear, since the second order derivatives $W_{y_iy_j}$ and W_s enter linearly.

Lemma 6.1. *If W_y is continuous on Q, then there exists a Borel measurable* $\mathbf{u}^*$ *from Q into U such that* (4.3) *holds almost all* $(s, y) \in Q$.

Proof. Since the dynamic programming equation now has the form (6.6), we can rewrite condition (4.3) as

$$
\begin{aligned}
W_y f^{\mathbf{u}^*} + L^{\mathbf{u}^*} &= \min_{v \in U} [W_y f(s, y, v) + L(s, y, v)] \\
&= H(s, y, W_y).
\end{aligned}
\tag{6.7}
$$

The function

$$\Theta(s, y, v) = W_y(s, y) f(s, y, v) + L(s, y, v)$$

is continuous on $Q \times U$, and U is compact by (6.2). By a selection theorem (Appendix B) there exists a Borel measurable $\mathbf{u}^*$ from Q into U such that

$$\Theta(s, y, \mathbf{u}^*(s, y)) = \min_{v \in U} \Theta(s, y, v). \quad \square$$

We now state the following result.

Theorem 6.1. *Let* (6.1)–(6.3) *hold. Then* (6.6) *has a unique solution in Q satisfying the boundary data* (4.2), *such that W is in $C^{1,2}(Q)$ and continuous in $\bar{Q}$.*

This theorem is a special case of general theorems about nonlinear parabolic equations. Let us outline another proof of Theorem 6.1, of independent interest. It is based on results about linear parabolic equations. The technical details of the argument are presented in Appendix E.

We define sequences of functions $W^1, W^2, \ldots$ on $\bar{Q}$ and of bounded Borel measurable feedback control laws $\mathbf{u}^0, \mathbf{u}^1, \ldots$ as follows. Choose $\mathbf{u}^0$ arbitrarily. Then W^{m+1} is the solution of the following boundary problem:

(6.8) (a)

$$0 = W_s^{m+1} + \mathscr{A}^{\mathbf{u}^m}(s) W^{m+1} + L^{\mathbf{u}^m}, \qquad m = 0, 1, 2, \ldots \text{ with } W^{m+1} = \Psi \text{ on } \partial^* Q;$$

and for almost all $(s, y) \in Q$, $m = 1, 2, \ldots$,

(b)

$$
\begin{aligned}
\mathscr{A}^{\mathbf{u}^m}(s) W^m + L^{\mathbf{u}^m} &= \min_{v \in U} [\mathscr{A}^v(s) W^m + L(s, y, v)] \\
&= \tfrac{1}{2} \sum_{i,j=1}^{n} a_{ij}(s, y) W^m_{y_i y_j} + H(s, y, W^m_y).
\end{aligned}
$$

The functions W^m need not be in $C^{1,2}(Q)$; however, they belong to the space $\mathscr{E}^{\lambda 0}(Q)$ defined in V.11 for $1 < \lambda < \infty$ and W^m_y is continuous on $\bar{Q}$. It can be shown, using a form of the maximum principle for parabolic equations, that

$$W^1 \geqq W^2 \geqq \cdots.$$

The desired solution W of (6.6) is the limit of W^m as $m \to \infty$.

Remarks. (1) Uniqueness of W with the properties in Theorem 6.1 follows from Corollary 4.1, Lemma 6.1 and the existence of the transition density $p^{\mathbf{u}^*}$. In fact, $W(s, x) = W^0(s, x)$ is the minimum expected performance, as in formula (4.4). The same remark holds true regarding the solution W obtained in Theorem 6.2 below.

(2) More can be said about the continuity of partial derivatives at points of ∂Q. Actually, W_y is continuous in Q; and W_s, $W_{y_i y_j}$ are continuous on $\bar{Q}-\partial^* Q$. If ∂G, $\Psi(T,x)$, and $\tilde{\Psi}$ are of class C^3, then these partial derivatives are continuous in $\bar{Q}-\{T\}\times\partial G$. This follows from Hölder estimates for partial derivatives cited in Appendix E. On the other hand, it can be shown by more elaborate methods Ladyzhenskaya-Solonnikov-Ural'seva [1, Chap. VI] that a solution W with the properties in Theorem 6.1 exists if ∂G merely satisfies an exterior sphere property and Ψ is continuous.

(3) The method of defining the kind of monotone sequence W^1, $W^2, \ldots$ in the proof of Theorem 6.1 was originally suggested by Bellman [1], as a formal general procedure in dynamic programming. He called the technique approximation in policy space.

We turn now to the case $Q=Q^0$, when J has the form (2.2). This is the case of fixed stopping time $\tau\equiv T$. Since Q^0 is unbounded the assumptions in Theorem 6.1 must be strengthened to include the behavior of $\sigma(t,x)$, $f(t,x,u)$, etc. for $|x|$ large. We now suppose, in addition to (6.2)

(6.9) (a) $f(t,x,u)=\tilde{b}(t,x)+\sigma(t,x)\theta(t,x,u)$.

(b) $\tilde{b}$, σ are in $C^{1,2}(\bar{Q}^0)$; moreover, σ, σ^{-1}, σ_x, $\tilde{b}_x$ are bounded on $\bar{Q}^0$, while θ is in $C^1(\bar{Q}^0\times U)$, with θ, θ_x bounded.

(c) $L(t,x,u)$ is in $C^1(\bar{Q}^0\times U)$; and L, L_x satisfy a polynomial growth condition.

(d) $\Psi(x)$ is in $C^2(E^n)$; and Ψ, Ψ_x satisfy a polynomial growth condition.

Every Borel measurable $\mathbf{u}$ (with values in the bounded set U) is admissible. This follows from (iii) §3. Moreover, there is a transition density $p^{\mathbf{u}}(s,x,t,y)$ as in the Note preceding (6.4). In particular, if W is a solution of (6.6) with W_y continuous, the control $\mathbf{u}^*$ defined by Lemma 6.1 is admissible.

To extend Theorem 6.1 we need an a priori estimate, on any bounded set, for a solution W to (6.6) and for its gradient W_y in the space-like variables. The proof uses stochastic control methods. Lemma 7.2 below is a similar result, under different assumptions.

Lemma 6.2. *Let $B\subset E^n$ be bounded, and W a solution of* (6.6) *in $C^{1,2}(Q^0)$ with W continuous in $\bar{Q}^0$ and $W(T,x)=\Psi(x)$. Then there exists a constant M_B, such that*

$$|W|\leq M_B, \quad |W_y|\leq M_B \quad \textit{for all } y\in B,\ T_0\leq s<T.$$

We defer the proof of Lemma 6.2 to Appendix E.

Theorem 6.2. *Let $Q=Q^0$. Assume* (6.2) *and* (6.9). *Then Eq.* (6.6) *with the Cauchy data $W(T,y)=\Psi(y)$ has a unique solution W in $C_p^{1,2}(Q)$ with W continuous in $\bar{Q}^0$.*

Under sufficiently strong assumptions about the behavior of $f(t,x,u)$, $L(t,x,u)$, and $\Psi(x)$ for $|x|$ large, the same proof as for Theorem 6.1 is valid. Using an approximation argument, we shall give a proof of the Theorem in Appendix E.

Remarks. (4) Just as in Remark 2, W_y is actually continuous in $\bar{Q}^0$. If Ψ is C^3, then W is in $C_p^{1,2}(\bar{Q}^0)$.

(5) By (6.2), U is compact. For an extension of Theorem 6.2 to cases where U is unbounded see Fleming [8], Lee [1].

Existence of Optimal Feedback Controls. From the existence of W in Theorem 6.1 or 6.2 we can easily deduce the existence of an optimal $\mathbf{u}^*$. Recall that, under the assumptions in the present section, any Borel measurable $\mathbf{u}$ with values in the bounded set U is admissible.

Theorem 6.3. *Under the assumptions of either Theorem* 6.1 *or Theorem* 6.2, *there exists an optimal admissible feedback control law* $\mathbf{u}^*$, *satisfying* (4.3) *almost everywhere in* Q.

Proof. Define $\mathbf{u}^*(s, y)$ in Q by Lemma 6.1; when $Q \neq Q^0$, define $\mathbf{u}^*(s, y)$ arbitrarily outside Q so that $\mathbf{u}^*$ is Borel measurable with values in Q. The theorem follows immediately from Corollary 4.1. □

This result is in contrast with the deterministic situation, treated in Chap. IV. For the latter problem, there is no general theorem about existence of optimal feedback control laws. In the deterministic case, there is even a problem of defining solutions to the system equations corresponding to a discontinuous feedback control law $\mathbf{u}$.

The optimal $\mathbf{u}^*$ in Theorem 6.3 may be discontinuous. This is easily seen from examples, in which discontinuities of $\mathbf{u}^*$ occur at points (s, y) where $\Theta = W_y f + L$ does not have a unique minimum on U. If $\Theta(s, y, \cdot)$ has a unique minimum on U for every $(s, y) \in Q$, then it follows using compactness of U that $\mathbf{u}^*$ is continuous on Q. Sufficient additional conditions for such a unique minimum are convexity of U and strict convexity of $\Theta(s, y, \cdot)$.

Under somewhat stronger conditions, we next prove a local Lipschitz condition for $\mathbf{u}^*(s, \cdot)$.

Lemma 6.3. *Let U be compact and convex; and $Q' = (T_0, T') \times G'$. Let $\Theta(s, y, u)$ be a function satisfying:*

(i) *$\Theta(s, y, \cdot)$ is C^2 for each $(s, y) \in \bar{Q}'$;*

(ii) *$\Theta_u(s, \cdot, u)$ satisfies on $\bar{G}'$ a Lipschitz condition, uniformly with respect to s, u; and*

(iii) *The characteristic values of the matrices Θ_{uu} are bounded below by $\gamma > 0$.*

Let $\mathbf{u}^*(s, y)$ be the unique $v \in U$ at which $\Theta(s, y, v)$ is minimum on U. Then $\mathbf{u}^*(s, \cdot)$ satisfies a Lipschitz condition on $\bar{G}'$, uniformly for $T_0 \leqq s \leqq T'$.

Proof. Assumption (iii) implies that $\Theta(s, y, \cdot)$ is a strictly convex function; and hence it has a unique minimum on U as asserted. Given $s, y_1, y_2 \in G'$, let

$$v_1 = \mathbf{u}^*(s, y_1), \qquad v_2 = \mathbf{u}^*(s, y_2).$$

From (i), (iii) and Taylor's formula,

$$\Theta(s, y_1, v_2) - \Theta(s, y_1, v_1) \geqq \Theta_u(s, y_1, v_1) \cdot (v_2 - v_1) + \frac{\gamma}{2} |v_2 - v_1|^2.$$

Since $\Theta(s, y_1, \cdot)$ is minimum on the convex set U at v_1, the first term on the right side is nonnegative. We apply the integral form of the mean-value theorem to the left side, to get

$$(*)\qquad \int_0^1 \Theta_u(P_1(\lambda))\cdot(v_2-v_1)\,d\lambda \geqq \frac{\gamma}{2}|v_2-v_1|^2,$$
$$P_1(\lambda)=(s, y_1, v_1+\lambda(v_2-v_1)).$$

Similarly, by interchanging the roles of y_1 and y_2,

$$(**)\qquad -\int_0^1 \Theta_u(P_2(\lambda))\cdot(v_2-v_1)\,d\lambda \geqq \frac{\gamma}{2}|v_2-v_1|^2,$$
$$P_2(\lambda)=(s, y_2, v_1+\lambda(v_2-v_1)).$$

By adding (*) and (**) we get

$$\int_0^1 [\Theta_u(P_1(\lambda))-\Theta_u(P_2(\lambda))]\cdot(v_2-v_1)\,d\lambda \geqq \gamma|v_2-v_1|^2.$$

We use Cauchy's inequality inside the integral and cancel a factor $|v_2-v_1|$, obtaining

$$\int_0^1 |\Theta_u(P_1(\lambda))-\Theta_u(P_2(\lambda))|\,d\lambda \geqq \gamma|v_2-v_1|.$$

By assumption (ii), and $|P_1(\lambda)-P_2(\lambda)|=|y_1-y_2|$,

$$|\Theta_u(P_1(\lambda))-\Theta_u(P_2(\lambda))| \leqq M|y_1-y_2|,$$

where M is a Lipschitz constant for $\Theta_u(s,\cdot,v)$. Then $M|y_2-y_1|\geqq\gamma|v_2-v_1|$; in other words

$$(6.10)\qquad |\mathbf{u}^*(s,y_2)-\mathbf{u}^*(s,y_1)| \leqq \gamma^{-1}M|y_2-y_1|$$

for $y_1, y_2 \in G'$. Thus $\gamma^{-1}M$ is a Lipschitz constant for $\mathbf{u}^*(s,\cdot)$ on G'. □

Let us now take $\Theta = W_y f + L$ as in the proof of Lemma 6.1 and assume that f is linear in the control variables:

$$(6.11)\qquad f(t,x,u)=\alpha(t,x)+\beta(t,x)u.$$

If the indicated partial derivatives exist and are continuous,

$$\Theta_u = W_y\,\beta + L_u, \qquad \Theta_{uu} = L_{uu},$$
$$\Theta_{uy} = W_{yy}\,\beta + W_y\,\beta_x + L_{ux}.$$

Condition (ii) in Lemma 6.3 will be satisfied if Θ_{uy} is bounded on $\bar{Q}'$, while (iii) is equivalent to the same assumption on L_{uu}. We, therefore, have:

Theorem 6.4. *Besides the assumptions of Theorem* 6.1 (*or Theorem* 6.2) *let us assume:*

(a) *U is convex;*

(b) *f has the form* (6.11);

(c) *$L(t,x,u)$ is C^2, and the characteristic values of the matrices L_{uu} are bounded below by $\gamma>0$.*

Let $Q'=(T_0, T')\times G'$, where G' is bounded and W is in $C^{1,2}(\bar{Q}')$. Then $\mathbf{u}^*(s,\cdot)$ satisfies a Lipschitz condition on $\bar{G}'$, uniformly for $T_0\leqq s\leqq T'$.

In Theorem 6.4 for Q' we may take any bounded cylinder with $\bar{Q}'\subset\bar{Q}-\partial^*Q$. Under slightly stronger (class C^3) boundary assumptions, as in Remarks 2, 4 we may take any Q' with $\bar{Q}'\subset\bar{Q}-\{T\}\times\partial G$. In that case, there is an extension of $\mathbf{u}^*$ to $\bar{Q}^0-\bar{Q}$ satisfying conditions (i) §3 for admissibility. (Recall that feedback controls are defined on $\bar{Q}^0$, not merely on $\bar{Q}$, according to our definition.) This can be done as follows. We may assume that $0\in U$. Let $\nu(z)$ denote the exterior unit normal at z to the C^3 manifold ∂G. Let $N=N_{r_0}$ denote the r_0-neighborhood of $\bar{G}$. For r_0 sufficiently small, every $y\in N-\bar{G}$ is uniquely representable in the form $y=z+r\,\nu(z)$, with $z\in\partial G$, $0<r<r_0$. Let $y'=z-r\,\nu(z)$ be the point in G symmetrical to y; and let α be a C^1 function with $\alpha(y)=1$ on G, $\alpha(y)=0$ outside N, and $0\leqq\alpha\leqq 1$. Define $\mathbf{u}^*$ on $\bar{Q}^0-\bar{Q}$ by

$$\begin{aligned}\mathbf{u}^*(s,y)&=\alpha(y)\,\mathbf{u}^*(s,y'), && \text{if } y\in N-\bar{G},\\ &=0 && \text{if } y\in E^n-N,\end{aligned}$$

and for $T_0\leqq s<T$, with $\mathbf{u}^*(T,y)$ arbitrary. Since U is convex and 0, $\mathbf{u}^*(s,y')\in U$, $0\leqq\alpha\leqq 1$, we have $\mathbf{u}^*(s,y)\in U$. Moreover, $\mathbf{u}^*(s,\cdot)$ satisfies a uniform Lipschitz condition for $T_0\leqq s\leqq T'<T$.

Existence Results in the Autonomous Case. Eq. (4.1^a) is now uniformly elliptic, semilinear:

$$0=\tfrac{1}{2}\sum_{i,j=1}^{n} a_{ij}(y)\,W_{y_iy_j}+H(y,W_y). \tag{6.6^a}$$

We consider it in an open set G, with the data $W=\Psi$ on ∂G.

Theorem 6.1^a. *Let G be bounded with ∂G a class C^2 manifold, and with Ψ of class C^2. Assume* (6.2) *and the autonomous version of* (6.3)(a), (b). *Then* (6.6^a) *has a unique solution in G satisfying $W(y)=\Psi(y)$ for $y\in\partial G$, such that W is in $C^2(G)$ and in $C^1(\bar{G})$.*

A proof very similar to that for Theorem 6.1 can be given. One can also quote a general existence theorem for elliptic equations Ladyzenskaya-Ural'seva [1, p. 373]. Remarks about boundary behavior of W like Remark 2 hold.

From Theorem 6.1^a follows the existence of an optimal autonomous Borel measurable feedback control $\mathbf{u}^*$, just as in Theorem 6.3. If ∂G and Ψ are C^3 and hypotheses (a), (b), (c) of Theorem 6.4 hold, then $\mathbf{u}^*$ is Lipschitz on $\bar{G}$. By the argument following Theorem 6.4, $\mathbf{u}^*$ has a Lipschitz extension to E^n, with values in U.

§ 7. Dependence of Optimal Performance on y and σ

For the existence results in §6, the assumption about σ^{-1} in (6.3a) was crucial. This assumption guaranteed uniform parabolicity of the dynamic programming equation (6.6). However, uniform parabolicity does not hold for many systems of interest. In §'s 7, 8, 9 we outline a number of results which do not assume that σ^{-1}

exists. The method for getting these results involves approximating the noise coefficient σ by σ^ε which has an inverse $(\sigma^\varepsilon)^{-1}$ for each $\varepsilon>0$.

To simplify the exposition in §'s 7, 8, 9 we shall make the following assumptions (7.1)–(7.4). These assumptions can be weakened in various ways, at the expense of complicating the discussion; see Fleming [3][7][10].

(7.1) Fixed terminal time T $(Q=Q^0)$.

(7.2) U is compact.

(7.3) $f(t,x,u)$, $L(t,x,u)$, $\Psi(x)$ are in $C^\infty(\bar{Q}^0\times U)$, $C^\infty(E^n)$ and satisfy a uniform Lipschitz condition in (x,u) for $x\in E^n$, $u\in U$, $(|f_x|, |f_u|, |L_x|, |L_u|, |\Psi_x|$ all bounded by some constant D.)

(7.4) $\sigma=\sigma(t)$ is in $C^\infty(\mathscr{T}_0)$.

Assumption (7.4) that the noise coefficient is not state dependent allows us to use standard results about ordinary differential equations instead of corresponding results for stochastic differential equations. The matrices $a(t)=\sigma(t)\,\sigma(t)'$ are nonnegative definite, but not necessarily positive definite. In particular, we do not exclude the deterministic problem $(\sigma=0)$ already discussed in earlier chapters.

By adding an additional system equation

$$d\xi_{n+1}=L(t,\xi_1,\ldots,\xi_n,u)\,dt,\qquad \xi_{n+1}(s)=0$$

and replacing $\Psi(x_1,\ldots,x_n)$ by

$$\tilde{\Psi}(x_1,\ldots,x_n,x_{n+1})=x_{n+1}+\Psi(x_1,\ldots,x_n),$$

we can reduce the performance criterion J in (2.7) to the special (Mayer) form with $L=0$. In this section we shall thus suppose that

$$J(s,y;u)=E_{sy}\,\Psi[\xi(T)].$$

This simplifies various calculations.

In making the estimates below, it is essential to make use of nonanticipative control processes. Given an initial time s, consider the class $\mathscr{U}(s)$ of admissible nonanticipative controls u defined in § 4. By (7.2), u is automatically bounded.

Given initial data (s,y) consider the optimal expected system performance, using nonanticipative controls:

(7.5) $$V(s,y)=\inf_{\mathscr{U}(s)} J(s,y;u).$$

We call $u^*\in\mathscr{U}(s)$ optimal, for the initial data (s,y), if $V(s,y)=J(s,y,u^*)$.

If the conditions of the Verification Theorem 4.1 are satisfied, then $V=W=W^0$ (formula (4.4)). In that case, $u^*(t)=\mathbf{u}^*(t,\xi^*(t))$ is optimal as already noted in § 4.

In the present section we shall study V as a function of the initial state y. The results are presented as five lemmas, needed in §'s 8, 9.

Given $u\in\mathscr{U}(s)$ let us write the system equation (3.1) in integrated form:

(7.6) $$\xi(t)=y+\int_s^t f(r,\xi(r),u(r))\,dr+\int_s^t \sigma(r)\,dw.$$

The process ξ is a solution of a stochastic differential equation with random, nonanticipative coefficient (see end of Sect. V.4). In the notation there $\sigma=\sigma(t)$ and

$$b(t, x, \omega)=f(t, x, u(t, \omega)),$$

temporarily indicating explicitly dependence on $\omega\in\Omega$. Sometimes we need to indicate dependence of the solution ξ of (7.6) on the initial data by writing $\xi=\xi(t; s, y)$. By (7.4) the term $\int_s^t \sigma\, dw$ can be treated as a continuous forcing term, not involving the initial state y. By standard results about ordinary differential equations $\xi(t; s, \cdot)$ is C^1. The $n\times n$ matrix $\xi_y=(\partial\xi_i/\partial y_j)$ of partial derivatives satisfies, from (7.6) and the fact that $f_x=b_x$

$$d\xi_y=f_x\,\xi_y\,dt, \qquad s\leqq t\leqq T, \tag{7.7}$$

with $\xi_y(s; s, y)=$identity. By (7.3), ξ_y is bounded; in fact $\|\xi_y\|\leqq \exp D(T-s)$.

Remark 7.1. This estimate for $\|\xi_y\|$ is the basis for Lemma 7.2. The simplicity of the estimate shows the technical advantage of nonanticipative controls. For a feedback $\mathbf{u}$ an additional term $f_u\,\mathbf{u}_x\,\xi_y\,dt$ would be needed in (7.7), and a Lipschitz condition on $\mathbf{u}(t, \cdot)$ would have to be imposed to get an estimate for $\|\xi_y\|$. By using nonanticipative controls we have avoided this difficulty.

For state dependent σ, the formula corresponding to (7.7) would be $d\xi_y=f_x\,\xi_y\,dt+\sigma_x\,\xi_y\,dw$; see Gikhman-Skorokhod [2, § 8].

Lemma 7.1 ($L=0$). *The gradient of J with respect to the initial state y exists, and satisfies*

$$J_y(s, y, u)=E_{sy}\{\Psi_x[\xi(T)]\,\xi_y(T)\}. \tag{7.8}$$

Proof. We have

$$J(s, y, u)=E\,\Psi[\xi(T; s, y)],$$

where we are now indicating dependence on s, y by writing $\xi(\cdot\,; s, y)$ instead of E_{sy}. Since Ψ_x is bounded by assumption (7.3), we get (7.8) by differentiating with respect to y inside the expectation. □

From (7.8) we have the bound $M=D\exp D(T-T_0)$ for $|J_y|$, when $T_0\leqq s\leqq T$. Thus $J(s, \cdot, u)$ satisfies the uniform Lipschitz condition

$$|J(s, y', u)-J(s, y, u)|\leqq M\,|y'-y|,$$

for every $u\in\mathscr{U}(s)$. From the definition (7.5) of V, we then have the same Lipschitz condition for V:

Lemma 7.2. *Let $M=D\exp D(T-T_0)$, with D as in (7.3). Then*

$$|V(s, y')-V(s, y)|\leqq M\,|y'-y|$$

for $y, y'\in E^n$, $T_0\leqq s\leqq T$.

Remark 7.2. In Lemma 7.2 we have used the boundedness of L_x and Ψ_x assumed in (7.3). Under the less restrictive assumptions (6.9)(c), (d) on L, Ψ a local estimate of this type was already proved in Lemma 6.2.

We next note that V_y also satisfies (7.8), provided it exists and there is an optimal u^*:

Lemma 7.3. *Suppose that $V_y(s, y)$ exists and $u^* \in \mathscr{U}(s)$ is optimal for initial data (s, y). Then $V_y(s, y) = J_y(s, y, u^*)$.*

Proof. For any y',

$$J(s, y', u^*) - V(s, y') \geqq 0.$$

Equality holds when $y' = y$. Hence the gradient of $J(s, \cdot, u^*) - V(s, \cdot)$ is 0 at y, provided it exists there. □

Since $V(s, \cdot)$ is Lipschitz Rademacher's theorem, Federer [2, p. 216], implies that $V(s, \cdot)$ is differentiable almost everywhere in E^n for each s. If $V(s, \cdot)$ is differentiable at y, then the gradient V_y exists there and Lemma 7.3 applies. Moreover, $|V_y| \leqq M$.

We now turn to some estimates of the dependence of V on σ. Let us replace σ by another noise coefficient σ', and let ξ' be the corresponding solution of the system equations:

$$\xi'(t) = y + \int_s^t f(r, \xi'(r), u(r))\, dr + \int_s^t \sigma'(r)\, dw. \tag{7.6'}$$

If we subtract (7.6) from (7.6'), recall (7.3), and use Gronwall's inequality, we get

$$\begin{gathered} \|\xi' - \xi\|_t \leqq \|\zeta\|_t \exp D(t-s), \\ \zeta(t) = \int_s^t [\sigma'(r) - \sigma(r)]\, dw, \qquad \|\zeta\|_t = \max_{s \leqq r \leqq t} |\zeta(r)|. \end{gathered} \tag{7.9}$$

Let us also write $J'(s, y, u)$, $V'(s, y)$ to mean that σ is replaced by σ'. Since $|\Psi_x| \leqq D$, Ψ is Lipschitz with constant D. Thus, for any $u \in \mathscr{U}(s)$,

$$\begin{aligned} |J'(s, y, u) - J(s, y, u)| &\leqq E\,|\Psi[\xi'(T)] - \Psi[\xi(T)]| \\ &\leqq D \exp D(t-s)\, E\,\|\zeta\|. \end{aligned}$$

Here $\|\zeta\| = \|\zeta\|_T$; for its expected value we have by Cauchy-Schwarz and V.3.7(a) the estimate

$$E\,\|\zeta\| \leqq (E\,\|\zeta\|^2)^{1/2} \leqq 2 \left(\int_s^T |\sigma' - \sigma|^2\, dt \right)^{1/2}$$

for $T_0 \leqq s \leqq T$. Let

$$C_1 = 2(T - T_0)^{1/2}, \qquad M_1 = C_1 M$$

with M as in Lemma 7.2. Then

$$E\,\|\zeta\| \leqq C_1\, \|\sigma' - \sigma\|, \tag{7.10}$$

$$|J'(s, y, u) - J(s, y, u)| \leqq M_1\, \|\sigma' - \sigma\|. \tag{7.11}$$

Since this is true for every u, we have:

Lemma 7.4. *For all $(s, y) \in \bar{Q}^0$,*

$$|V'(s, y) - V(s, y)| \leqq M_1\, \|\sigma' - \sigma\|.$$

Let us now suppose that $\sigma' = \sigma^\varepsilon$ depending on $\varepsilon > 0$, and that $\|\sigma^\varepsilon - \sigma\| \to 0$ as $\varepsilon \to 0$. We write correspondingly J^ε for the expected system performance, and V^ε for its infimum in (7.5). According to Lemma 7.4, V^ε tends uniformly on $\bar{Q}^0$ to V as $\varepsilon \to 0$. This is a rather weak statement; in §9 we shall outline some stronger results in case $\sigma = 0$. In preparation for those results we state another lemma. In this lemma we let $u^{\varepsilon *}$ denote a nonanticipative process optimal for noise coefficient σ^ε and initial data (s, y) (thus $V^\varepsilon(s, y) = J^\varepsilon(s, y, u^{\varepsilon *})$ and u^* a nonanticipative process optimal for the original σ and the same initial data.

Lemma 7.5. *Suppose that $V_y(s, y)$, $V_y^\varepsilon(s, y)$ exist and that $\int_s^T |u^{\varepsilon *}(t) - u^*(t)|\, dt \to 0$ in probability as $\varepsilon \to 0$. Then $V_y^\varepsilon(s, y) \to V_y(s, y)$ as $\varepsilon \to 0$.*

Proof. Let us write $\xi^{\varepsilon *}$, ξ^* for the solutions of the system equations corresponding to $u^{\varepsilon *}$, u^* respectively, with initial data (s, y). By Lemma 7.3

$$V_y^\varepsilon(s, y) - V_y(s, y) = E_{sy}\{\Psi_x[\xi^{\varepsilon *}(T)]\, \xi_y^{\varepsilon *}(T) - \Psi_x[\xi^*(T)]\, \xi_y^*(T)\}.$$

Now Ψ_x is bounded and continuous; since f_x is bounded, $\xi_y^{\varepsilon *}$ is uniformly bounded by (7.7). To prove the lemma, it suffices to show that $\xi^{\varepsilon *}(T) \to \xi^*(T)$ and $\xi_y^{\varepsilon *}(T) \to \xi_y^*(T)$ as $\varepsilon \to 0$, the convergence being in probability. This will follow if we show $\|\xi^{\varepsilon *} - \xi^*\|$ and $\|\xi_y^{\varepsilon *} - \xi_y^*\|$ tend to 0 in probability.

Define η as the solution of

$$\eta(t) = y + \int_s^t f(t, \eta(r), u^{\varepsilon *}(r))\, dr + \int_s^t \sigma(r)\, dw.$$

Then

$$\xi^{\varepsilon *}(t) - \eta(t) = \int_s^t [f(r, \xi^{\varepsilon *}(r), u^{\varepsilon *}(r)) - f(r, \eta(r), u^{\varepsilon *}(r))]\, dr$$

$$+ \int_s^t [\sigma^\varepsilon(r) - \sigma(r)]\, dw.$$

By (7.9)

$$\|\xi^{\varepsilon *} - \eta\|_t \leq \|\zeta^\varepsilon\|_t \exp D(t-s)$$

where D is as in (7.3) and $\zeta^\varepsilon(t) = \int_s^t (\sigma^\varepsilon - \sigma)\, dw$. By (7.10) with $\sigma' = \sigma^\varepsilon$, $\|\xi^{\varepsilon *} - \eta\| \to 0$ in probability as $\varepsilon \to 0$. Next,

$$\eta(t) - \xi^*(t) = \int_s^t [f(r, \eta(r), u^{\varepsilon *}(r)) - f(r, \xi^*(r), u^*(r))]\, dr;$$

by using again (7.3) and Gronwall's inequality

$$\|\eta - \xi^*\|_t \leq D\, e^{D(t-s)} \int_s^t |u^{\varepsilon *}(r) - u^*(r)|\, dr.$$

Using the hypothesis of the lemma, $\|\eta - \xi^*\| \to 0$ in probability as $\varepsilon \to 0$. Therefore, $\|\xi^{\varepsilon *} - \xi^*\| \to 0$ in probability as $\varepsilon \to 0$.

Let $\chi^\varepsilon = \xi_y^{\varepsilon *}$, $\chi = \xi_y^*$; by (7.7) $d\chi^\varepsilon = A^\varepsilon \chi^\varepsilon$, $d\chi = A\chi$, where

$$A^\varepsilon(t) = f_x(t, \xi^{\varepsilon *}(t), u^{\varepsilon *}(t)), \qquad A(t) = f_x(t, \xi^*(t), u^*(t))$$

and $\chi^\varepsilon = \chi =$ identity when $t=s$. Then

$$\chi^\varepsilon - \chi = \int_s^t A(r)[\chi^\varepsilon(r) - \chi(r)]\, dr + \int_s^t [A^\varepsilon(r) - A(r)]\, \chi^\varepsilon(r)\, dr.$$

Since A and χ^ε are uniformly bounded on $[s, T]$, Gronwall's inequality implies

$$\|\chi^\varepsilon - \chi\| \leqq C \int_s^T |A^\varepsilon(r) - A(r)|\, dr$$

for some C. From what has already been shown,

$$\lim_{\varepsilon \to 0} E_{sy} \int_s^T |A^\varepsilon(r) - A(r)|\, dr = 0.$$

Therefore, $\|\chi^\varepsilon - \chi\| \to 0$ in probability. □

Remark. A different kind of result about convergence of partial derivatives $V^\varepsilon_{y_i}$ to V_{y_i} for certain i will be proved in Lemma 8.2(c).

§ 8. Generalized Solutions of the Dynamic Programming Equation

In many systems of interest, noise enters only certain components of the system equations (3.1). In such cases the matrices $a = \sigma\sigma'$ are singular and the dynamic programming equation (4.1) [or (6.6)] is degenerate (not parabolic). We no longer know that a $C^{1,2}$ solution W exists, as required in the Verification Theorem 4.1.

In this section we show that the function V in (7.5) is a kind of generalized solution to (4.1), for an interesting class of systems which we now describe. This is Theorem 8.2. Theorem 8.1 takes the place of Theorem 4.1.

Let us assume that noise enters the last $n-v$ system components, $1 \leqq v < n$. Moreover, control is applied to the system only through these same components. Thus, we assume that Eqs. (3.1) now have the form $(1 \leqq v < n)$:

$$\text{(8.1)} \quad \begin{aligned} d\xi_i &= \tilde{b}_i(t, \xi(t))\, dt, & i &= 1, \ldots, v, \\ d\xi_i &= [\tilde{b}_i(t, \xi(t)) + g_i(t, \xi(t), u(t))]\, dt + \sum_{j=v+1}^{n} \sigma_{ij}(t)\, dw_j, & i &= v+1, \ldots, n. \end{aligned}$$

In vector-matrix notation, let us write $\hat{x} = (x_{v+1}, \ldots, x_n)$ for the last $n-v$ components of a vector x. Moreover, let

$$g = \begin{pmatrix} 0 \\ \hat{g} \end{pmatrix}, \qquad \sigma = \begin{pmatrix} 0 & 0 \\ 0 & \hat{\sigma} \end{pmatrix},$$

where $\hat{g}$ has components $g_{v+1}, \ldots, g_n$, $\hat{\sigma} = (\sigma_{ij})$, $i, j = v+1, \ldots, n$. Then (8.1) has the form (3.1), with $f = \tilde{b} + g$. Let us assume besides the general assumptions (7.1), (7.2), and on L, Ψ in (7.3):

(8.2) $\tilde{b}(t,x)$ is in $C^\infty(\bar{Q}^0)$ and satisfies a uniform Lipschitz condition in x ($|\tilde{b}_x| \leqq D$).

(8.3) $\hat{g}(t, x, u) = \hat{\sigma}(t)\, \hat{\theta}(t, x, u)$, where $\hat{\theta}$ is in $C^\infty(\bar{Q}^0 \times U)$ and $\hat{\theta}$, $\hat{\theta}_x$ are bounded.

(8.4) The $(n-\nu)\times(n-\nu)$ matrices $\hat{\sigma}(t)$ are nonsingular, and in $C^\infty(\mathcal{T}_0)$.

(8.5) The transition distributions corresponding to the system $d\tilde{\xi}=\tilde{b}(t,\tilde{\xi}(t))\,dt+\sigma(t)\,dw$ have a density $\tilde{p}$ satisfying, for some $\tilde{\mu}>1$,

$$\int_{Q_{s'T}} |\tilde{p}(s,y,t,x)|^{\tilde{\mu}}\,dt\,dx<\infty, \qquad T_0\leqq s<s'<T.$$

Under the assumptions (8.2)–(8.4), any Borel measurable feedback control law $\mathbf{u}$ is admissible (by (7.2) $\mathbf{u}$ is automatically bounded). For any such $\mathbf{u}$, the function $\hat{\Theta}^{\mathbf{u}}(t,x)=\hat{\theta}(t,x,\mathbf{u}(t,x))$ has properties V(10.6). A solution ξ of (8.1) with given initial data exists and is unique in probability law, by Theorems V.10.2 and V.10.3. (Assumption (8.5) will be used below to allow us to apply Theorem V.11.1.)

Example. Suppose that the system equations are linear, with $\tilde{b}(t,x)=A(t)x$, $\hat{g}(t,u)=\hat{B}(t)\,u$. In (8.3) we take $\hat{\theta}(t,u)=\hat{\sigma}^{-1}(t)\,\hat{B}(t)\,u$. If the pair (A,σ) is completely controllable at any s, then by Theorem V.9.1, the transition densities $\tilde{p}(s,y,t,x)$ are gaussian. Then (8.5) holds for any $\tilde{\mu}>1$.

For a system of the form (8.1), the dynamic programming equation (4.1) or (6.6) has the form

$$\begin{aligned} 0&=W_s+\tfrac{1}{2}\sum_{i,j=\nu+1}^{n} a_{ij}(s)\,W_{y_i y_j}+W_y\,\tilde{b}+H(s,x,W_{\hat{y}}),\\ a(s)&=\hat{\sigma}(s)\,\hat{\sigma}(s)', \qquad W_{\hat{y}}=(W_{y_{\nu+1}},\ldots,W_{y_n}),\\ H(s,y,\hat{p})&=\min_{v\in U}\,[\hat{p}'\,\hat{g}(s,y,v)+L(s,y,v)]. \end{aligned} \tag{8.6}$$

Eq. (8.6) is not parabolic, since the matrix of coefficients a_{ij} has rank $n-\nu$ rather than full rank n. Such an equation is often called *degenerate parabolic.* Instead of merely solutions of (8.6) in the classical sense ($C^{1,2}$ functions W), we shall now admit certain solutions W which satisfy (8.6) almost everywhere. We recall the definition of spaces $\mathcal{E}^{\lambda\nu}(Q)$ in Chap. V.11.

Definition. A function $W(s,y)$ is a *generalized solution* of (8.6) in Q if W is in $\mathcal{E}^{\lambda\nu}(Q)$ with $\lambda^{-1}+\mu^{-1}=1$, $1<\mu<\tilde{\mu}$, and W satisfies (8.6) at almost all points $(s,y)\in Q$.

The main results of Sects. 4 and 6 were the verification theorems, and the existence of solutions to (6.6) and of an optimal control law $\mathbf{u}^*$. The verification theorems extend with no difficulty; in the proof of Theorem 4.1 (or Theorem 4.2) one uses Theorem V.11.1 in place of Theorem V.5.1. Recall also Corollary 4.1.

Theorem 8.1. *Under assumptions* (8.2)–(8.5), *Theorem* 4.1 *remains true if W is a generalized solution of* (4.1). *In Theorem* 4.1(b), *we now require that* (4.3) *holds for almost all* $(s,y)\in Q$.

In the remainder of this section we shall prove that the function V defined by (7.5) is a generalized solution, for the case of fixed terminal time ($Q=Q^0$). From the verification theorem, it will then follow that the minimum of J is the same among feedback laws $\mathbf{u}$ and among nonanticipative controls u. (We do not know whether these two minima are the same when our assumptions are violated.)

For $\varepsilon>0$, let us replace $\sigma(t)$ by the nonsingular $n\times n$ matrix $\sigma^\varepsilon(t)$ defined by

$$\sigma^\varepsilon=\begin{pmatrix}(2\varepsilon)^{1/2}I & 0\\ 0 & \hat{\sigma}\end{pmatrix},$$

where I is the identity $\nu\times\nu$ matrix. Let $V^\varepsilon(s,x)$ denote the corresponding infimum in (7.5). From Theorem 4.1, the discussion of nonanticipative controls following it, Theorems 6.2, 6.3, and Remark 4 following Theorem 6.2, we know that V^ε is in $C^{1,2}(\bar{Q}^0)$ and satisfies the dynamic programming equation

$$(8.6^\varepsilon)\qquad \begin{aligned}0=V_s^\varepsilon+\varepsilon\sum_{i=1}^{\nu}V^\varepsilon_{y_iy_i}+\sum_{i,j=\nu+1}^{n}a_{ij}(s)\,V^\varepsilon_{y_iy_j}\\ +V_y^\varepsilon\,\tilde{b}(s,y)+H(s,y,V_{\hat{y}}^\varepsilon)=0,\end{aligned}$$

with the Cauchy data $V^\varepsilon(T,y)=\Psi(y)$. Note that (8.6^ε) is uniformly parabolic when $\varepsilon>0$.

Lemma 8.1. *Let* $0<\varepsilon<1$. *Then:*

(a) $|V_y^\varepsilon|\leqq M$, *where* M *is as in Lemma* 7.2.

(b) *For any bounded set* $Q\subset Q^0$, *and* $1<\lambda<\infty$ *there exists* $M_{Q,\lambda}$ *such that*

$$\int_Q\Big[|V_s^\varepsilon|^\lambda+\sum_{i,j=\nu+1}^{n}|V^\varepsilon_{y_iy_j}|^\lambda\Big]\,dy\,ds\leqq M_{Q,\lambda}.$$

Proof. (a) follows at once from Lemma 7.2. To prove (b), let us put

$$\chi^\varepsilon(s,y)=V_y^\varepsilon(s,y)\,\tilde{b}(s,y)+H(s,y,V_{\hat{y}}^\varepsilon(s,y)).$$

From part (a) we know that V_y^ε, and hence also χ^ε is uniformly bounded on any compact set. It follows from (8.2)–(8.4), V(4.6) and (7.5) that $V^\varepsilon(s,y)$ is uniformly bounded on any compact set. Let us write

$$\bar{y}=(y_1,\dots,y_\nu),\qquad \hat{y}=(y_{\nu+1},\dots,y_n).$$

Given $\bar{y}_0$, let us make the substitution

$$\begin{aligned}\varepsilon^{1/2}z_i&=y_i-y_{0i}, && i=1,\dots,\nu;\\ z_i&=y_i, && i=\nu+1,\dots,n;\\ \psi^\varepsilon(s,z)&=V^\varepsilon(s,y).\end{aligned}$$

In the variables (s,z) we have (recall (8.6^ε))

$$(8.7)\qquad \begin{aligned}&\psi^\varepsilon_{z_i}=\varepsilon^{1/2}\,V^\varepsilon_{y_i},\quad i\leqq\nu,\qquad \psi^\varepsilon_{z_i}=V^\varepsilon_{y_i},\quad i>\nu,\\ &0=\psi_s^\varepsilon+\sum_{i=1}^{\nu}\psi^\varepsilon_{z_iz_i}+\tfrac{1}{2}\sum_{i,j=\nu+1}^{n}a_{ij}\,\psi^\varepsilon_{z_iz_j}+\chi^\varepsilon.\end{aligned}$$

Moreover, ψ^ε, ψ_z^ε are uniformly bounded on any compact set. Eq. (8.7) is uniformly parabolic. From local estimates for solutions of parabolic partial differential equations (Appendix E), given any bounded set $\hat{K} \subset E^{n-\nu}$ and $K = \{|\bar{z}| \leqq 1\} \times \hat{K}$,

$$\int_{T_0}^{T} \int_{K} \left[|\psi_s^\varepsilon|^\lambda + \sum_{i,j=1}^{n} |\psi_{z_i z_j}^\varepsilon|^\lambda \right] dz\, ds \leqq C_1$$

where C_1 depends on K and λ. Upon returning to the original variables, this implies

$$\int_{T_0}^{T} \int_{|\bar{y}-\bar{y}_0| \leqq \varepsilon^{1/2}} \int_{\hat{K}} \left[|V_s^\varepsilon|^\lambda + \sum_{i,j=\nu+1}^{n} |V_{y_i y_j}^\varepsilon|^\lambda \right] d\hat{y}\, d\bar{y}\, ds \leqq C_1 \varepsilon^{\nu/2}.$$

Given Q bounded, we have $Q \subset [T_0, T] \times \bar{D} \times \hat{K}$, where $\bar{D}$ is the union of fewer than $N \varepsilon^{-\nu/2}$ nonoverlapping ν-dimensional cubes of side $2\varepsilon^{1/2}$ and N does not depend on ε. From this we get (b), with $M_{Q,\lambda} = N C_1$. □

Lemma 8.2. *As* $\varepsilon \to 0$:

(a) $V^\varepsilon \to V$ *uniformly on* $\bar{Q}^0$;

(b) *The partial derivatives* V_s^ε, $V_{y_i}^\varepsilon$, $i = 1, \ldots, \nu$, *and* $V_{y_i y_j}^\varepsilon$, $i, j = \nu+1, \ldots, n$ *converge to the corresponding* V_s, V_{y_i}, $V_{y_i y_j}$ *weakly in* $L^\lambda(Q)$, *for* $1 < \lambda < \infty$ *and bounded subsets* Q *of* Q^0.

(c) *For* $i = \nu+1, \ldots, n$, $V_{y_i}^\varepsilon$ *converges to* V_{y_i} *pointwise for every* $(s, y) \in Q^0$.

Proof. (a) follows from Lemma 7.4, with $\sigma' = \sigma^\varepsilon$.

Part (b) follows from the estimates in Lemma 8.1 and (a).

To prove (c), fix $\bar{y}_0 = (y_{01}, \ldots, y_{0\nu})$ and introduce new variables $z_1, \ldots, z_n$ as in the proof of Lemma 8.1. For notational simplicity we take $\bar{y}_0 = \bar{0}$. Then

$$\psi^\varepsilon(s, z) - V(s, \bar{0}, \hat{y}) = V^\varepsilon(s, \varepsilon^{1/2}\bar{z}, \hat{y}) - V^\varepsilon(s, \bar{0}, \hat{y}) + V^\varepsilon(s, \bar{0}, \hat{y}) - V(s, \bar{0}, \hat{y}).$$

By Lemma 8.1(a),

$$|V^\varepsilon(s, \varepsilon^{1/2}\bar{z}, \hat{y}) - V^\varepsilon(s, \bar{0}, \hat{y})| \leqq \varepsilon^{1/2} |\bar{z}|.$$

Hence by part (a) of the present lemma, $\psi^\varepsilon(s, z)$ converges to $V(s, \bar{0}, \hat{y})$ uniformly as $\varepsilon \to 0$. Since ψ^ε satisfies the uniformly parabolic Eq. (8.7), with ψ^ε, ψ_z^ε and χ^ε uniformly bounded on any compact set, $\psi_{z_i}^\varepsilon$ satisfies for $i = 1, \ldots, n$ a uniform Holder condition on any compact set (see Appendix (E.9)). Therefore, $\psi_{z_i}^\varepsilon$ also converges uniformly on every compact set. For $i = \nu+1, \ldots, n$ we have $z_i = y_i$; and this implies in particular (c). □

Theorem 8.2. *Under the assumptions* (8.2)–(8.5), *the function* V *in* (7.5) *is a generalized solution of the dynamic programming Eq.* (8.6) *in* Q^0.

Proof. Let V^ε be as above. Given a C^∞ function ψ with compact support in Q^0, we multiply (8.6^ε) by ψ and integrate by parts the terms involving $V_{y_i y_i}^\varepsilon$, $i \leqq \nu$:

$$0 = -\varepsilon \int_{Q^0} \sum_{i=1}^{\nu} V_{y_i}^\varepsilon \psi_{y_i}\, dy\, ds + \int_{Q^0} \left[V_s^\varepsilon + \tfrac{1}{2} \sum_{i,j=\nu+1}^{n} a_{ij} V_{y_i y_j}^\varepsilon + V_y^\varepsilon \tilde{b} \right] \psi\, dy\, ds$$
$$+ \int_{Q^0} H(s, y, V_{\hat{y}}^\varepsilon) \psi\, dy\, ds.$$

As $\varepsilon \to 0$ the first integral tends to 0, by Lemma 8.1(a). By Lemma 8.2(b) the second integral tends to the corresponding expression with V^ε replaced by V. The same is true for the third integral by Lemmas 8.1(a) and 8.2(c). Thus

$$0=\int_{Q^0}\left[V_s+\tfrac{1}{2}\sum_{i,j=\nu+1}^{n} a_{ij}V_{y_i y_j}+V_y\tilde{b}+H(s,y,V_{\hat{y}})\right]\psi\,dy\,ds.$$

Since this holds for all such ψ, the function V satisfies (8.6) almost everywhere in Q^0. By Lemma 8.2(a), V is continuous on $\bar{Q}^0$. By Lemma 7.2, V then satisfies a linear growth condition $|V(s,y)|\leq M_1(1+|y|)$ for some M_1. Thus V is in $\mathscr{E}^{\lambda\nu}(Q^0)$. □

Theorem 8.3. *Under the assumptions of Theorem* 8.2 *there exists an optimal feedback control law* $\mathbf{u}^*$, *satisfying* (4.3) *almost everywhere in* Q^0.

Proof. With the aid of Lusin's theorem, Lemma 6.1 remains true when W_y is measurable on $Q(=Q^0$ here). The proof then follows that for Theorem 6.3, using Theorem 8.1 with $W=V$ instead of Theorem 4.1. □

§ 9. Stochastic Approximation to the Deterministic Control Problem

Let us now consider the case when the system Eqs. (3.1) are deterministic ($\sigma=0$):

(9.1) $$d\xi=f(t,\xi(t),u(t))\,dt,\qquad t\geq s,$$

with $\xi(s)=y$, and with performance criterion

(9.2) $$J=\int_s^T L(t,\xi(t),u(t))\,dt+\Psi[\xi(T)].$$

(No expectation occurs in (9.2) since the problem is deterministic.) Necessary conditions for a minimum (Pontryagin's principle) were given for this problem in Chap. II, and conditions for existence of a minimum in Chap. III. As deterministic controls we admit all Borel measurable functions $u(\cdot)$ from $[s,T]$ into U; recall that U is bounded by (7.2). Let $\mathscr{U}_d(s)$ denote the set of all such $u(\cdot)$. Since no noise enters the system, the controller could do no better using non-anticipative controls in $\mathscr{U}(s)$. Thus, the function V in (7.5) now satisfies

$$V(s,y)=\inf_{u\in\mathscr{U}_d(s)} J(s,y,u).$$

If the problem is rewritten in Mayer form (with $L=0$), $V(s,y)$ is the same as in IV(3.1). By Theorems IV.4.1 and IV.4.2, V is a locally Lipschitz function which satisfies almost everywhere in Q^0 the dynamic programming equation

(9.3) $$0=V_s+H(s,y,V_y)$$

with $V(T,y)=\Psi(y)$. In certain portions N of Q^0, which we call below regions of strong regularity, V is C^1 and agrees in N with a solution of (9.3) computed by the method of characteristics.

Let us now make the problem stochastic by adding a white noise term, with small coefficients $(2\varepsilon)^{1/2}$, in each component of (9.1). The perturbed system equations are

$$(9.1^\varepsilon) \qquad d\xi^\varepsilon = f(t, \xi^\varepsilon(t), u(t))\,dt + (2\varepsilon)^{1/2}\,dw;$$

in the notation of §7, $\sigma^\varepsilon = (2\varepsilon)^{1/2} I$ where I is the identity matrix. We again denote by J^ε the expected system performance; for $\varepsilon = 0$, $J^0 = J$ as in (9.2). For each $\varepsilon > 0$, the condition (6.3) for uniform parabolicity is satisfied. The function V^ε satisfies, according to (4.4) and Theorems 6.2, 6.3:

$$V^\varepsilon(s, y) = \min_{\mathscr{V}} J^\varepsilon(s, y, \mathbf{u}) = \min_{\mathscr{U}(s)} J^\varepsilon(s, y, u)$$

(the right side equals $V^\varepsilon(s, y)$ by definition (7.5)). Moreover, V^ε is a $C^{1,2}$ solution of the dynamic programming equation:

$$(9.3^\varepsilon) \qquad 0 = V_s^\varepsilon + \varepsilon \Delta_y V^\varepsilon + H(s, y, V_y^\varepsilon),$$

with the Cauchy data

$$(9.4) \qquad V^\varepsilon(T, y) = V(T, y) = \Psi(y).$$

Here Δ_y is the Laplace operator in the variables $y = (y_1, \ldots, y_n)$.

By Lemma 7.2, $V^\varepsilon(s, \cdot)$ satisfies a uniform Lipschitz condition; in fact, $|V_y^\varepsilon| \leq M$. By Lemma 7.4, V^ε tends to V uniformly as $\varepsilon \to 0$. Thus the optimal performance V^ε for the stochastically perturbed problem tends to that for the deterministic problem. This is rather a weak result. It does not, for instance, say anything about convergence of optimal feedback controls; nor does it tell how one might find an approximate solution to the perturbed stochastic problem, in terms of quantities computable from a solution to the deterministic one. We turn now to these matters.

To begin with, let us give conditions for convergence V_y^ε to V_y. We shall suppose from now on that an optimal $u^* \in \mathscr{U}_d(s)$ exists for each $(s, y) \in Q^0$; recall the existence Theorem III.4.1 and its corollary.

Definition. Let us call (s, y) a *regular point* if the following holds: for every $\rho > 0$ there exists $\delta > 0$ such that

$$J(s, y, u) < J(s, y, u^*) + \delta \quad \text{implies} \quad \int_s^T |u(t) - u^*(t)|\,dt < \rho.$$

Clearly regularity of (s, y) implies that u^* is the unique optimal control in $\mathscr{U}_d(s)$, for these initial data. As usual, controls agreeing almost everywhere on $[s, T]$ are regarded as the same. Under some further assumptions, uniqueness of u^* is equivalent to regularity. Let us write ξ^* for the solution of (9.1) corresponding to u^*, with $\xi^*(s) = y$; let P^* be the solution of the adjoint equations (Problem II.3):

$$\frac{dP^*}{dt} = -P^{*\prime} f_x(t, \xi^*(t), u^*(t)) + L_x(t, \xi^*(t), u^*(t))$$

$$h(t, u) = P^*(t)' f(t, \xi^*(t), u) - L(t, \xi^*(t), u).$$

The terminal condition is $P^*(T)' = -\Psi_x[\xi^*(T)]$. The multiplier $\lambda_1 \leq 0$ cannot be 0; we have taken $\lambda_1 = -1$. For almost all $t \in [s, T]$, $h(t, \cdot)$ is maximum on U at $u^*(t)$.

Besides the general assumptions (7.1)–(7.4), let us suppose in this section that

(9.5) (a) U is convex;

(b) $f(t, x, u) = \alpha(t, x) + \beta(t, x)u$;

(c) $L(t, x, \cdot)$ is convex.

Theorem 9.1. *Assume that:* (i) *u^* is the unique optimal control in $\mathscr{U}_d(s)$ for initial data (s, y); and* (ii) *for almost all $t \in [s, T]$, $h(t, v) < h(t, u^*(t))$ whenever $v \neq u^*(t)$. Then (s, y) is a regular point.*

Note that the conditions for Corollary III.4.1 to the existence Theorem III.4.1 are satisfied. If we strengthen (9.5c) to require that $L(t, x, \cdot)$ is a strictly convex function, then (ii) is automatically satisfied since $-h(t, \cdot)$ is then also strictly convex.

Proof of Theorem 9.1. Regularity of (s, y) is equivalent to the following statement: any minimizing sequence $u^1, u^2, \ldots$ in $\mathscr{U}_d(s)$ converges to u^* in $L^1[s, T]$. Since $u^r(t) \in U$, and U is bounded, it suffices to prove instead that u^r tends to u^* in measure as $r \to \infty$.

Let u^r be such a minimizing sequence, and ξ^r the corresponding solution of (9.1). By uniqueness of u^* and Lemma III.3.2, $\|\xi^r - \xi^*\|$ tends to 0 (uniform norm), while u^r tends to u^* weakly* in $L^\infty[s, T]$ as $r \to \infty$. Using (9.5b)

$$\lim_{r \to \infty} \int_s^T P^*(t)' f(t, \xi^*(t), u^r(t))\, dt = \int_s^T P^*(t)' f(t, \xi^*(t), u^*(t))\, dt.$$

Since $\Psi[\xi^r(T)]$ tends to $\Psi[\xi^*(T)]$ and $J(s, y, u^r)$ tends to $J(s, y, u^*)$, we have using (9.2)

$$\lim_{r \to \infty} \int_s^T L(t, \xi^*(t), u^r(t))\, dt = \int_s^T L(t, \xi^*(t), u^*(t))\, dt.$$

Therefore

$$\lim_{r \to \infty} \int_s^T h(t, u^r(t))\, dt = \int_s^T h(t, u^*(t))\, dt.$$

Since $h(t, u^r(t)) \leq h(t, u^*(t))$ almost everywhere, $h(t, u^r(t))$ tends to $h(t, u^*(t))$ in measure on $[s, T]$.

Let K be a compact subset of $[s, T]$, such that the restriction to K of u^* is continuous and $h(t, \cdot)$ has a unique maximum on U at $u^*(t)$ for each $t \in K$. Given $\alpha > 0$ there exists $\beta = \beta(\alpha)$, tending to 0 as $\alpha \to 0$, such that

$$(\#) \qquad h(t, u^*(t)) - h(t, v) < \alpha \Rightarrow |v - u^*(t)| < \beta,$$

provided $t \in K$, $v \in U$. See Problem 5. Then convergence in measure of $h(t, u^r(t))$ to $h(t, u^*(t))$ implies convergence in measure of u^r to u^* on K.

By (ii) and Lusin's theorem the measure of $[s, T] - K$ can be made arbitrarily small. Hence u^r tends to u^* in measure on $[s, T]$. □

Theorem 9.2. *If (s, y) is a regular point and $V_y(s, y)$ exists, then $V_y^\varepsilon(s, y)$ tends to $V_y(s, y)$ as $\varepsilon \to 0$.*

Proof. By Lemma 7.5 it suffices to prove that $\int_s^T |u^{\varepsilon *}(t)-u^*(t)|\,dt$ tends to 0 in probability, where $u^{\varepsilon *}\in\mathscr{U}(s)$ is optimal for the stochastic problem. (By Theorem 6.3 we can take $u^{\varepsilon *}(t)=\mathbf{u}^{\varepsilon *}(t,\xi^{\varepsilon *}(t))$, where $\mathbf{u}^{\varepsilon *}$ is an optimal control policy and $\xi^{\varepsilon *}$ the corresponding solution of (9.1) with $\xi^\varepsilon(s)=y$.) Define a process η^ε by

$$d\eta^\varepsilon = f(t,\eta^\varepsilon(t),u^{\varepsilon *}(t))\,dt,$$

with $\eta^\varepsilon(s)=y$. Let us apply (7.11), with $u=u^{\varepsilon *}$, $\sigma'=\sigma^\varepsilon$, $\sigma=0$:

$$|J^\varepsilon(s,y,u^{\varepsilon *})-E_{sy}J(s,y,u^{\varepsilon *})|\leqq M_1(2\varepsilon)^{1/2}.$$

(For J as in (9.2), $J(s,y,u^{\varepsilon *})$ is a random variable; in (7.11) its expectation appears.) Now $J^\varepsilon(s,y,u^{\varepsilon *})=V^\varepsilon(s,y)$, which tends to $V(s,y)$ as $\varepsilon\to 0$. On the other hand,

$$J(s,y,u^{\varepsilon *})\geqq V(s,y)$$

with probability 1, since $u^{\varepsilon *}(\cdot,\omega)$ is with probability 1 in $\mathscr{U}_d(s)$. Then

$$E_{sy}J(s,y,u^{\varepsilon *})\geqq V(s,y),$$

and the left side tends to $V(s,y)$ as $\varepsilon\to 0$. Therefore,

$$\lim_{\varepsilon\to 0} P_{sy}[J(s,y,u^{\varepsilon *})>J(s,y,u^*)+\delta]=0$$

for each $\delta>0$. Since (s,y) is regular, for each $\rho>0$

$$\lim_{\varepsilon\to 0} P_{sy}\left[\int_s^T |u^{\varepsilon *}-u^*|\,dt>\rho\right]=0. \quad \square$$

Remark. In some cases, regularity of (s,y) is equivalent to the differentiability of V at (s,y). For the simplest problem of calculus of variations this was shown by Kuznetzov and Siskin [1]. See also Fleming [7, §3].

We proceed now to stronger results which hold if the idea of regularity is strengthened. The treatment follows Fleming [10]. We begin with some definitions and notation. Recall the definition (6.4) of H. Let Γ be an open subset of E^{2n+1} with the following property:

(9.6) For $(s,y,p)\in\Gamma$, $p'f(s,y,u)+L(s,y,u)$ is minimum on U at a unique $u=G(s,y,p)$; moreover, G is C^∞ on Γ.

Since $H(s,y,p)=p'f(s,y,G)+L(s,y,G)$, H is also C^∞ on Γ.

Example. Suppose $L=L(s,y)$ and $f(s,y,u)$ is as in Theorem 9.1. Let $U=\{|u|\leqq 1\}$. Then $\Gamma=\{(s,y,p)\colon p'\beta(s,y)\neq 0,\ G(s,y,p)=-|p'\beta|^{-1}p'\beta$.

The terminal hyperplane is now denoted by $M=\{T\}\times E^n$; in the notation of §6, $M=\partial^*Q$. We call $\gamma^*=\{(t,\xi^*(t)),\ s\leqq t\leqq T\}$ an *optimal trajectory* when ξ^* is the solution of (9.1) corresponding to an optimal $u^*\in\mathscr{U}_d(s)$. A C^∞ hypersurface in E^{n+1} is a manifold of dimension n and class C^∞. Recall that ψ of class $C^k(S)$ means that ψ agrees on S with a function of class $C^k(\mathscr{O})$, where $\mathscr{O}$ is some open set containing S.

Definition. A set N is a *region of strong regularity* if:

(1) $N\subset Q^0\cup\Sigma_1$, where Σ_1 is a relatively open subset of the terminal hyperplane M.

(2) Each $(s, y) \in N - \Sigma_1$ is regular; moreover, $\gamma^* \subset N$, where $\gamma^* = \gamma^*(s, y)$ is the unique optimal trajectory starting at (s, y).

(3) V is of class $C^1(\bar{N})$.

(4) There exist disjoint open subsets $N_1, \ldots, N_m$ of N such that V is of class $C^\infty(\bar{N}_j)$, $j = 1, \ldots, m$.

(5) There exist C^∞ hypersurfaces $\Sigma_2, \ldots, \Sigma_m$ disjoint from each other and from Σ_1 such that:

(a) $N - (\Sigma_1 \cup \cdots \cup \Sigma_m) = N_1 \cup \cdots \cup N_m$.

(b) $\partial N_j \subset \Sigma_j \cup \Sigma_{j+1} \cup \partial N$ for $j = 1, \ldots, m-1$, $\partial N_m \subset \Sigma_m \cup \partial N$.

(c) For $(s, y) \in N_j$, $\gamma^*(s, y)$ meets Σ_i nontangentially at a single point (s_i, y_i) for $i \leqq j$ and $\gamma^*(s, y)$ does not meet Σ_i for $i > j$.

(6) (a) For $(s, y) \in N_j$, $j = 1, \ldots, m$ and $(s, y) \in \Sigma_1$, $(s, y, V_y(s, y)) \in \Gamma$.

(b) $\mathbf{u}^*(s, y) = G(s, y, V_y(s, y))$ agrees on each N_j with a function of class $C^\infty(\bar{N}_j)$.

(c) The unique optimal control $u^* \in U_d(s)$, corresponding to initial data $(s, y) \in N - \Sigma_1$, satisfies $u^*(t) = \mathbf{u}^*(t, \xi^*(t))$ except for the finitely many times t when γ^* meets Σ_j, $j = 2, \ldots, m$.

In the terminology of Chap. IV.6, each N_j has the same role as a cell of maximum dimension $n+1$. For $j > 1$, Σ_j has the role of a cell of type II and dimension n. There are no lower dimensional cells.

We call $\mathbf{u}^*$ *the optimal feedback control in* N, for the deterministic problem.

The C^∞ extensions of $\mathbf{u}^*$ from N_{j-1} and N_j to their closures are not required to agree on Σ_j. When $\mathbf{u}^*$ is discontinuous across Σ_j, we call Σ_j a *switching surface.* If $\mathbf{u}^*$ is continuous across Σ_j but its first order partial derivatives are discontinuous there, then we call Σ_j a *transition surface.* Across a switching surface, the first order partials of V are continuous while second order derivatives generally have jumps. Across a transition surface, the second order derivatives are continuous while the third order derivatives generally have jumps.

The classical technique for constructing regions of strong regularity is the method of characteristics; see Fleming [10, Appendix]. The method of characteristics constructs piecewise smooth solutions to (9.3) with the data $V = \Psi$ on in certain portions of Q^0, working backward in time from M. At some points (s, y) there may be several solutions to (9.3), corresponding to families of characteristic trajectories leading from (s, y) to different portions of M. The function V is the C^1 and piecewise C^∞ solution to (9.3) in N, with the data on M, which gives the absolute minimum value to $J(s, y, u)$. Other solutions of (9.3) constructed by the method of characteristics correspond to local minima of J.

As before let $\mathbf{u}^{\varepsilon *}$ be an optimal feedback control law for the stochastic problem with system Eqs. (9.1$^\varepsilon$), and $V^\varepsilon(s, y)$ the optimal expected performance. In the following result a stronger convexity condition than (9.5 c) is imposed on $L(s, y, \cdot)$.

Theorem 9.3. *In addition to the previous assumptions, suppose that the characteristic values of the matrices* L_{uu} *are bounded below by* $c > 0$. *Then, uniformly for* (s, y) *in any compact subset of* N:

(a) $V_y^\varepsilon = V_y + \varepsilon (Z_1)_y + o(\varepsilon)$;

(b) $\lim_{\varepsilon \to 0} \mathbf{u}^{\varepsilon *}(s, y) = \mathbf{u}^*(s, y)$;

(c) $V^\varepsilon = V + \varepsilon Z_1 + \varepsilon^2 Z_2 + o(\varepsilon^2)$, *where* Z_1, Z_2 *satisfy the linear first order equations*

$$(Z_1)_s + (Z_1)_y f^*(s, y) + \Delta_y V = 0, \tag{9.7}$$

$$(Z_2)_s + (Z_2)_y f^*(s, y) + \tfrac{1}{2}(Z_1)_y H^*_{pp}(Z_1)'_y + \Delta_y Z_1 = 0, \tag{9.8}$$

$$Z_1(T, y) = Z_2(T, y) = 0 \quad \text{for } y \in \Sigma_1. \tag{9.9}$$

Here $f^*(s, y) = f(s, y, \mathbf{u}^*(s, y))$, $H^*_{pp}(s, y) = H_{pp}(s, y, V_y(s, y))$, *and* Δ_y *is the Laplace operator.*

For a proof of this theorem we refer to Fleming [10, §6]. The argument there becomes somewhat simpler for the present case of fixed final time T. The first order linear partial differential Eqs. (9.7), (9.8) can be solved for Z_1 and Z_2 by the method of characteristics.

Theorem 9.3 is best possible in the sense that V^ε cannot always be expanded up to terms of order ε^3, nor V^ε_y up to terms of order ε^2, at points of the transition surfaces $\Sigma_2, \ldots, \Sigma_m$. Such transition surfaces often occur as boundary hypersurfaces between two regions in E^{n+1} such that $\mathbf{u}^*$ obeys some control constraint in one region but not in the other.

When $L(s, y, \cdot)$ is merely convex, as in (9.5c), there may be switching surfaces. In that case the results are less complete. It is known that V^ε_y tends to V_y as $\varepsilon \to 0$, uniformly on any compact $N^1 \subset N$ Fleming [10, Corollary 5.6]. This implies that $\mathbf{u}^{\varepsilon *}$ tends to $\mathbf{u}^*$ uniformly on any compact subset of $N - (\Sigma_2 \cup \cdots \cup \Sigma_m)$. Perhaps the expansions (a), (c) hold on compact subsets of $N - (\Sigma_2 \cup \cdots \cup \Sigma_m)$, but this has not been proved. It appears that a condition $(1, f(s, y, u))$ not tangent to Σ_j for all $u \in U$ should be imposed at points (s, y) of a switching surface Σ_j.

If the strong convexity condition on L in Theorem 9.3 holds and there are no control constraints ($U = E^m$), then there are neither transition nor switching surfaces. In this case stronger results have been obtained, under some further technical assumptions on f and L. In fact, V^ε, V^ε_y, $\mathbf{u}^{\varepsilon *}$ have in this case asymptotic expansions in powers $\varepsilon, \varepsilon^2, \ldots, \varepsilon^m$ for any finite m. See Fleming [10, §7]. Included is the stochastically perturbed simplest problem of calculus of variations for which $f(u) = u$ and $U = E^n$.

This result can be interpreted as a result in the theory of first order partial differential equations, using the following device. Consider a Cauchy problem for some first order equation

$$V_s + F(s, y, V_y) = 0, \quad T_0 < s < T, \tag{9.10}$$

with $V(T, y) = \Psi(y)$. Under some assumptions Fleming [10, (7.9)] including a strong concavity condition on $F(s, y, \cdot)$ one can introduce the function L dual to F in the sense of the classical canonical transformation:

$$L(s, y, u) = \max_{p \in E^n} (F - p'u)$$

$$F(s, y, p) = \min_{u \in E^n} (L + p'u).$$

If we now take the simplest problem $f(u) = u$, $U = E^n$, then $F = H$ in (6.4). Thus (9.10) becomes the dynamic programming (or Hamilton-Jacobi) Eq. (9.3). The second order equation obtained by adding a term $\varepsilon \Delta_y V$ in (9.10) becomes the dynamic programming Eq. (9.3$^\varepsilon$) for the corresponding stochastic control problem with

$$\xi^\varepsilon(t) = y + \int_s^t u(r)\,dr + (2\varepsilon)^{1/2}[w(t) - w(s)].$$

For literature on the relation between these partial differential equations and a nonlinear conservation law (when $n=1$) see Fleming [7].

§ 10. Problems with Partial Observations

In the preceding sections we have supposed that the controller knows the state $\xi(t)$ at each time t. Let us now formulate a problem in which the controller has only partial observations. Generally speaking, such problems turn out to be more difficult. In this section we merely indicate some literature dealing with them. In § 11, we shall find that for a special class of problems, with linear system and observation equations, there is a separation principle by which they can be reduced to ones with completely observed states. The separation principle gives, in particular, a solution to the partially observed linear regulator problem.

Open Loop Control. Suppose that the controller has no information, except the distribution of the initial state $\xi(s)$. The control must then be a deterministic function of time, namely $u \in \mathscr{U}_d(s)$. Such controls are called open loop. If $\sigma = \sigma(t)$, then a necessary condition rather similar to Pontryagin's principle was obtained by Kushner [3]. A necessary condition which applies to problems where $\sigma = \sigma(t, \xi)$ may be state dependent was later obtained by Warfield [1]. She used McShane's approach to stochastic integration. For small noise coefficient $\sigma = (2\varepsilon)^{1/2} I$, Holland [2][3] obtained expansions in powers of ε of the optimal open loop system performance and (under further restrictions) of the optimal open loop control. There are some similarities with the results in § 9 for the complete observation case, but the methods are quite different.

Control for Partially Observable States. Here one must specify the kind of information on which the controller can base his decision at each time t. It is frequently assumed that all data obtained at times $\leqq t$ is available at time t. This is the "classical" information structure for partially observable problems. It was for this kind of information structure that a separation principle will be obtained in § 11. General necessary and sufficient dynamic programming conditions for problems with classical information structure were given in Rishel [2]. The solution there of the dynamic programming equation is a function of the observations and controls up to each time t. A frequently used model for the data process is a solution η of some system of stochastic differential equations

$$(10.1) \qquad d\eta = g(t, \xi(t), \eta(t))\,dt + \sigma_1(t, \xi(t), \eta(t))\,dw_1,$$

with $\eta(s) = 0$. The data available at time t are the observations $\eta(r)$ for $s \leqq r \leqq t$. In that case, more complete results were obtained in Davis-Varaiya [2]; it was shown there that a "relative completeness" condition of Rishel [2] always holds.

Another possibility is for the controller to use only currently observed data ("zero memory" controls). For some results about this problem, see Fleming [6]. A counterexample of Witsenhausen [1] shows that for a zero memory information structure, a separation principle need not hold even for simple linear systems with quadratic performance criterion.

More general kinds of information structures are treated in Witsenhausen [2] and Ho-Chu [1].

§ 11. The Separation Principle

Let us now suppose that the system Eqs. (3.1) and observation Eqs. (10.1) have the following (linear) form:

$$d\xi = [A(t)\xi(t) + B(t)u(t)]\,dt + \sigma(t)\,dw, \tag{11.1}$$

$$d\eta = H(t)\xi(t) + \sigma_1(t)\,dw_1. \tag{11.2}$$

The initial states are a gaussian random vector $\xi(s)$, and $\eta(s)=0$; w_1 is a standard brownian motion independent of w and $\xi(s)$. At time t the controller knows the observations $\eta(r)$ for $s \leqq r \leqq t$. The control value $u(t)$ is to be dependent on this data.

To express the way in which this dependence will be formulated, in analogy with the definition of nonanticipative control in § 4, let $\mathscr{F}_t$ be an increasing family of σ-algebras such that $\xi(s)$ is $\mathscr{F}_s$ measurable and the brownian motions w and w_1 are adapted to $\{\mathscr{F}_t\}$ and define $\mathscr{U}$ to be the class of controls which are stochastic processes such that:

(1) $u(t) \in U$ for $s \leqq t \leqq T$

(2) $E \int_s^T |u(t)|^k\,dt < \infty$ for every $k > 0$

(3) u is nonanticipative with respect to $\{\mathscr{F}_t\}$.

For each control u in $\mathscr{U}$, (11.1) (11.2) with initial conditions $\xi(s)$, 0 will have a unique solution ξ, η satisfying

$$\sup_{s \leqq t \leqq T} E\{|\xi(t)|^k\} < \infty.$$

Let

$$\mathscr{G}_t^u = \mathscr{F}\{\eta(r), s \leqq r \leqq t\}$$

where η, ξ are the processes which are solutions of (11.1) (11.2) corresponding to the control u.

Let us say that the control u depends on the measurements if, for each t in $[s, T]$, $u(t)$ is $\mathscr{G}_t^u$ measurable. This will not be the case for all controls in $\mathscr{U}$. Let $\mathscr{Y}$ be a subset of controls of $\mathscr{U}$ which depend on the measurements in this sense. Later we take a particular choice for $\mathscr{Y}$. For technical reasons we cannot define $\mathscr{Y}$ merely as the set of all controls in $\mathscr{U}$ depending on the measurements.

For controls u in $\mathscr{Y}$ consider the problem of minimizing

$$J(u) = E\left\{\int_s^T L(t, \xi(t), u(t))\,dt + \Psi[\xi(T)]\right\}. \tag{11.3}$$

(In (11.3) we have not indicated the dependence of J on s and the distribution of $\xi(s)$. We keep s and T fixed throughout the discussion to follow.)

To begin the discussion of the separation principle, let us first consider the uncontrolled system, with $u(t)=0$ in (11.1):

$$d\xi^0 = A(t)\,\xi^0(t)\,dt + \sigma(t)\,dw, \tag{11.1°}$$

and the corresponding observations

$$d\eta^0 = H(t)\,\xi^0(t)\,dt + \sigma_1(t)\,dw_1. \tag{11.2°}$$

The initial data are the same: $\xi^0(s)=\xi(s)$, $\eta^0(s)=0$. We make the same assumptions as in V.9 regarding the Kalman-Bucy filter model. The mean square optimal estimate of $\xi^0(t)$ given the observations $\eta^0(r)$, $s\leq r\leq t$, is the conditional expectation

$$\hat{\xi}^0(t) = E\{\xi^0(t)\,|\,\mathscr{G}_t^0\},$$
$$\mathscr{G}_t^0 = \mathscr{F}(\eta^0(r),\, r\leq t).$$

By Theorem V.9.2, $\hat{\xi}^0$ obeys the stochastic differential equation:

$$d\hat{\xi}^0 = A(t)\,\hat{\xi}^0(t)\,dt + F(t)\,[d\eta^0 - H(t)\,\hat{\xi}^0(t)\,dt] \tag{11.4}$$

with initial data $\hat{\xi}^0(s) = E\,\xi(s)$. As in V(9.8)

$$F(t) = R(t)\,H'(t)\,[a_1(t)]^{-1}, \tag{11.5}$$

where the covariance matrix $R(t)$ of the error process $\tilde{\xi}^0 = \xi^0 - \hat{\xi}^0$ obeys the Riccati-type matrix differential equation V(9.9). Moreover, by Theorem V.9.2(c),

$$\eta^0(t) - \int_s^t H(r)\,\hat{\xi}^0(r)\,dr = \int_s^t \sigma_1(r)\,d\hat{w}(r), \tag{11.6}$$

where $\hat{w}$ is a standard k-dimensional brownian motion adapted to $\{\mathscr{G}_t^0\}$.

Let us define processes ξ^+, η^+ as solutions of the ordinary differential equations

$$d\xi^+ = [A(t)\,\xi^+(t) + B(t)\,u(t)]\,dt, \tag{11.7}$$

$$d\eta^+ = H(t)\,\xi^+(t)\,dt, \tag{11.8}$$

with $\xi^+(s) = \eta^+(s) = 0$. Then

$$\xi = \xi^0 + \xi^+, \qquad \eta = \eta^0 + \eta^+. \tag{11.9}$$

Lemma 11.1. *If $u(t)$ is $\mathscr{G}_t^0$ measurable and the equality $\mathscr{G}_t^0 = \mathscr{G}_t^u$ holds between the σ-fields of measurements of the uncontrolled and controlled processes, then:*

(a) *The conditional distribution of $\xi(t)$ given $\mathscr{G}_t^u$ is gaussian, with mean $\hat{\xi}(t) = \hat{\xi}^0(t) + \xi^+(t)$ and covariance matrix $R(t)$.*

(b) *The conditional mean $\hat{\xi}(t)$ obeys the stochastic differential equation*

$$d\hat{\xi} = [A(t)\,\hat{\xi}(t) + B(t)\,u(t)]\,dt + F(t)\,\sigma_1(t)\,d\hat{w}, \tag{11.10}$$

with $\hat{w}$ the brownian motion in (11.6).

Proof. From the equality of $\mathcal{G}_t^u$ and $\mathcal{G}_t^0$ conditional expectations with respect to $\mathcal{G}_t^u$ and with respect to $\mathcal{G}_t^0$ are the same. Moreover, $u(t)$ is $\mathcal{G}_t^u=\mathcal{G}_t^0$ measurable; and hence $\xi^+(t), \eta^+(t)$ are also $\mathcal{G}_t^0$ measurable. Part (a) follows from (11.9) and Remark V.9.2 regarding the Kalman-Bucy filter model. To get (b) we add Eqs. (11.4) and (11.7). From (11.6), $d\eta^0 - H\hat{\xi}^0\,dt = \sigma_1\,d\hat{w}$. □

We note that

$$d\eta^0 - H\hat{\xi}^0\,dt = d(\eta-\eta^+) - H(\hat{\xi}-\xi^+)\,dt = d\eta - H\hat{\xi}\,dt.$$

Hence, $\sigma_1\,d\hat{w} = d\eta - H\hat{\xi}\,dt$.

Let us set

$$g(t,x) = (2\pi)^{-\frac{n}{2}}(\det R(t))^{-\frac{1}{2}}\exp[-\tfrac{1}{2}x' R(t)^{-1}x].$$

Under the conditions of Lemma 11.1, the conditional density of $\xi(t)$ given $\mathcal{G}_t^u$ is $g(t, x-\hat{\xi}(t))$. Let

$$\hat{L}(t,\hat{x},u) = \int_{E^n} L(t,x,u)\,g(t,x-\hat{x})\,dx, \qquad \hat{\Psi}(\hat{x}) = \int_{E^n} \Psi(x)\,g(T,x-\hat{x})\,dx,$$

(11.11)

$$\hat{J}(u) = E\left\{\int_s^T \hat{L}(t,\hat{\xi}(t),u(t))\,dt + \hat{\Psi}[\hat{\xi}(T)]\right\}.$$

Lemma 11.2. *Under the conditions of Lemma* 11.1 $J(u)=\hat{J}(u)$.

Proof. By definition of $\hat{L}$ and $\mathcal{G}_t^0$ measurability of $u(t)$

$$E\{L(t,\xi(t),u(t))\,|\,\mathcal{G}_t^0\} = \hat{L}(t,\hat{\xi}(t),u(t))$$

with probability 1. Hence

$$E\int_s^T L(t,\xi(t),u(t))\,dt = E\int_s^T E\{L\,|\,\mathcal{G}_t^0\}\,dt = E\int_s^T \hat{L}\,dt.$$

Similarly, $E\Psi[\xi(T)] = E\hat{\Psi}[\hat{\xi}(T)]$. □

We can now explain the idea of the separation principle. The partially observed stochastic control problem is converted into a completely observed control problem in the following way. Take $\hat{\xi}(t)$ as the state of the system. The estimate $\hat{\xi}(t)$ satisfies (11.10) which is taken as the new system equation. The new expected performance is taken to be $\hat{J}(u)$.

For the separated problem, a natural class of controls is

$$\mathcal{Y}_0 = \{u\in\mathcal{U}: u(t) \text{ is } \mathcal{G}_t^0 \text{ measurable}, s\leq t\leq T\}.$$

The problem of minimizing $\hat{J}(u)$ on $\mathcal{Y}_0$ is of the type considered in previous sections. In view of Lemmas 11.1 and 11.2 let us consider the subset

$$\mathcal{Y} = \{u\in\mathcal{Y}_0: \mathcal{G}_t^0 = \mathcal{G}_t^u \text{ for } s\leq t\leq T\}$$

If the minimum of $\hat{J}(u)$ on $\mathcal{Y}_0$ occurs at some $u^*\in\mathcal{Y}$, then u^* also minimizes $J(u)$ on $\mathcal{Y}$ by Lemma 11.2. We next show following Wonham [3], that $\mathcal{Y}$ contains the class $\mathscr{Y}_1$ of controls u obtained below via (11.12) from Lipschitz feedback functions of past observations.

Let $\mathscr{C}^k = \mathscr{C}^k[s, T]$ denote the space of continuous functions from $[s, T]$ into E^k, with the sup norm $\| \quad \|$. Let Γ denote the space of functions γ from $[s, T] \times \mathscr{C}^k$ into the control set U which satisfy:

(a) If $g(r) = h(r)$ for $s \leqq r \leqq t$, then $\gamma(t, g) = \gamma(t, h)$.

(b) For each $\gamma \in \Gamma$ there exists a constant K such that $|\gamma(t, g) - \gamma(t, h)| \leqq K \|g - h\|$ for all $g, h \in \mathscr{C}^k$ and $s \leqq t \leqq T$.

(c) $\gamma(\cdot, \cdot)$ is Borel measurable.

(d) $\gamma(t, 0)$ is bounded.

For each $\gamma \in \Gamma$, and given $w, w_1, \xi(s), \eta(s) = 0$, Eqs. (11.1), (11.2) have a solution ξ, η if we set

$$u(t) = \gamma(t, \eta), \qquad s \leqq t \leqq T. \tag{11.12}$$

The proof is by the same method as for Theorem V.4.1. Moreover, there is (pathwise) uniqueness of ξ, η in the sense of that theorem.

Lemma 11.3. *For each $\gamma \in \Gamma$, and $s \leqq t \leqq T$, $u(t)$ is $\mathscr{G}_t^0$ measurable and $\mathscr{G}_t^u = \mathscr{G}_t^0$.*
In the proof of Lemma 11.3 we use the following:

Lemma 11.4. *Let Φ be an operator in $\mathscr{C}^k$ such that, for some constant K,*

$$\|\Phi g - \Phi h\| \leqq K \|g - h\|, \qquad \text{all } g, h \in \mathscr{C}^k.$$

Then, given $h \in \mathscr{C}^k$, the equation

$$g(t) = h(t) + \int_s^t (\Phi g)(r)\, dr, \qquad s \leqq t \leqq T, \tag{$*$}$$

has a unique solution $g \in \mathscr{C}^k$. The sequence of successive approximations $g_0 = 0$,

$$g_{l+1}(t) = h(t) + \int_s^t (\Phi g_l)(r)\, dr, \qquad l = 0, 1, 2, \ldots$$

converges in norm to g.

Proof of Lemma 11.4. Define the operator B on $\mathscr{C}^k$ by

$$(B g)(t) = \int_s^t (\Phi g)(r)\, dr.$$

The n-th iterate B^n of B satisfies

$$\|B^n g - B^n g'\| \leqq \frac{K^n (T-s)^n}{n!} \|g - g'\|,$$

for any $g, g' \in \mathscr{C}^k$. If g and g' are both solutions of (*), then $\|g-g'\| = \|B^n g - B^n g'\|$, which tends to 0 as $n \to \infty$. This shows uniqueness. Now

$$\begin{aligned}\|g_{l+m} - g_l\| &\leqq \sum_{j=1}^{m} \|g_{l+j} - g_{l+j-1}\| \\ &\leqq \sum_{j=1}^{m} \frac{K^{l+j-1}(T-s)^{l+j-1}}{(l+j-1)!} \|g_1 - g_0\|.\end{aligned}$$

Comparison with the exponential series shows that the sequence g_l converges in norm. The limit g satisfies (*). □

Proof of Lemma 11.3. By definition, $\eta(r)$ is $\mathscr{G}_t^u$ measurable for $s \leqq r \leqq t$. Let

$$\begin{aligned}\eta_t(r) &= \eta(r), \qquad \text{if } r \leqq t, \\ \eta_t(r) &= \eta(t), \qquad \text{if } r > t.\end{aligned}$$

By (a) of the definition of the space Γ,

$$u(t) = \gamma(t, \eta_t).$$

It follows that $u(t)$ is $\mathscr{G}_t^u$ measurable. From (11.7) and (11.8)

$$\eta^+(t) = \int_s^t H(\tau) \int_s^\tau W(\tau, \rho) B(\rho) u(\rho) \, d\rho \, d\tau,$$

where $W(s,t)$ is the fundamental solution matrix for $A(t)$:

$$\begin{aligned}dW &= A(t) W dt, \qquad s \leqq t, \\ W(s,s) &= \text{identity}.\end{aligned}$$

For $r \leqq t$, $\eta^+(r)$ is $\mathscr{G}_t^u$ measurable, and hence so is $\eta^0(r) = \eta(r) - \eta^+(r)$. This shows that $\mathscr{G}_t^0 \subset \mathscr{G}_t^u$.

Define an operator Φ on $\mathscr{C}^k$ by

$$(\Phi g)(t) = H(t) \int_s^t W(t, \rho) B(\rho) \gamma(\rho, g) \, d\rho.$$

By (b) of the definition of the space Γ, Φ satisfies the hypothesis of Lemma 11.4. Since $\eta = \eta^0 + \eta^+$, we have

$$\eta(t) = \eta^0(t) + \int_s^t (\Phi \eta)(\tau) \, d\tau.$$

Define $\eta_0 = 0$ and

$$\eta_{l+1}(t) = \eta^0(t) + \int_s^t (\Phi \eta_l)(\tau) \, d\tau, \qquad l = 0, 1, 2, \ldots.$$

By induction on l, $\eta_l(t)$ is $\mathscr{G}_t^0$ measurable; while by Lemma 11.4, $\|\eta_l - \eta\| \to 0$ as $l \to \infty$. Hence $\eta(r)$ is $\mathscr{G}_t^0$ measurable for $r \leqq t$. This implies that $u(t)$ is $\mathscr{G}_t^0$ measurable and $\mathscr{G}_t^u \subset \mathscr{G}_t^0$. □

The Linear Regulator Problem. Let us apply these results to the problem of the partially observed linear regulator. We now take $U = E^m$, L, Ψ quadratic as in § 5. The initial data $\xi(s)$ has given gaussian distribution. Wetake the class $\mathscr{Y}_1$ of control processes of the form (11.12) where $\gamma \in \Gamma$. Among them we shall show the optimal control is of the form $u(t)=G(t)\,\hat{\xi}(t)$, where $G=-N^{-1}B'K$ just as for the complete observation case in § 5. In order to show that such a control is in the class, we associate with any G of class $C^1[s,T]$ a corresponding $\gamma_G \in \Gamma$. Let W_G be the following fundamental solution matrix:

$$dW_G=(A+BG-FH)\,W_G\,dt, \qquad t \geqq s,$$

with $W_G(s,s)=$identity. Let

$$\gamma_G(t,g)=G(t)\left[W_G(t,s)\,E[\xi(s)]+\int_s^t W_G(t,r)\,F(r)\,dg(r)\right].$$

Here g denotes a generic element of $\mathscr{C}^k$, and the integral is in the Riemann-Stieltjes sense. Since $W_G F$ is C^1, integration by parts gives (see Graves [1, p. 265])

$$\int_s^t W_G F\,dg(r)=W_G F g|_s^t-\int_s^t \frac{\partial}{\partial r}(W_G F)\,g(r)\,dr.$$

Therefore, $\gamma_G \in \Gamma$.

To compute $\hat{J}(u)$ for the linear regulator problem, we have

$$\hat{L}(t,\hat{x},u)=\int_{E^n} x'M(t)\,x\,g(t,x-\hat{x})\,dx+u'N(t)\,u$$

$$\hat{\Psi}(\hat{x})=\int_{E^n} x'D\,x\,g(T,x-\hat{x})\,dx.$$

After the change of variables $\tilde{x}=x-\hat{x}$ in these integrals, we get

$$\hat{L}=\hat{x}'M(t)\,\hat{x}+u'N(t)\,u+Z_1(t)$$

$$\hat{\Psi}=\hat{x}'D\,\hat{x}+Z_2,$$

$$Z_1(t)=\int_{E^n}\tilde{x}'M(t)\,\tilde{x}\,g(t,\tilde{x})\,d\tilde{x},$$

$$Z_2=\int_{E^n}\tilde{x}'D\,\tilde{x}\,g(T,\tilde{x})\,d\tilde{x}.$$

Here we have used the fact that the gaussian density $g(t,\tilde{x})$ satisfies

$$1=\int_{E^n} g(t,\tilde{x})\,d\tilde{x}, \qquad 0=\int_{E^n}\tilde{x}\,g(t,\tilde{x})\,d\tilde{x}.$$

Thus, for the linear regulator

$$\text{(11.13)}\qquad \begin{aligned} \hat{J}(u) &\doteq E\left\{\int_s^T [\hat{\xi}(t)'M(t)\,\hat{\xi}(t)+u(t)'N(t)\,u(t)]\,dt+\hat{\xi}(T)'D\,\hat{\xi}(T)\right\}+Z,\\ Z &= E\left\{\int_s^T \tilde{\xi}(t)'M(t)\,\tilde{\xi}(t)\,dt+\tilde{\xi}(T)'D\,\tilde{\xi}(T)\right\}, \end{aligned}$$

where $\tilde{\xi}=\xi-\hat{\xi}$ is the estimation error. The (gaussian) distribution of $\tilde{\xi}(t)$ is unaffected by the control. Hence Z is also unaffected by the choice of $\gamma \in \Gamma$. Except for Z, $\hat{J}(u)$ has the same form as $J(u)$. However, ξ is to be replaced by $\hat{\xi}$, and the system Eq. (11.1) by (11.10).

In stating the separation principle for the linear regulator, we again let $K(t)$ be the matrix solution of (5.6) with $K(T)=D$. $F(t)$, $R(t)$ are defined by (11.5), V(9.9) with $R(s)=E\,\xi(s)\,\xi(s)'$; and $\hat{y}=E\,\xi(s)$. We denote an optimal control by u^*, and the corresponding solutions of (11.1), (11.2), (11.10) by $\xi^*, \eta^*, \hat{\xi}^*$.

Theorem 11.1. (Separation Principle for the Partially Observed Linear Regulator.) *An optimal control for the partially observed linear regulator problem is*

$$u^*(t) = -N^{-1}(t)\,B(t)'\,K(t)\,\hat{\xi}^*(t);$$

$\hat{\xi}^*$ *satisfies the stochastic differential equation*

$$d\hat{\xi}^* = [A(t)\,\hat{\xi}^*(t) + B(t)\,u^*(t)]\,dt + F(t)[d\eta^* - H(t)\,\hat{\xi}^*(t)\,dt]$$

with initial data $\hat{\xi}^*(s) = E[\xi^*(s)]$.

Proof. The control u^* corresponds to $\gamma_{G^*}\in\Gamma$, for $G^* = -N^{-1}B'K$. By Theorem 5.1, Corollary 4.2, and Lemmas 11.1, 11.2, 11.3, $J(u^*)\leqq J(u)$ for any $u\in\mathscr{Y}$. □

Let us conclude with a result corresponding to Theorem 11.1 when $J(u)$ has the more general form (11.3). We now regard $\hat{\xi}(t)$ as the completely observed state of a system obeying (11.10), with expected performance criterion (11.11). We now seek an optimal control in the class $\mathscr{Y}_1$ of controls of the form (11.12). Let $\mathscr{V}_1$ denote the class of all Borel measurable functions $\hat{\mathbf{u}}(t,\hat{x})$ such that $\hat{\mathbf{u}}(t,0)$ is bounded and

$$|\hat{\mathbf{u}}(t,\hat{x}) - \hat{\mathbf{u}}(t,\hat{y})| \leqq M\,|\hat{x}-\hat{y}|$$

for $s\leqq t\leqq T$ and all $\hat{x}, \hat{y}\in E^n$. Recall that s is fixed throughout the present section.

To each $\hat{\mathbf{u}}\in\mathscr{V}_1$ we associate $\gamma_{\hat{\mathbf{u}}}\in\Gamma$ as follows. Let Y be the fundamental solution matrix for

$$dY = (A - FH)\,Y\,dt, \qquad Y(s,s) = \text{identity}.$$

Let $\zeta(t)$ be defined by

$$\zeta(t) = Y(t,s)\,\xi(s) + \int_s^t Y(t,r)\,F(r)\,d\eta(r) + \int_s^t Y(t,r)\,B(r)\,\hat{\mathbf{u}}(r,\zeta(r))\,dr.$$

It follows that $\zeta(t)$ is of the form $\zeta(t)=\phi(t,\eta)$ where $\phi(t,g)$ is continuous on $[s,T]\times\mathscr{C}^k$, $\phi(t,\cdot)$ is Lipschitz uniformly in t, and $\phi(t,g)=\phi(t,\tilde{g})$ if $g(r)=\tilde{g}(r)$ for $s\leqq r\leqq t$. See Problem 12. We take

$$\gamma(t,g) = \hat{\mathbf{u}}(t,\phi(t,g)).$$

We then find in the same way as for Theorem 11.1:

Theorem 11.2. *Suppose that there exists an optimal feedback control law* $\hat{\mathbf{u}}^*\in\mathscr{V}_1$ *for problem* (11.10), (11.11), *in which* $\hat{\xi}(t)$ *is regarded as a completely observed state. Then* $u^*(t)=\hat{\mathbf{u}}^*(t,\hat{\xi}^*(t))$ *is an optimal control in the class* $\mathscr{Y}$ *for the partially observed problem which has* (11.1), (11.2) *as system and observation equations and expected performance criterion* (11.3).

Under sufficiently severe restrictions, the existence of $\mathbf{u}^*$ with the properties required in Theorem 11.2 can be proved. See Problem 13.

We note that the existence of a $C_p^{1,2}$ solution $W(s, \hat{y})$ to the dynamic programming equation for (11.10)–(11.11) requires that $F(t)\,\sigma_1(t)$ be $n \times n$ nonsingular, and in particular that state and observation vectors have the same dimension $k=n$. When $k<n$ Eqs. (11.10) do not split in the manner of (8.1). It is an open question to find results like Theorems 8.2 and 8.3 for the problem (11.10) to (11.11).

Other Results on the Separation Principle. We mention Davis-Varaiya [1], Lindquist [2], Lee [1], Balakrishnan [3][5], Bensoussan-Viot [1]. These results are concerned with equality of the data σ-algebras $\mathscr{G}_t^u$ and $\mathscr{G}_t^0$, with the separation principle for infinite dimensional problems (e.g., systems with time delays and distributed parameter systems) and with related matters.

Problems—Chapter VI

(1) Let $n=1$, $d\xi = u\,dt + \sigma\,dw$, with $\sigma>0$ a constant. Take $L=1$ and $U=[\alpha, \beta]$ with $\alpha<0$, $\beta>0$. Look for a solution $W(y)$ of (4.1^a) on the interval $0 \leqq y \leqq 1$ with $W''(y)<0$ and $W(0)=W(1)=0$. Derive a transcendental equation for the point in the interval $0<y<1$ where the optimal feedback control law $\mathbf{u}^*(y)$ switches from α to β. Use Theorem 4.2 to verify that $W(y)$ is the minimum expected time for $\xi(t)$ to reach 0 or 1 starting from $\xi(0)=y$.

(2) (Linear regulator with state dependent noise.) Let $n=1$, $d\xi=(A\xi+Bu)\,dt+(C\xi+D)\,dw$, with $U=E^1$ and $J(s, y, u)$ as in (5.2). Here A, B, C, D are scalar-valued functions of time. Find a solution of (4.1) of the form $W(s, y)=K(s)\,y^2+p(s)\,y+q(s)$, with $W(T, y)=0$. Show that $\dot{K}=-C^2 K-2KA+N^{-1}K^2B^2-M$, $\dot{p}=-2CK-pA+N^{-1}KB^2p$, $\dot{q}=-D^2K+(4N)^{-1}p^2B^2$, $\mathbf{u}^*=-N^{-1}BKy-(2N)^{-1}Bp$. Can you extend this result to more dimensions?

(3) Let $n=1$, $d\xi=u(t)\,dt+\sigma\,dw$, $\sigma>0$ constant, $U=(-\infty, \infty)$. Consider the open loop stochastic control problem of minimizing $\int_0^1 [\xi(t)^4+u(t)^2]\,dt$ with $\xi(0)=y$ given. Using the separation principle for the open loop case, reduce this problem to the following calculus of variations problem: minimize $\int_0^1 [6\sigma^2 t\,x(t)^2+x(t)^4+\dot{x}(t)^2]\,dt$ with $x(0)=y$ and $x(1)$ free.

(4) Prove inequality (**), in the proof of Lemma (6.2).

(5) Prove assertion (#), in the proof of Theorem 9.1.

(6) Suppose that hypotheses (6.1), (6.2), (6.3) hold. Let $\mathbf{u}^1, \mathbf{u}^2, \ldots$ be a sequence of admissible feedback controls such that $\mathbf{u}^j(s, y)$ tends to $\mathbf{u}(s, y)$ as $j \to \infty$, almost everywhere with respect to Lebesgue measure on Q. Show that $J(s, y; \mathbf{u}^j)$ tends to $J(s, y, \mathbf{u})$ as $j \to \infty$. *Hint.* $\psi^j(s, y)=J(s, y; \mathbf{u}^j)$ satisfies a linear parabolic equation of the form Appendix (E.1) for each j. Use the estimates (E.8), (E.9).

(7) Suppose that (3.2), (3.6) hold and that U is compact. Let τ_R = exit time of $(t, \xi(t))$ from the cylinder $Q_R = (T_0, T) \times (|x| < R)$, and $J_R(s, y, u) = E_{sy}\{\int_s^{\tau_R} L(t, \xi(t), u(t))\, dt + \Psi[\xi(\tau_R)]\}$. Let $J(s, y, u)$ be the corresponding expression with τ_R replaced by T. Show that $|J_R - J| \leqq \beta_R$, where $\beta_R \to 0$ as $R \to \infty$ and β_R depends only on U, T_0, T, the constants in (3.2), (3.6), and a bound for $|y|$. *Hint.* Inequality V(4.7).

(8) Let $(\Omega, \mathscr{F}, P)$, $\{\mathscr{F}_t\}$, and a standard brownian motion w adapted to $\{\mathscr{F}_t\}$ be given. Let $\tilde{\mathscr{U}}(s)$ be the class of bounded stochastic processes on $(\Omega, \mathscr{F}, P)$, nonanticipative with respect to $\{\mathscr{F}_t\}$. Let $\tilde{V}(s, y) = \inf J(s, y, u)$, the infimum being over $\tilde{\mathscr{U}}(s)$; the solutions of (3.1) are with respect to the given w. Show that $\tilde{V}(s, y) = V(s, y)$, with V as in (7.5), under the hypotheses of either §6 or §7. *Hints.* First consider the case of a bounded cylinder Q, with assumptions (6.1)–(6.3). In Problem 6 take each $\mathbf{u}^j$ Lipschitz. Use Theorem V.4.1 and Corollary 4.2. Reduce the case $Q = Q^0$ to the bounded one, using Problem 7 and Lemma 7.4.

(9) Let $d\xi = [A(t)\,\xi(t) + B(t)\,u(t)]\,dt + \sigma(t)\,dw(t)$, $\xi(s) = y$. Let J have the form (2.2) with T fixed. Let U be convex, and $L(t, \cdot, \cdot)$, $\Psi(\cdot)$ be convex functions.

(a) Show that $J(s, \cdot, \cdot)$ is a convex function on $E^n \times \tilde{\mathscr{U}}(s)$, with $\tilde{\mathscr{U}}(s)$ as in Problem 8.

(b) Show that $V(s, \cdot)$ is convex, using (a).

(c) Show that $W(s, \cdot)$ in Theorem 6.2 is convex if in addition $\sigma^{-1}(t)$ exists and L, Ψ satisfy (6.9)(c)(d).

(10) In Problem 9, let $n = 1$, $B(t) = 1$, $\sigma(t) > 0$, $U = [-1, 1]$. Let $J = E_{sy}\,\Psi[\xi(\tau)]$, where Ψ is convex and Ψ together with its first derivative Ψ' satisfy a polynomial growth condition. Show that an optimal feedback control law is $\mathbf{u}^*(s, y) = 1$ for $-\infty < y \leqq c(s)$, $\mathbf{u}^*(s, y) = -1$ for $c(s) < y < \infty$, where $c(s) = \inf\{y;\, W_y(s, y) > 0\}$.

(11) Let U be compact and $H(s, y, p)$ as in (6.4). Show that:

(a) H is Lipschitz on any compact set; and

(b) $H(s, y, p)(1 + |p|)^{-1}$ is bounded for (s, y) in any compact set. *Note.* For further properties of H see Fleming [10, p. 479].

(12) (a) Consider the following operator B on the space $\mathscr{C}^n = \mathscr{C}^n[s, T]$:

$$Bf(t) = \int_s^t \beta(r, f(r))\, dr.$$

Assume that β is Borel measurable with $\beta(t, 0)$ bounded and $\beta(t, \cdot)$ satisfying a uniform Lipschitz condition. Show that $(I - B)^{-1} = I + B + B^2 + \cdots$; moreover, $(I - B)^{-1}$ is Lipschitz on $\mathscr{C}^n$. *Hint.* See the proof of Lemma 11.4.

(b) Consider the integral equation

$$f(t) = h(t) + \int_s^t \alpha(r)\, dg(r) + \int_s^t \beta(r, f(r))\, dr,$$

with $g \in \mathscr{C}^k$, $h \in \mathscr{C}^n$, α $(n \times k)$-matrix valued of class C^1, and β as above. Using (a) show that there is a unique solution (for fixed h) $f(t) = \phi(t, g)$. Moreover, $\phi(t, g)$ is continuous on $[s, T] \times \mathscr{C}^k$, $\phi(t, \cdot)$ is Lipschitz uniformly in t, and $\phi(t, g) = \phi(t, \tilde{g})$ if $g(r) = \tilde{g}(r)$ for $s \leqq r \leqq t$.

(13) The dynamic programming equation for (11.10)–(11.11) is

$$0 = W_s + \min_{v \in V} [\hat{\mathscr{A}}^v W + \hat{L}(s, \hat{y}, v)],$$

$$\hat{\mathscr{A}}^v = \tfrac{1}{2} \sum_{i,j=1}^{n} \hat{a}_{ij}(s) \frac{\partial^2}{\partial \hat{y}_i \partial \hat{y}_j} + (A(s)\hat{y} + B(s)v) \frac{\partial}{\partial \hat{y}},$$

$$\hat{a}(s) = F(s)\sigma_1(s)\sigma_1'(s)F'(s).$$

Suppose that $W(s, \hat{y})$ is a solution in $C_p^{1,2}(\bar{Q}^0)$ with $\dot{W}_{\hat{y}\hat{y}}$ bounded and $W(T, \hat{y}) = \hat{\Psi}$. Moreover, suppose that: (i) U is compact and convex; (ii) L is C^2 with characteristic values of L_{uu} bounded below by $\gamma > 0$; and (iii) L_{ux} is bounded for $u \in U$. Show that there exists an optimal $\mathbf{u}^*$ with the property required in Theorem 11.2. *Hints.* Show that the partial derivatives of $\hat{L}$ have corresponding properties, by differentiating under the integral sign. Use Lemma 6.3 with $\Theta = W_{\hat{y}} Bu + \hat{L}$, and Theorem 4.1.

Appendices

A. Gronwall-Bellman Inequality

In this short appendix we prove the following result, which we call the Gronwall-Bellman inequality:

Let $m(t)$ be continuous and satisfy

$$0 \leqq m(t) \leqq h(t) + \int_s^t g(r)\, m(r)\, dr, \qquad s \leqq t \leqq T, \tag{A.1}$$

where $g(t) \geqq 0$ on $[s, T]$, $\int_s^T g(t)\, dt < \infty$, and $h(t)$ is bounded on $[s, T]$. Then for $s \leqq t \leqq T$

$$m(t) \leqq h(t) + \int_s^t g(r)\, h(r) \exp\left[\int_r^t g(u)\, du\right] dr. \tag{A.2}$$

Proof. Let $z(t) = \int_s^t g(r)\, m(r)\, dr$. Then (A.1) implies after multiplying by $g(t)$

$$\dot{z}(t) - g(t)\, z(t) \leqq g(t)\, h(t).$$

We multiply by the integrating factor $\exp[-\int_s^t g(u)\, du]$ and integrate both sides, obtaining

$$z(t) \leqq \int_s^t g(r)\, h(r) \exp\left[\int_r^t g(u)\, du\right] dr.$$

By (A.1),

$$m(t) - h(t) \leqq z(t).$$

Hence, comparing these two inequalities gives (A.2). □

Corollary. *If $0 \leqq m(t) \leqq D + C \int_s^t m(r)\, dr$ for $s \leqq t \leqq T$, where C, D are nonnegative constants, then*

$$m(t) \leqq D e^{C(t-s)}, \qquad s \leqq t \leqq T. \tag{A.3}$$

Proof. In (A.2) take $g(t) = C$, $h(t) = D$, and perform the integrations. □

Inequality (A.3) is called *Gronwall's inequality*.

B. Selecting a Measurable Function

In this appendix we prove a lemma, which is needed in Chap. III and VI. We use the following notation. Given positive integers p and m, we denote points of E^{p+m} by (z, u) with

$$z=(z_1, \ldots, z_p), \qquad u=(u_1, \ldots, u_m).$$

For a set $D \subset E^{p+m}$, let

$$D^z=\{u: (z, u) \in D\}, \qquad \Delta=\{z: D^z \text{ is not empty}\}.$$

We call D σ-compact if $D=D_1 \cup D_2 \cup \cdots$ where $D_1, D_2, \ldots$ are compact sets. For instance, every open set is σ-compact, and also every closed set. By "almost all z" we mean except for a set of p-dimensional Lebesgue measure 0. A vector valued function $u(z)=(u_1(z), \ldots, u_m(z))$ is measurable if and only if each component u_i is measurable (Lebesgue). By changing a Lebesgue measurable function u on a set of p-dimensional measure 0, we can arrange that u is Borel measurable.

Lemma B. *If D is σ-compact, then there exists a measurable function $u=u(z)$ with $(z, u(z)) \in D$ for almost all $z \in \Delta$.*

Proof. Suppose first that D is compact. Then Δ is also compact, since Δ is the image of D under the projection sending (z, u) into z. Also each set D^z is compact. Let us use induction on m. For $m=1$, let $u(z)$ be the least element of D^z. In this case $u(z)$ is lower semicontinuous on Δ, hence measurable. Assuming the result true in dimension $m-1$, let $\tilde{D}$ be the projection of D onto $(z, u_1, \ldots, u_{m-1})$ space E^{p+m-1}. Then there exist measurable $u_1, \ldots, u_{m-1}$ such that $(z, u_1(z), \ldots, u_{m-1}(z)) \in \tilde{D}$ for almost all $z \in \Delta$. Let $u_m(z)$ be the least v, such that $(z, u_1(z), \ldots, u_{m-1}(z), v) \in D$. By Lusin's theorem, there exists a sequence of compact sets $K_1 \subset K_2 \subset \cdots$ such that $\Delta-(K_1 \cup K_2 \cup \cdots)$ has measure 0 and the restriction of $u_1, \ldots, u_{m-1}$ to K_l is continuous, for each $l=1, 2, \ldots$. The restriction of u_m to K_l is then lower semicontinuous, which implies that u_m is measurable. This proves the lemma for compact D.

For D σ-compact, we have

$$D=D_1 \cup D_2 \cup \cdots, \qquad D_1 \subset D_2 \subset \cdots, \qquad D_l \text{ compact for } l=1, 2, \ldots.$$

For the projection Δ_l of D_l we have

$$\Delta=\Delta_1 \cup \Delta_2 \cup \cdots, \qquad \Delta_1 \subset \Delta_2 \subset \cdots.$$

By what we have shown, there is a measurable u^l with $(z, u^l(z)) \in D_l$ for almost all $z \in \Delta_l$, $l=1, 2, \ldots$. Set

$$u(z)=u^l(z), \qquad \text{for } z \in \Delta_l-\Delta_{l-1}. \quad \square$$

In Chap. III we apply Lemma B twice, then once in Chap. VI, as follows:

(1) For Lemma III.3.4. Since $\dot{x}^*$ is a measurable function on $[T_0, T_1]$, there exist by Lusin's theorem compact sets $\Delta_1 \subset \Delta_2 \subset \cdots$ such that

$$[T_0, T_1]-(\Delta_1 \cup \Delta_2 \cup \cdots)$$

has measure 0 and the restriction of $\dot{x}^*$ to each Δ_l is continuous. Let

$$D_l=\{(t,u)\colon t\in\Delta_l,\ \dot{x}^*(t)=f(t,x^*(t),u),\ u\in U\},\qquad l=1,2,\ldots,$$
$$D=D_1\cup D_2\cup\cdots.$$

Here $p=1$. By Lemma II.3.3, D^t is not empty for $t\in\Delta=\Delta_1\cup\Delta_2\cup\cdots$. By Lemma B there is a measurable function u^* such that, for almost all $t\in[T_0,T_1]$,

$$\dot{x}^*(t)=f(t,x^*(t),u^*(t)),\qquad u^*(t)\in U.$$

(2) For Lemma III.5.6. The reasoning is very similar to that for Lemma III.3.4. In the notation in III.5, let $\tilde{u}=(u,v)$ and $\tilde{U}_l=\{\tilde{u}\in\tilde{U}\colon |\tilde{u}|\leqq l\}$,

$$D_l=\{(t,\tilde{u})\colon t\in\Delta_l,\ \dot{\tilde{x}}^*=\tilde{f}(t,x^*(t),\tilde{u}),\ \tilde{u}\in U_l\}.$$

Now Δ_l is a compact set such that $\dot{\tilde{x}}^*=(\dot{x}^*,\dot{Z}^*)$ is continuous on Δ_l and $[T_0,T_1]-(\Delta_1\cup\Delta_2\cup\cdots)$ has measure 0.

(3) For Lemma VI.6.1. Take $p=n+1$, $z=(s,y)$. Write $Q=K_1\cup K_2\cup\cdots$, where $K_1\subset K_2\subset\cdots$ and each K_l is compact. Let

$$D=\{(s,y,u)\colon (s,y)\in Q,\ W_y(s,y)f(s,y,u)+L(s,y,u)=H(s,y,W_y(s,y))\},$$

with H defined by VI(6.4).

Then $D=D_1\cup D_2\cup\cdots$, where D_l is defined in the same way but with $(s,y)\in K_l$. Each D_l is compact. We apply Lemma B to find $\mathbf{u}^*(s,y)$ with the property required in Lemma VI.6.1.

C. Convex Sets and Convex Functions

Let us review the definitions, and a few elementary properties used in this book. For deeper results in the theory of convexity see for instance Eggleston [1], Valentine [1].

Definition. Let $\mathscr{V}$ be a vector space. A set $\mathscr{K}\subset\mathscr{V}$ is *convex* if $u_1,u_2\in\mathscr{K}$ and $0\leqq\varepsilon\leqq 1$ imply $(1-\varepsilon)u_1+\varepsilon u_2\in\mathscr{K}$.

One says that u is a convex combination of $u_1,\ldots,u_m$ if

$$u=\sum_{j=1}^m \varepsilon_j u_j,\qquad \varepsilon_j\geqq 0\ \text{ for } j=1,\ldots,m,\qquad \sum_{j=1}^m \varepsilon_j=1.$$

If $\mathscr{K}$ is convex, then any convex combination u of points of $\mathscr{K}$ is also in $\mathscr{K}$.

Consider now the special case $\mathscr{V}=E^n$. Let $K\subset E^n$ be convex, and for $a>0$ let K_a be the a-neighborhood of K. Then $x\in K_a$ if and only if $|x-y|<a$ for some $y\in K$. The sets K_a are convex for each $a>0$. This is immediate from the definitions and the triangle inequality.

The following is an extension of the fact about convex combinations mentioned above. Let K be convex, and let $G=(G_1,\ldots,G_n)$ be a function from an interval

$[t_0, t_1]$ into K, with the components G_i integrable for $i=1, \ldots, n$. Then

$$(t_1 - t_0)^{-1} \int_{t_0}^{t_1} G(t)\,dt \in \bar{K}, \tag{C.1}$$

where $\bar{K}$ is the closure of K.

To prove this, if G is a step function (with finitely many values $z_1, \ldots, z_m$ in K), then the left side of (C.1) is a convex combination of $z_1, \ldots, z_m$ and hence belongs to K since K is convex. Next suppose that $G(t)$ is bounded. There is a sequence $G^1, G^2, \ldots$ of step functions tending to G in measure on $[t_0, t_1]$. Moreover, given $a>0$ we may assume that $G^r(t) \in K_a$ for each $r=1, 2, \ldots$. (To do this, we arbitrarily redefine $G^r(t) = z^*$ for some $z^* \in K$, when $G^r(t) \notin K_a$.) Then

$$(t_1 - t_0)^{-1} \int_{t_0}^{t_1} G^r(t)\,dt \in K_a$$

for each $r=1, 2, \ldots$. Hence the left side of (C.1) is in $\bar{K}_a$. Since this is true for each $a>0$ we get (C.1) when G is bounded. For unbounded G, we approximate G in L^1 norm by bounded functions with values in K.

Definition. Let $\mathcal{K}$ be a convex set. A real valued function J is *convex on* $\mathcal{K}$ if $u_1, u_2 \in \mathcal{K}$ and $0 \leqq \varepsilon \leqq 1$ imply

$$J[(1-\varepsilon)u_1 + \varepsilon u_2] \leqq (1-\varepsilon) J(u_1) + \varepsilon J(u_2). \tag{C.2}$$

The function J is strictly convex if strict inequality holds in (C.2) when $u_1 \neq u_2$ and $0 < \varepsilon < 1$.

For a function $\phi(x)$, $x=(x_1, \ldots, x_n)$, of class C^2 on an open convex set $K \subset E^n$, a necessary and sufficient condition for convexity is that the matrix $\phi_{xx} = (\phi_{x_i x_j})$, $i, j = 1, \ldots, n$, of second order partial derivatives be non-negative definite for all $x \in K$. Positive definiteness of the matrices ϕ_{xx} is sufficient for strict convexity. See Fleming [1, p. 56].

A linear function maps convex sets into convex sets. To show this suppose that $f(u) = \alpha u + \beta$ is from $U \subset E^m$ into E^n. Let $F = f(U) = \{f(u)\colon u \in U\}$, as in Chap. III.2. Then

$$f[(1-\varepsilon)u_1 + \varepsilon u_2] = (1-\varepsilon) f(u_1) + \varepsilon f(u_2).$$

Hence F is convex if U is convex. If $f(u) = \alpha u + \beta$, U is convex, and $L(u)$ is convex on U, as in Chap. III.4, let

$$\tilde{F} = \{\tilde{z} = (z, z_{n+1})\colon\ z = f(u),\ z_{n+1} \geqq L(u),\ u \in U\}.$$

Then $\tilde{F}$ is convex. To show this, let

$$z_j = f(u_j), \qquad z_{n+1,j} \geqq L(u_j), \quad j = 1, 2,$$
$$z = (1-\varepsilon) z_1 + \varepsilon z_2, \qquad z_{n+1} = (1-\varepsilon) z_{n+1,1} + \varepsilon z_{n+1,2}.$$

For $0 \leqq \varepsilon \leqq 1$, $u = (1-\varepsilon) u_1 + \varepsilon u_2$, we have

$$z = f(u), \qquad z_{n+1} \geqq (1-\varepsilon) L(u_1) + \varepsilon L(u_2) \geqq L(u).$$

Thus $\tilde{z} = (z, z_{n+1})$ is in $\tilde{F}$, which shows that $\tilde{F}$ is convex.

D. Review of Basic Probability

Here we review some concepts used in Chap. V and VI. Some books which the reader may consult regarding them are Breiman [1], Doob [1], Feller [1], Gikhman-Skorokhod [1], Loeve [1], Meyer [1].

Let $(\Omega, \mathscr{F}, P)$ be a probability space; Ω is a set, $\mathscr{F}$ a σ-algebra of subsets of Ω, and P a probability measure on $\mathscr{F}$ with $P(\Omega)=1$. Sets $A\subset\mathscr{F}$ are called events. An event A is called P-almost sure if $P(A)=1$; we also say that A occurs with probability 1. A random variable η is a real valued, $\mathscr{F}$-measurable function on Ω. Its expectation is

$$E\eta = \int_{\Omega} \eta(\omega)\, P(d\omega),$$

provided $E|\eta|<\infty$.

Random Vectors. Let Σ be a complete separable metric space, which we call a state space. A function from Ω into Σ will be called a random vector if the event $A=\{\omega\in\Omega\colon \xi(\omega)\in B\}$ is in $\mathscr{F}$ whenever B is an open subset of Σ. As is customary in works on probability, we shall denote such a set (for notational brevity) as $A=\{\xi\in B\}$. (Among mathematicians another common notation, which we shall not use, is $A=\xi^{-1}(B)$.) Let $\mathscr{B}(\Sigma)$ denote the least σ-algebra containing all open subsets of Σ; a set $B\in\Sigma$ is called a Borel subset of Σ. Then ξ is a random vector if and only if $\{\xi\in B\}$ is in $\mathscr{F}$ for every $B\in\mathscr{B}(\Sigma)$. In measure theoretic terminology, a random vector is simply a function from Ω into Σ measurable with respect to the pair of σ-algebras $(\mathscr{F}, \mathscr{B}(\Sigma))$.

We recall some standard notions of convergence for a sequence ξ^j, $j=1, 2, \ldots$, of random vectors. Let $d(x, y)$ denote the distance between $x, y\in\Sigma$. Then ξ^j tends to ξ as $j\to\infty$:

(i) *P-almost surely* if $d[\xi^j(\omega), \xi(\omega)]\to 0$ as $j\to\infty$, with probability 1;

(ii) In *probability* if, for any $\varepsilon>0$, $P(d(\xi^j, \xi)>\varepsilon)\to 0$ as $j\to\infty$.

(iii) In mean of order p if $E\,d(\xi^j, \xi)^p\to 0$ as $j\to\infty$.

Examples. The following are of particular interest in Chap. V and VI.

(1) Let $\Sigma=E^n$, with the usual euclidean distance $d(x, y)=|x-y|$. Then $\xi=(\xi_1, \ldots, \xi_n)$, where $\xi_1, \ldots, \xi_n$ are random variables, is called an *n-dimensional* random vector.

(2) Let $\Sigma=\mathscr{C}^n(\mathscr{T})$, where $\mathscr{T}$ is a compact interval of E^1 and $\mathscr{C}^n(\mathscr{T})$ denotes the space of continuous functions from $\mathscr{T}$ into E^n. We give $\mathscr{C}^n(\mathscr{T})$ the usual sup norm metric:

$$d(g, h) = \|g-h\| = \max_{t\in\mathscr{T}} |g(t)-h(t)|.$$

Distribution of Random Vectors. If ξ is any random vector, its *probability distribution measure* is the measure on $\mathscr{B}(\Sigma)$ defined by

$$P^*(B)=P(\xi\in B), \qquad B\in\mathscr{B}(\Sigma).$$

In measure theoretic terms, P^* is simply the measure into which P is transformed by the $(\mathscr{F}, \mathscr{B}(\Sigma))$ measurable function ξ. Therefore, by the transformation formula for integrals

(D.1) $$EG(\xi)=\int_{\Omega} G[\xi(\omega)]\, P(d\omega)=\int_{\Sigma} G(x)\, P^*(dx)$$

provided G is real valued, Borel measurable and either integral exists.

Joint Distributions. Let $\xi^1, \ldots, \xi^m$ be random vectors, defined on the same $(\Omega, \mathscr{F}, P)$ but with values in possibly different spaces $\Sigma^1, \ldots, \Sigma^m$. The m tuple $\xi=(\xi^1, \ldots, \xi^m)$ defines a random vector on the cartesian product $\Sigma=\Sigma^1 \times \cdots \times \Sigma^m$. Its distribution measure P^* is called the *joint distribution* of the random vectors $\xi^1, \ldots, \xi^m$.

In particular, let $\xi=(\xi_1, \ldots, \xi_n)$ be an n-dimensional random vector. In this case $m=n$ and each Σ^j is the real line E^1. If P^* is absolutely continuous with respect to n-dimensional Lebesgue measure, then on the right side of (D.1) one can put $P^*(dx)=f(x)\, dx$, where

$$f(x) \geqq 0, \qquad \int_{E^n} f(x)\, dx=1.$$

Here $x=(x_1, \ldots, x_n)$ and dx denotes integration with respect to Lebesgue measure. The function f is called the *density* of ξ.

Example 3. An n-dimensional random vector ξ is called *non-degenerate gaussian* if

$$f(x)=(2\pi)^{-\frac{n}{2}}(\det R)^{-\frac{1}{2}} \exp[-\tfrac{1}{2}(x-\mu)' R^{-1}(x-\mu)],$$

where the vector $\mu=(\mu_1, \ldots, \mu_n)$ and positive symmetric matrix $R=(\rho_{ij})$ turn out to consist of the means and covariances of the components $\xi_1, \ldots, \xi_n$:

$$\mu_i=E\,\xi_i, \qquad \rho_{ij}=E(\xi_i-\mu_i)(\xi_j-\mu_j), \quad i, j=1, \ldots, n.$$

A random vector ξ is *degenerate gaussian* if some linear change of coordinates in E^n transforms ξ into a vector $(\eta_1, \ldots, \eta_m, 0, \ldots, 0)$ with $0 \leqq m<n$ and $(\eta_1, \ldots, \eta_m)$ non-degenerate gaussian in dimension m.

Independence. We recall that events $A_1, \ldots, A_m$ are called independent if

$$\prod_{j \in \mathscr{C}} P(A_j)=P\Big(\bigcap_{j \in \mathscr{C}} A_j\Big)$$

for any finite subcollection $\mathscr{C}$ of these events. Random vectors $\xi^1, \ldots, \xi^m$ are called (mutually) independent if the events $\xi^1 \in B_1, \ldots, \xi^m \in B_m$ are independent for any $B_1, \ldots, B_m$ with $B_j \in \mathscr{B}(\Sigma^j)$. In measure theoretic terms this means that the joint distribution is a product measure: $P^*=(P^1)^* \times \cdots \times (P^m)^*$, where $(P^j)^*$ is the distribution of ξ^j. If $\mathscr{F}_1$ is a σ-algebra, $\mathscr{F}_1 \subset \mathscr{F}$, a random vector ξ is said to be independent of $\mathscr{F}_1$ if the events A, $\xi \in B$ are independent for any $A \in \mathscr{F}_1$, $B \in \mathscr{B}(\Sigma)$.

Conditional Probabilities and Expectations. Let us mention two concepts of conditional expectation, and the relation between them. Conditional probabilities can be treated as a special case, by using indicator functions (formula (D.5)). As references we mention Breiman [1, Chap. IV], Gikhman-Skorokhod [1, Chap. III].

First, suppose that ζ is a random vector with state space Σ_0 (a complete separable metric space). Let π be the distribution measure of ζ. If η is a random variable with $E|\eta|<\infty$, then the conditional expectation $E(\eta|\zeta=z)$ is defined by

(D.2) $$\int_D E(\eta|\zeta=z)\,\pi(dz)=\int_{\zeta\in D}\eta(\omega)\,P(d\omega),$$

for all $D\in\mathscr{B}(\Sigma_0)$. Existence and uniqueness π-almost surely of this kind of conditional expectation follow from the Radon-Nikodym Theorem. For brevity, we often write $E(\eta|\zeta=z)$ as $E(\eta|z)$.

For many purposes it is convenient to define a more general conditional expectation than the one above. Let $\mathscr{G}$ be a σ-algebra with $\mathscr{G}\subset\mathscr{F}$, and η a random variable with $E|\eta|<\infty$. The *conditional expectation* $E(\eta|\mathscr{G})$ is defined as any $\mathscr{G}$-measurable random variable such that

$$\int_A E(\eta|\mathscr{G})(\omega)\,P(d\omega)=\int_A\eta(\omega)\,P(d\omega)$$

for every $A\in\mathscr{G}$. More precisely, any such random variable is called a version of the conditional expectation. The Radon-Nikodym theorem again insures that $E(\eta|\mathscr{G})$ exists, and is uniquely determined P-almost surely.

Some elementary properties are as follows. If $\mathscr{G}\subset\mathscr{G}'$, then P-almost surely

(D.3) $$E\{E(\eta|\mathscr{G}')|\mathscr{G}\}=E(\eta|\mathscr{G}).$$

If $E|\zeta|$, $E|\eta|$, and $E|\zeta\,\eta|$ are all finite and ζ is $\mathscr{G}$-measurable, then P-almost surely

(D.4) $$E(\zeta\,\eta|\mathscr{G})=\zeta\,E(\eta|\mathscr{G}).$$

Given $A\in\mathscr{G}$, its conditional probability with respect to $\mathscr{G}$ is defined by

(D.5) $$P(A|\mathscr{G})=E(\chi_A|\mathscr{G}),$$

where χ_A is the indicator function ($\chi_A(\omega)=1$ for $\omega\in A$, $\chi_A(\omega)=0$ otherwise.) In particular, let ξ be a random vector with values in a complete separable metric space Σ. For $B\in\mathscr{B}(\Sigma)$ let

$$\hat{P}(B|\mathscr{G})=P(\xi\in B|\mathscr{G}).$$

From this formula, $\hat{P}(B|\mathscr{G})$ is determined up to a set Γ_B with $P(\Gamma_B)=0$. However, it can be shown that a version of $\hat{P}(B|\mathscr{G})$ exists for all $B\in\mathscr{B}(\Sigma)$, such that $\hat{P}(\cdot|\mathscr{G})$ is a probability measure for each $\omega\in\Omega$. Such a family of probability measures $\hat{P}(\cdot|\mathscr{G})$ is called a *regular conditional distribution* for the random vector ξ given $\mathscr{G}$. If $\eta=G(\xi)$ with G Borel measurable, then conditional expectation can be computed by integrating with respect to the conditional distribution:

(D.6) $$E[G(\xi)|\mathscr{G}]=\int_\Sigma G(x)\,\hat{P}(dx|\mathscr{G}),$$

provided $E|G(\xi)|<\infty$.

Given a random vector ζ as above, let $\mathscr{F}(\zeta)$ denote the σ-algebra consisting of all sets $A=\{\zeta\in D\}$, $D\in\mathscr{B}(\Sigma_0)$. On the left side of (D.2) let us substitute $z=\zeta(\omega)$ and use the transformation formula for integrals. Then (D.2) becomes

(D.2′) $$\int_A E(\eta|\zeta=\zeta(\omega))\,P(d\omega)=\int_A E(\eta|\mathscr{F}(\zeta))\,P(d\omega)$$

for every $A \in \mathscr{F}(\eta)$. If we write $E(\eta|\zeta)$ for the random variable whose value at $z = \zeta(\omega)$ is $E(\eta|z)$, then (D.2′) shows that $E(\eta|\zeta)$ is a version of $E(\eta|\mathscr{F}(\zeta))$. This connects the two definitions of conditional expectation. Moreover, a version $\hat{P}(\cdot|\zeta)$ of the regular conditional distribution $\hat{P}(\cdot|\mathscr{F}(\zeta))$ can be defined. In this case, (D.6) is equivalent to

(D.6′) $$E[G(\xi)|\zeta = z] = \int_{\Sigma} G(x)\,\hat{P}(dx|z),$$

π-almost surely.

E. Results About Parabolic Equations

To begin with, we review the definition of generalized partial derivatives and some of their properties. Then we collect some results about second order parabolic partial differential equations, needed in Chap. V, VI.

Let $\psi(s, y)$ be a real valued function defined on an open set Q. Here $s \in \mathscr{T}_0 = [T_0, T]$ and $y = (y_1, \ldots, y_n) \in E^n$. We say that ϕ_i is the *generalized partial derivative* of ψ in the variable y_i, and write $\phi_i = \psi_{y_i}$, if ψ and ϕ_i are integrable over any compact subset of Q and

$$\int_Q \phi_i\,\beta\,ds\,dy = -\int_Q \psi\,\beta_{y_i}\,ds\,dy$$

for every C^1 function $\beta(s, y)$ with compact support in Q ($\beta(s, y) = 0$ for (s, y) not in some compact subset of Q).

The generalized partial derivative ψ_s is defined similarly; and $\psi_{y_i y_j}$ is the generalized partial derivative of ψ_{y_i} in the variable y_j.

Let $L^\lambda(K)$ denote the space of λ-th power integrable functions on $K \subset Q$, with $\|\ \|_{\lambda, K}$ the norm in $L^\lambda(K)$. The following is easily shown: if ψ^l, $l = 1, 2, \ldots$, is a sequence tending to ψ in $L^\lambda(K)$ and $\|\psi^l_{y_i}\|_{\lambda, K}$ is bounded, $1 < \lambda < \infty$, then $\psi^l_{y_i}$ tends to ψ_{y_i} weakly in $L^\lambda(K)$ as $l \to \infty$. Similar remarks hold about weak convergence of other generalized partial derivatives, of various orders $\geqq 1$, when they exist.

Now suppose that $\psi \in L^\lambda(K)$ and has support in K, with K compact. For $l = 1, 2, \ldots$, define ψ^l by the convolution

$$\psi^l(s, y) = \int_{E^{n+1}} \alpha^l(t, x)\,\psi(s - t, y - x)\,dt\,dx$$

where

$$\alpha^l \geqq 0, \qquad \int_{E^{n+1}} \alpha^l\,dt\,dx = 1,$$

$$\alpha^l(t, x) = 0 \quad \text{for } t^2 + |x|^2 \geqq l^{-2}$$

and α^l is C^∞. The following are standard results. For each l, ψ^l is C^∞ and has support in the l^{-1}-neighborhood of K. Moreover, $\|\psi^l - \psi\|_{\lambda, E^{n+1}} \to 0$ as $l \to \infty$; if ψ is continuous the convergence is uniform. Finally,

$$(\psi^l)_{y_i} = (\psi_{y_i})^l, \qquad (\psi^l)_s = (\psi_s)^l$$

when the generalized partial derivatives ψ_{y_i} or ψ_s exist. It follows that those generalized partial derivatives of ψ which are in $L^\lambda(E^{n+1})$ are the limits in λ-norm of the corresponding partial derivatives of ψ^l as $l\to\infty$. This gives in particular Lemma V.11.1.

Let us next turn to a summary of results about linear partial differential equations of the form

$$\text{(E.1)}\qquad \begin{gathered}\psi_s+\mathscr{A}(s)\,\psi+\Lambda(s,y)=0,\\ \mathscr{A}(s)\psi=\tfrac{1}{2}\sum_{i,j=1}^{n}a_{ij}(s,y)\,\psi_{y_iy_j}+\sum_{i=1}^{n}b_i(s,y)\,\psi_{y_i}+c(s,y)\,\psi.\end{gathered}$$

Basic references are Ladyzhenskaya-Solonnikov-Ural'seva [1], and Freidman [1]. Let us consider (E.1) in a cylindrical region $Q=(T_0,T)\times G$, with the boundary data

$$\text{(E.2)}\qquad \psi(s,y)=\Psi(s,y),\qquad (s,y)\in\partial^*Q=([T_0,T]\times\partial G)\cup(\{T\}\times G).$$

For Chap. VI.6 the two cases of interest are Q bounded and $Q=Q^0=(T_0,T)\times E^n$. We quote two kinds of results—first about solutions of (E.1) in terms of generalized partial derivatives locally integrable to some power λ, second about solutions in $C^{1,2}(Q)$.

For $1<\lambda<\infty$ let $\mathscr{H}^\lambda(Q)$ denote the space of functions ψ such that ψ together with all the generalized partial derivatives ψ_s, ψ_{y_i}, $\psi_{y_iy_j}$, $i,j=1,\dots,n$ are in $L^\lambda(Q)$. In $\mathscr{H}^\lambda(Q)$ let us introduce the norm (of Sobolev type)

$$\text{(E.3)}\qquad \|\psi\|^{(2)}_{\lambda,Q}=\|\psi\|_{\lambda,Q}+\|\psi_s\|_{\lambda,Q}+\sum_{i=1}^{n}\|\psi_{y_i}\|_{\lambda,Q}+\sum_{i,j=1}^{n}\|\psi_{y_iy_j}\|_{\lambda,Q}.$$

Let us assume that

(E.4) G is open, ∂G is a compact manifold of class C^2.

If $\lambda>\dfrac{n+2}{2}$, then every $\psi\in\mathscr{H}^\lambda(Q)$ is continuous on $\bar{Q}$. For $\lambda>n+2$, ψ_y is also continuous on $\bar{Q}$. These facts follow from Hölder estimates mentioned below. For Q bounded and $\lambda>\dfrac{n+2}{2}$, we have in the notation of V.11 and VI.8, $\mathscr{H}^\lambda(Q)=\mathscr{E}^{\lambda 0}(Q)$.

About the coefficients a_{ij}, b_i, c in (E.1), and Λ, Ψ let us assume:

(E.5) $a_{ij}(s,y)$ is bounded and Lipschitz on $\bar{Q}$, $i,j=1,\dots,n$; $a_{ij}=a_{ji}$; moreover there exists $\gamma>0$ such that

$$\sum_{i,j=1}^{n}a_{ij}(s,y)\,v_i\,v_j\geqq\gamma|v|^2\qquad\text{for all } v\in E^n.$$

(E.6) b_i, c are bounded, Borel measurable on Q, $i=1,\dots,n$; $\|\Lambda\|_{\lambda,Q}<\infty$.

(E.7) $\Psi(T,y)$ is of class $C^2(\bar{G})$; and $\Psi(s,y)=\tilde{\Psi}(s,y)$ for $T_0\leqq s\leqq T$, $y\in\partial G$, where $\tilde{\Psi}$ is of class $C^{1,2}(\bar{Q})$. Moreover,

$$\|\Psi\|^{(2)}_{\lambda,\partial^*Q}=\sum_{i=1}^{n}\|\Psi_{y_i}(T,\cdot)\|_{\lambda,G}+\sum_{i,j=1}^{n}\|\Psi_{y_iy_j}(T,\cdot)\|_{\lambda,G}+\|\tilde{\Psi}\|^{(2)}_{\lambda Q}$$

is finite.

Existence Theorem in $\mathscr{H}^\lambda(Q)$. Assumption (E.4) implies that (E.1) is a uniformly parabolic equation. The following result is contained in Ladyzhenskaya-Solonnikov-Ural'seva [1, Theorem 9.1, Chap. IV]. The problem (E.1)–(E.2) has a unique solution in $\mathscr{H}^\lambda(Q)$. Moreover, the estimate

$$\|\psi\|_{\lambda,Q}^{(2)} \leqq M(\|\Lambda\|_{\lambda,Q} + \|\Psi\|_{\lambda,\partial^* Q}^{(2)}) \tag{E.8}$$

holds, with constant M depending only on bounds for b_i, c on Q, if Q and the second order coefficients a_{ij} in (E.1) are regarded as given.

Local Estimates. For unbounded Q, the finiteness of $\|\psi\|_{\lambda,Q}^{(2)}$ imposes rather stringent global integrability conditions on ψ and its partial derivatives. However, local estimates can be derived from (E.8) if it is known that ψ, ψ_y are locally bounded. Suppose that ψ, ψ_y are bounded on $Q'' = (T_0, T) \times G''$, and that $\bar{G}' \subset G''$. Let α be C^2 with

$$\alpha(y) = 1 \quad \text{for } y \in G', \qquad \alpha(y) = 0 \quad \text{for } y \notin G''.$$

Then $\psi' = \alpha\psi$ satisfies

$$\psi'_s + \mathscr{A}(s)\psi' = -\Lambda' = -\alpha\Lambda + \sum a_{ij}(\alpha_{y_j}\psi_{y_i} + \alpha_{y_i}\psi_{y_j}) + \left(\sum a_{ij}\alpha_{y_i y_j} + \alpha_y b\right)\psi.$$

Let $Q' = (T_0, T) \times G'$. By (E.8), $\|\psi\|_{\lambda,Q'}^{(2)}$ has a bound depending on bounds for b_i, c, ψ, ψ_y on Q''.

Hölder Estimates. Let us also consider, for $0 < \mu \leqq 1$, norms of Hölder type as follows. Let

$$\|\psi\|_Q = \sup_{(s,y)\in Q} |\psi(s,y)|,$$

$$|\psi|_Q^\mu = \|\psi\|_Q + \sup_{\substack{x,y\in \bar{G} \\ T_0 \leqq s \leqq T}} \frac{|\psi(s,x) - \psi(s,y)|}{|x-y|^\mu} + \sup_{\substack{T_0 \leqq s,t \leqq T \\ y \in \bar{G}}} \frac{|\psi(s,y) - \psi(t,y)|}{|s-t|^{\mu/2}}.$$

Finiteness of $|\psi|_Q^\mu$ means that ψ is bounded and satisfies a Hölder condition; in particular, ψ is continuous on $\bar{Q}$. For $\lambda > \frac{n+2}{2}$, $\|\psi\|_{\lambda,Q}^{(2)} < \infty$ implies $|\psi|_Q^\mu < \infty$ for some $\mu > 0$ Ladyzhenskaya-Solonnikov-Ural'seva [1, p. 80].

We also let

$$|\psi|_Q^{1+\mu} = |\psi|_Q^\mu + \sum_{i=1}^n |\psi_{y_i}|_Q^\mu,$$

$$|\psi|_Q^{2+\mu} = |\psi|_Q^{1+\mu} + |\psi_s|_Q^\mu + \sum_{i,j=1}^n |\psi_{y_i y_j}|_Q^\mu.$$

The estimate (E.8) implies $|\psi|_Q^{1+\mu} < \infty$ for some $\mu > 0$, if λ is large enough. In fact, if $\lambda > n+2$

$$|\psi|_Q^{1+\mu} \leqq M_1 \|\psi\|_{\lambda,Q}^{(2)}, \qquad \mu = 1 - \frac{n+2}{\lambda}, \tag{E.9}$$

where the constant M_1 depends on Q and λ. See Ladyzhenskaya-Solonnikov-Ural'seva [1, p. 80, p. 342].

Under stronger assumptions on the coefficients of $\mathscr{A}(s)$ and Λ, in the form of Hölder conditions, there are Hölder estimates for all partial derivatives of ψ which appear in (E.1). In fact, suppose that Q', Q'' are bounded open subsets of Q with $\bar{Q}' \subset Q''$. Then

(E.10) $$|\psi|_{Q'}^{2+\mu} \leqq M_2(|\Lambda|_{Q''}^{\mu} + \|\psi\|_{Q''}),$$

where the constant M_2 depends on Q', Q'', and $|b_i|_{Q''}^{\mu}$, $i, j = 1, \ldots, n$, $|c|_{Q''}^{\mu}$. See Ladyzhenskaya-Solonnikov-Ural'seva [1, Theorem 10.1, Chap. IV]. If b_i, c, Λ satisfy a Hölder condition on Q, then we may take $Q'' = Q$ and Q' any open subset with compact closure $\bar{Q}'$ contained in Q. The solution ψ is then in $C^{1,2}(Q)$; in other words, ψ is a solution of (E.1) in the classical sense.

If ∂G, $\Psi(T, x)$, and $\tilde{\Psi}$ are of class C^3, then we may take any Q' with $\bar{Q}' \subset \bar{Q} - \{T\} \times \partial G$. This is also contained in Ladyzhenskaya-Solonnikov-Ural'seva [1, Theorem 10.1, Chap. IV].

A Maximum Principle. We also need a form of the maximum principle for linear second order parabolic equations. A standard result Friedman [1, p. 39] states that if $\psi \in C^{1,2}(Q)$, $\psi_s + \mathscr{A}(s)\psi \geqq 0$ in Q, ψ is continuous in $\bar{Q}$ and $\psi \leqq 0$ on $\partial^* Q$, then $\psi \leqq 0$ in Q. This result extends to $\psi \in \mathscr{H}^{\lambda}(Q)$ with $-\Lambda = \psi_s + \mathscr{A}(s)\psi$ in $L^{\lambda}(Q)$ as follows. For b_i, c satisfying (E.6) take sequences b_i^m, c^m, of C^1 functions $m = 1, 2, \ldots$, uniformly bounded and tending to b_i, c almost everywhere in Q. Denote the corresponding operators by $\mathscr{A}^m(s)$. Suppose that $\Lambda \leqq 0$ in Q and $\psi \leqq 0$ on $\partial^* Q$. Take $\Lambda^m \leqq 0$ of class C^1 tending to Λ in $L^{\lambda}(Q)$ as $m \to \infty$. The solution ψ^m of

$$\psi_s^m + \mathscr{A}^m(s)\psi^m + \Lambda^m = 0 \quad \text{in } Q,$$

$$\psi^m = \Psi \leqq 0 \quad \text{on } \partial^* Q$$

has $\|\psi^m\|_{\lambda,Q}^{(2)} \leqq M$ for some constant M, by (E.8). Moreover, ψ^m tends to ψ weakly in $L^{\lambda}(Q)$. Since $\psi^m \leqq 0$ in Q for each $m = 1, 2, \ldots$, we have $\psi \leqq 0$ almost everywhere in Q as required.

Let us now complete the proofs of Theorems VI.6.1 and VI.6.2. These were only sketched in Chap. VI.

Proof of Theorem VI.6.1. Eq. VI(6.8a) with $W^{m+1} = \Psi$ on $\partial^* Q$ has a unique solution in $\mathscr{H}^{\lambda}(Q)$. Note that $b^m = f^{\mathbf{u}^m}$, $\Lambda^m = L^{\mathbf{u}^m}$ are bounded on Q, with bound not depending on m since Q and U are bounded. We apply (E.8) with $\psi = W^{m+1}$, $c^m = 0$, to conclude that $\|W^{m+1}\|_{\lambda,Q}^{(2)}$ is bounded for any $\lambda > 1$. By VI(6.8b)

$$W_s^m + \mathscr{A}^{\mathbf{u}^m}(s) W^m + L^{\mathbf{u}^m} \leqq W_s^m + \mathscr{A}^{\mathbf{u}^{m-1}}(s) W^m + L^{\mathbf{u}^{m-1}} = 0.$$

By subtracting this from VI(6.8a) we get

$$(W^{m+1} - W^m)_s + \mathscr{A}^{\mathbf{u}^m}(s)(W^{m+1} - W^m) \geqq 0 \quad \text{in } Q$$

with $W^{m+1} - W^m = 0$ on $\partial^* Q$. By the form of the maximum principle above, $W^{m+1} - W^m \leqq 0$ in Q. The sequence $W^1, W^2, \ldots$ is nonincreasing. Let W be its limit as $j \to \infty$. Since $\|W^m\|_{\lambda,Q}^{(2)}$ is bounded for $1 < \lambda < \infty$, by (E.9) with $\lambda > n+2$, $\|W^n\|_Q^{1+\mu}$ is bounded for some $\mu > 0$. As $m \to \infty$, W^m, W_y^m converge to W, W_y uniformly on $\bar{Q}$, while W_s^m, $W_{y_i y_j}^m$ converge weakly in $L^{\lambda}(Q)$ to W_s, $W_{y_i y_j}$, $i, j = 1, \ldots, n$.

By VI(6.8 b)

$$W_s^m + \mathscr{A}^{\mathbf{u}}(s)\, W^m + L^{\mathbf{u}} \geqq W_s^m + \mathscr{A}^{\mathbf{u}^m}(s)\, W^m + L^{\mathbf{u}^m}$$
$$= (W^m - W^{m+1})_s + \mathscr{A}^{\mathbf{u}^m}(s)(W^m - W^{m+1})$$

for any admissible $\mathbf{u}$. As $m \to \infty$, the left side tends weakly in $L^\lambda(Q)$ to $W_s + \mathscr{A}^{\mathbf{u}}(s)\, W + L^{\mathbf{u}}$, and the right side weakly to 0. Hence

$$0 \leqq W_s + \mathscr{A}^{\mathbf{u}}(s)\, W + L^{\mathbf{u}},$$

almost everywhere in Q. Define $\mathbf{u}^*$ by Lemma VI.6.1. Then

$$W_s + \mathscr{A}^{\mathbf{u}^*} W + L^{\mathbf{u}^*} \leqq W_s + \mathscr{A}^{\mathbf{u}^m}(s)\, W + L^{\mathbf{u}^m}$$
$$= (W - W^{m+1})_s + \mathscr{A}^{\mathbf{u}^m}(s)(W - W^{m+1}).$$

Again, the right side tends weakly to 0 as $m \to \infty$.

We have shown that

$$0 = W_s + \mathscr{A}^{\mathbf{u}^*}(s)\, W + L^{\mathbf{u}^*} = W_s + \tfrac{1}{2} \sum_{i,j=1}^{n} a_{ij} W_{y_i y_j} + H(s, y, W_y)$$

almost everywhere in Q. Thus W is a solution of VI(6.6) in $\mathscr{H}^\lambda(Q)$. Since H is locally Lipschitz and $|W_y|_Q^\mu < \infty$ for some $\mu > 0$, $|H(s, y, W_y)|_Q^\mu < \infty$. By using (E.10), W is in $C^{1,2}(Q)$. □

Proof of Lemma VI.6.2. We recall from (4.4) that $W(s, y)$ is the infimum of $J(s, y, u)$ among nonanticipative controls in $\mathscr{U}(s)$. Let k be as in the polynomial growth condition (3.6). Since U is compact, (3.2) together with V(4.6) imply that $E_{sy}|\xi(t)|^k$ is bounded uniformly with respect to $s \in [T_0, T)$, $y \in B$, and $u \in \mathscr{U}(s)$. It follows from (3.6) that $J(s, y, u)$ is uniformly bounded. Hence $|W(s, y)| \leqq M_B$ for suitable constant M_B.

To show the corresponding estimate for W_y it suffices to show that

$$|W(s, y') - W(s, y)| \leqq M_B |y' - y|$$

for all $s \in [T_0, T)$ and $y, y' \in B$. This will follow if we show that, for all such s, y, y' and $u \in \mathscr{U}(s)$

$$(*) \qquad |J(s, y', u) - J(s, y, u)| \leqq M_B |y' - y|,$$

for suitable constant M_B. Given $u \in \mathscr{U}(s)$, let ξ, ξ' be the corresponding solutions of (3.1) with $\xi(s) = y$, $\xi'(s) = y'$. Then

$$\xi'(t) - \xi(t) = y' - y + \int_s^t [f(r, \xi'(r), u(r)) - f(r, \xi(r), u(r))]\, dr$$
$$= + \int_s^t [\sigma(r, \xi'(r)) - \sigma(r, \xi(r))]\, dw.$$

By standard estimates (Problem VI.4)

$$(**) \qquad E|\xi'(t) - \xi(t)|^2 \leqq K |y' - y|^2, \qquad s \leqq t \leqq T,$$

where the constant K depends only on the bounds $|\sigma_x| \leqq C$, $|f_x| \leqq C$ in (3.2) and on $T-T_0$. By the mean value theorem,

$$\Psi[\xi'(T)] - \Psi[\xi(T)] = \int_0^1 \Psi_x[\xi_\lambda(T)] \cdot [\xi'(T) - \xi(T)]\, d\lambda,$$

$$\xi_\lambda = \lambda \xi + (1-\lambda)\xi'.$$

Upon taking expectations and using Cauchy-Schwarz together with (**), we get

$$E|\Psi[\xi'(T)] - \Psi[\xi(T)]|^2 \leqq K|y'-y|^2 \int_0^1 E\Psi_x^2[\xi_\lambda(T)]\, d\lambda.$$

By (6.9d), Ψ_x satisfies for some constants C_1, l

$$|\Psi_x| \leqq C_1(1+|x|)^l.$$

Since $|\xi_\lambda| \leqq |\xi| + |\xi'|$, $E|\xi_\lambda(T)|^{2l}$ is bounded (from V(4.6)) by a constant depending only on the constants in (3.2), l, and a bound for $|y|$, $|y'|$. In this way, we get for $y, y' \in B$

$$E|\Psi[\xi'(T)] - \Psi[\xi(T)]| \leqq (E|\Psi[\xi'(T)] - \Psi[\xi(T)]|^2)^{1/2} \leqq M_{1B}|y'-y|$$

for some M_{1B}. In a similar way, by using the polynomial growth condition

$$|L_x(t, x, u)| \leqq C_1(1+|x|)^l$$

for some C_1, l, we get

$$E\int_s^T |L(t, \xi'(t), u(t)) - L(t, \xi(t), u(t))|\, dt \leqq M_{2B}|y'-y|.$$

We take $M_B = M_{1B} + M_{2B}$. □

Note. The constant M_B depends only on B, $T-T_0$, the constants C, k in (3.2), (3.6), and C_1, l in the proof above.

Proof of Theorem 6.2. First, suppose that f is bounded and L, Ψ have compact support. Then the same proof as for Theorem VI.6.1 gives the desired solution W of (6.6), with $\|W\|_{\lambda, Q^0}^{(2)} < \infty$. To eliminate these restrictions, for $l = 1, 2, \ldots$, let α_l be a C^∞ function with $\alpha_l \geqq 0$,

$$\alpha_l(y) = 1 \quad \text{for } |y| \leqq l, \qquad \alpha_l(y) = 0 \quad \text{for } |y| \geqq l+1,$$

and $|\alpha_{l y}| \leqq 2$. Let W_l be the solution of

$$(*) \qquad (W_l)_s + \tfrac{1}{2} \sum_{i,j=1}^n a_{ij}(W_l)_{y_i y_j} + \alpha_l(y) H(s, y, (W_l)_y) = 0,$$

$$W_l(T, y) = \alpha_l(y)\Psi(y).$$

Now (*) has the form (6.6), in which f, L are replaced by $\alpha_l f$, $\alpha_l L$. By Lemma VI.6.2 and the Note following it, W_l and $(W_l)_y$ are uniformly bounded on any compact subset of $\bar{Q}^0$. By the local estimates above, $\|W_l\|_{\lambda, Q'}^{(2)}$ is bounded for any bounded $Q' \subset Q$. By (E.9), $(W_l)_y$ satisfies a uniform Holder condition on each such Q'. Take $Q' = (T_0, T) \times (|y| < l_0)$. For $l \geqq l_0$, W_l is a solution of (6.6) in Q' with $W_l(T, y) = \Psi(y)$ for $|y| \leqq l_0$. Since $H(s, y, p)$ satisfies a local Lipschitz condition, we find by using

(E.10) as in the proof of Theorem VI.6.1 that $(W_l)_s$, $(W_l)_{y_i y_j}$ also satisfy a uniform Hölder condition on any compact subset of Q'. We now use Ascoli's theorem to find a subsequence of l such that W_l tends to a limit W, uniformly on each compact subset of $\bar{Q}^0$, while $(W_l)_s$, $(W_l)_{y_i}$, $(W_l)_{y_i y_j}$ tend to W_s, W_{y_i}, $W_{y_i y_j}$ uniformly on any compact subset of Q^0. The function W is the desired solution of (6.6) with the Cauchy data $W(T, x) = \Psi(x)$. By VI(3.6) and compactness of U, $|W_l(s, y)| \leqq M(1+|y|)^k$ for some M not depending on l. Hence, W satisfies the polynomial growth condition $|W(s, y)| \leqq M(1+|y|)^k$. Uniqueness of W was already pointed out in Remark 1 following Theorem VI.6.1. □

F. A General Position Lemma

In this Appendix we shall prove Lemma IV.7.1. First a special case of the general co-area formula of Federer [1] p. 426, [2] p. 248 will be stated.

Let X be an m-dimensional manifold and Y a k-dimensional manifold each contained in a Euclidean space, $m \geqq k$. Let

$$g: X \rightarrow Y$$

be a C^1 mapping. Let Dg be the $k \times m$ matrix of partial derivatives of g, computed using orthonormal bases for the tangent spaces to X, Y at x, $g(x)$ respectively. Define $Jg(x)$ to be the Jacobian of g at x, that is the square root of the sum of the squares of the determinants of the $k \times k$ submatrices of $Dg(x)$. Let H^n denote n-dimensional Hausdorff measure.

Theorem: (co-area formula) [1] [2]

$$\int_X Jg(x)\, dH^m = \int_Y H^{m-k}\{g^{-1}(y)\}\, dH^k. \tag{F.1}$$

We shall show that Lemma 7.1 follows from F.1 applied to the mapping taking $(t, x(t; t_0, y))$ into its initial condition y.

First we shall define more precisely this mapping. Since there is a solution of II(3.1) on $[t_0, t_1]$ with $x(t_0) = x_0$, theorems from differential equations on dependence of solutions on initial conditions imply there is a neighborhood N of x_0 in E^n such that for each $y \in N$ there is a solution $x(t, t_0, y)$ of II.3.1, with initial condition $x(t_0) = y$, defined on the entire interval $[t_0, t_1]$. Consider the mapping T

$$T: [t_0, t_1] \times N \rightarrow E^{n+1}$$

defined by

$$T(t, y) = (t, x(t; t_0, y)).$$

Uniqueness theorems for differential equations imply that T has an inverse T^{-1}

$$T^{-1}: T\{[t_0, t_1] \times N\} \rightarrow [t_0, t_1] \times N.$$

Let $\tau_0 = t_0$, $\tau_q = t_1$ and $\tau_1, \ldots, \tau_{q-1}$ denote the times of discontinuities of the control u on $[t_0, t_1]$. By taking the neighborhood N smaller, if necessary, we may assume N is bounded and from the continuity of $f(t, x, u)$ and of $f_x(t, x, u)$ that $f(t, x, u(t))$ and $f_x(t, x, u(t))$ will be bounded and continuous on each of the sets

$$[\tau_j, \tau_{j+1}] \times E^n \cap T\{[t_0, t_1] \times N\} \quad \text{for } j = 0, \ldots, q-1.$$

The matrix of partial derivatives of T is given by

$$\begin{pmatrix} 1 & f(t, x(t; t_0, y), u(t)) \\ 0 & x_y(t; t_0, y) \end{pmatrix} \tag{F.2}$$

where $x_y(t; t_0, y)$ is a continuous solution of

$$\dot{x}_y(t) = f_x(t, x(t; t_0, y), u(t))\, x_y(t)$$

with $x_y(t_0) = I$ where I is the identity matrix. This implies that on each of the sets $(\tau_j, \tau_{j+1}) \times N$, for $j = 0, \ldots, q-1$, T has bounded continuous partial derivatives and the determinant of the Jacobian matrix of T, (F.2), is bounded away from zero. Hence T^{-1} has bounded continuous partial derivatives on $T\{(\tau_j, \tau_{j+1}) \times N\}$.

Let π be the mapping taking E^{n+1} into E^n defined by

$$\pi(t, y) = y.$$

Define the mapping F taking the solution of II(3.1) into its initial conditions by defining F on $T\{[t_0, t_1] \times N\}$ by

$$F(t, x) = \pi(T^{-1}(t, x)).$$

We are now in a position to deduce Lemma 7.1 by an application of (F.1). On the each set $T\{(\tau_j, \tau_{j+1}) \times N\}$, $j = 0, \ldots, q-1$, $F(t, x)$ is C^1. Notice that $\mathscr{S} \cap T\{(\tau_j, \tau_{j+1}) \times N\}$ is an n-dimensional manifold. Apply (F.1) with

$$X = \mathscr{S} \cap T\{(\tau_j, \tau_{j+1}) \times N\},$$

$Y = N$, and $g = F$ to obtain

$$\int_X JF(t, x)\, dH^n = \int_N H^0\{F^{-1}(\{y\})\}\, dH^n. \tag{F.3}$$

Our previous reasoning has shown that $F(t, x)$ has bounded continuous partial derivatives on $T\{(\tau_j, \tau_{j+1}) \times N\}$. Hence $JF(t, x)$ is bounded. Since $\mathscr{S}$ is an n-dimensional manifold given by the zero set of a continuously differentiable function for which the gradient does not vanish the n-dimensional Hausdorff measure of $\mathscr{S}$ intersected with a bounded set is finite. Hence the right side of (F.3) is finite. This implies, for H^n almost every $y \in N$, that $H^0(F^{-1}(y))$ is finite. Hausdorff zero dimensional measure of a set is just the number of points in a set. For subsets of E^n Hausdorff n-dimensional measure and Lebesgue measure agree. Now

$$F^{-1}(y) = \{(t, x)\colon x(t;\, t_0, y) = x, \tau_j < t < \tau_{j+1}, (t, x) \in \mathscr{S}\}.$$

Hence for Lebesgue n-dimensional measure almost every y the set of points of the trajectory $(t, x(t; t_0, y))$ for which $\tau_j < t < \tau_{j+1}$ and $(t, x(t; t_0, y)) \in \mathscr{S}$ is finite or empty. Since this is true for each $j = 0, \ldots, q-1$ and there are only finitely many points of the from $(\tau_j, x(\tau_j; t_0, y))$ which are not of the type mentioned above we conclude that for almost every $y \in N$ the set of points of the form $(t, x(t; t_0, y))$ with $t_0 \leqq t \leqq t_1$ which belong to $\mathscr{S}$ is finite or empty.

Bibliography

Athans, M. ed.: [1] Special issue on the linear-quadratic-gaussian estimation and control problem. IEEE Transactions on Automatic Control. **AC-16**, 527–847 (1971).

[2] Status of optimal control theory and applications for deterministic systems. IEEE Transactions on Automatic Control. **AC-11**, 580–596 (1966).

Athans, M., Falb, P. L.: [1] Optimal control. McGraw Hill, 1966.

Astrom, K.: [1] Introduction to stochastic control theory. Academic Press, 1970.

Auslander, L.: [1] Differential geometry. New York: Harper and Row, 1967.

Balakrishnan, A. V. ed.: [1] Control theory and calculus of variations. New York: Academic Press, 1967.

[2] Stochastic differential systems. Lecture Notes in Economics and Mathematical Systems, **84**, Springer-Verlag, 1973.

[3] Stochastic control: a function space approach. SIAM J. on Control, **10**, 285–297 (1972).

[4] Stochastic optimization theory in Hilbert spaces I. Internat. J. Applied Math. Optim. **1** (1974).

[5] A note on the structure of optimal stochastic controls. Internat. J. Appl. Math. and Optimiz. **1**, 87–94 (1974).

[6] Techniques of optimization. New York: Academic Press, 1971.

Balakrishnan, A. V., Neustadt, L.: [1] Computing methods in optimization problems. New York: Academic Press, 1964.

[2] Mathematical theory of control. New York: Academic Press, 1967.

Aoki, M.: [1] Optimization of stochastic systems. New York: Academic Press, 1967.

Bather, J. A.: [1] A continuous time inventory model. J. Appl. Prob. **3**, 538–549 (1966).

[2] A diffusion model for the control of a dam. J. Appl. Prob. **5**, 55–71 (1968).

Bellman, R.: [1] Adaptive control processes. A guided tour. Princeton, N. J.: Princeton University Press, 1961.

[2] Dynamic programming. Princeton, N. J.: Princeton University Press, 1957.

Bellman, R., Kalaba, R. E.: [1] Quasilinearization and nonlinear boundary-value problems. American Elsevier, 1965.

Beltrami, E. J.: [1] An algorithmic approach to nonlinear analysis and optimization. Math. in Sci. and Engr. **63**, Academic Press, 1970.

Benes, V. E.: [1] Existence of optimal stochastic control laws. SIAM J. on Control. **9**, 446–472 (1971).

[2] A homotopy method for proving convexity in certain optimal stochastic control problems, Proc. Internat. Symp. on Control, Numerical Methods, and Computer Systems Modelling, IRIA June 1974, Springer Lecture Notes in Econ. and Math. Systems **107** (1975).

Bensoussan, A.: [1] Filtrage optimal des systemes lineaires. Paris: Dunod, 1971.

Bensoussan, A., Lions, J. L.: [1] Nouvelles methodes en controle impulsionnel. Appl. Math. Optimiz.

[2] Optimal impulse and continuous control: Method of nonlinear variational inequalities.

[3] Problems de temp d'arret optimaux et de perturbations singulieres dans les inequations variationelles. Proc. Internat. Symp. on Control, Numerical Methods and Computer Systems Modelling, IRIA, June, 1974. Springer Lecture Notes in Econ. and Math. Systems **107** (1975).

Bensoussan, A., Viot, M.: [1] Optimal control of stochastic linear distributed parameter systems. SIAM J. on Control. **13**, 904–926 (1975).

Berkovitz, L. D.: [1] A Hamilton-Jocobi theory for a class of control problems. Colloque sur la theorie mathematique du control optimal, held Bruxelles, April 1969, Centre Belge de Recherches. Mathematiques, Vander Louvain, 1970.

[2] Necessary conditions for optimal strategies in a class of differential games and control problems. SIAM Journal on Control, **5**, 1–24 (1967).

[3] Variational methods in problems of control programming. Journal of Mathematical Analysis and Applications, **3**, 145–199 (1961).

[4] Optimal control theory. Applied Math. Sci. **12**, Springer-Verlag, 1974.

Billingsley, P.: [1] Convergence of probability measures. New York: Wiley, 1968.

Bismut, J. M.: [1] Analyse convexe et probabilities, These de Doctorat d'Etat. Univ. Paris VI, 1973.

[2] Conjugate convex functions in optimal stochastic control. J. Math. Anal. **44**, 384–404 (1973).

[3] Theorie du potentiel et controle des diffusions markoviennes. Proc. Internat. Symp. on Control, Numerical Methods and Computer Systems Modelling, IRIA, June 1974. Springer Lecture Notes in Econ. and Math. Systems. 107 (1975).

Bliss, G. A.: [1] Calculus of variations. Carus Math. Monographs No. 1, Math. Assoc. Amer. 1925.

[2] Lectures on the calculus of variations. University of Chicago Press, 1946.

Blumenthal, R. M., Getoor, R. K.: [1] Markov processes and potential theory. New York: Academic Press, 1968.

Boel, R.: [1] Optimal control of jump processes. Ph.D. Thesis, Dept. of Electrical Engr. and Computer Sci. Univ. of California, Berkeley, 1974.

Boltyanskii, V. G.: [1] Sufficient conditions for optimality and justification of the dynamic programming method. SIAM J. on Control **4**, 326–361 (1966).

[2] Mathematical methods of optimal control. New York: Holt, Rinehart, Winston, 1971.

Breakwell, J. V., Speyer, J. L., Bryson, A. E.: [1] Optimization and control of nonlinear systems using the second variation. SIAM J. Control **1**, 193–223 (1963).

Breiman, L.: [1] Probability. Addison-Wesley, Reading, Mass. 1968.

Brockett, R. W.: [1] Finite dimensional linear systems. Wiley, 1970.

Bryson, A. E., Denham, W. F.: [1] A steepest-ascent method for solving optimum programming problems. Journal of Applied Mechanics ser E **29**, 247–257 (1962).

Bryson, A. E., Ho, Y.-C.: [1] Applied optimal control. Blaisdell, 1969.

Bucy, R. S., Joseph, P. D.: [1] Filtering for stochastic processes with applications to guidance. New York: Interscience, 1968.

Burmeister, E., Dobell, A. R.: [1] Mathematical theories of economic growth. MacMillan, 1970.

Cainiello, E. R.: [1] Functional analysis and optimization. Academic Press, 1966.

Cannon, M. D., Cullum, C. D., Polak, E.: [1] Theory of optimal control and mathematical programming. McGraw-Hill, 1970.

Caratheodory, C.: [1] Calculus of variations and partial differential equations of first order. Translated by R. B. Dean and J. J. Brand Statten, Holden Day, 1965.

Cesari, L.: [1] Existence theorems for weak and usual optimal solutions in Lagrange problems with unilateral constraints I. Trans. Amer. Math. Soc. **124**, 369–412 (1966).

Chernoff, H.: [1] Optimal stochastic control. Sankhya, Ser. A **30**, 221–252 (1968).

Crow, J., Kimura, M.: [1] An introduction to population genetics theory. New York: Harper and Row, 1970.

Courant, R., Hilbert, D.: [1] Methods of mathematical physics, vol. II. Interscience, 1962.

Davis, M. H. A.: [1] On the existence of optimal policies in stochastic control. SIAM J. on Control **11**, 587–594 (1973).

Davis, M. H. A., Varaiya, P. P.: [1] Information states in linear stochastic systems. J. Math. Anal. Appl. **37**, 384–402 (1972).

[2] Dynamic programming conditions for partially observable stochastic systems. SIAM J. on Control **11**, 226–261 (1973).

Donsker, M. D.: [1] An invariance principle for certain probability limit theorems. Memoirs Amer. Math. Soc. **6** (1951).

Doob, J. L.: [1] Stochastic processes. New York: Wiley, 1953.

Dorato, P., Hsieh, C.-M., Robinson, Prentiss, N.: [1] Optimal bang-bang control of linear stochastic systems with a small noise parameter. IEEE Trans. Automatic Control **AC-12**, 682–689 (1967).

Dorato, P., Van Melaert, L.: [1] Numerical solution of an optimal control problem with a probability criterion. IEEE Trans. Auto. Control **17**, 543–546 (1972).

Duncan, T., Varaiya, P. P.: [1] On the solutions of a stochastic control system. SIAM J. on Control **9**, 354–371 (1971), Part II Memo. ERL-M406, Berkeley, 1973.

Dyer, P., McReynolds, S. R.: [1] The computation and theory of optimal control. Academic Press, 1970.

Dynkin, E.B.: [1] Theory of Markov processes. Englewood Cliffs, N.J.: Prentice-Hall, 1961.
[2] Markov processes. Berlin: Springer, 1965 [English transl.].
Eggleston, H.G.: [1] Convexity. Cambridge, England, 1950.
Falb, P.L., DeJong, J.L.: [1] Some successive approximation methods in control and oscillation theory. New York: Academic Press, 1969.
Federer, H.: [1] Curvature measures. Transactions of the American Mathematical Society **93**, 418–491 (1959).
[2] Geometric measure theory. Springer-Verlag, 1969.
Feller, W.: [1] An introduction to probability theory and its applications, vol. II. New York: Wiley, 1966.
Filippov, A.F.: [1] On certain questions in the theory of optimal control, Vestnik Moskov. Univ. Ser. Math. Mech. Astronom. **2**, 25–32 (1959) = SIAM J. Control **1**, 76–84 (1962).
Fleming, W.H.: [1] Functions of several variables. Wesley: Addison, 1965.
[2] Some markovian optimization problems. J. Math. and Mech. **12**, 131–140 (1963).
[3] The Cauchy problem for degenerate parabolic equations. J. Math. Mech. **13**, 987–1008 (1964).
[4] Duality and a priori estimates in Markovian optimization problems. J. Math. Anal. Appl. **16**, 254–279 (1966); Erratum, Ibid **19**, 204 (1966).
[5] Stochastic Lagrange multipliers, in Math. Theory of Control (A.V. Balakrishnan and L.W. Neustadt eds.), pp. 433–440. New York: Academic Press, 1967.
[6] Optimal control of partially observable diffusions. J. SIAM Control **6**, 194–214 (1968).
[7] The Cauchy problem for a nonlinear first-order partial differential equation. J. Differential Equations, **5**, 515–530 (1969).
[8] Controlled diffusions under polynomial growth conditions, in Control Theory and the Calculus of Variations (ed. A.V. Balakrishnan), pp. 209–234, Academic Press, 1969.
[9] Optimal continuous-parameter stochastic control. SIAM Review **11**, 470–509 (1969).
[10] Stochastic control for small noise intensities. SIAM J. on Control, **9**, 473–517 (1971).
[11] Stochastically perturbed dynamical systems. Rocky Mountain Math. J. **4**, 407–433 (1974).
Fleming, W.H., Nisio, M.: [1] On the existence of optimal stochastic controls. J. Math. and Mechanics, **15**, 777–794 (1966).
Florentin, J.J.: [1] Optimal control of continuous time Markov stochastic systems. J. Electron. Control **10**, 473–488 (1961).
[2] Partial observability and optimal control. J. Electron. Control **13**, 263–279 (1962).
Freidlin, M.I.: [1] A priori estimates of solutions of degenerating elliptic equations. Dokl. Akad. Nauk SSSR **158**, 281–283 (1964); Soviet Math. Doklady **5**, 1231–1234.
[2] On the factorization of nonegative matrices. Theory Probability Appl. **13**, 354–358 (1968).
Friedman, A.: [1] Partial differential equations of parabolic type. Englewood Cliffs: Prentice Hall, 1964.
[2] Differential games. New York: Wiley, 1971.
[3] Stochastic differential equations. New York: Academic Press, 1975.
Fujisaki, M., Kallianpur, G., Kunita, H.: [1] Stochastic differential equations for the non linear filtering problem. Osaka Journal of Mathematics, **9**, 19–40 (1972).
Gabasov, R.: [1] One problem in the theory of optimal processes. Automation and Remote Control **28**, 1085–1094 (1967).
Gabasov, R., Kirillova, F.M.: [1] High order necessary conditions for optimality. SIAM Journal on Control **10**, 127–168 (1972).
Gikhman, I.I., Skorohod, A.V.: [1] Introduction to the theory of random processes. Philadelphia: Saunders, 1969.
[2] Stochastic differential equations. Springer-Verlag, 1972.
Girsanov, I.V.: [1] On transforming a certain class of stochastic processes by absolutely continuous substitution of measures. Theory of Prob. and its Appl. **5**, 285–301 (1960).
[2] Lectures on the mathematical theory of extremum problems. Lecture Notes in Economics and Mathematical Systems, No. 67, Springer-Verlag, 1972.
Goh, B.S.: [1] Necessary conditions for singular extremals involving multiple control variables. SIAM Journal on Control **4** 716–731 (1966).
[2] The second variation for the singular Bolza problem. SIAM Journal on Control **4**, 309–325 (1966).
Goldstein, S.: [1] Classical mechanics. Addison Wesley, 1959.

Graves, L.M.: [1] The theory of functions of real variables. McGraw-Hill, 1946.

Gurley, J.G.: [1] Optimal-thrust trajectories in an arbitrary gravitational field. SIAM Journal on Control, Vol. **2**, 423–432 (1965).

Halkin, H.: [1] A generalization of La Salle's bang-bang principle. SIAM J. Control **2**, 199–202 (1965).
[2] Optimal control as programming in infinite dimensional spaces. Centro Internazionale Mathematico Estivo (C.I.M.E.) Bressanone, Italy, 1966.

Halkin, H., Hendricks, E.C.: [1] Subintegrals of set-valued functions with semianalytic graphs. Proc. Nat. Acad. Sci. USA, **59**, 365–367 (1968).

Hausmann, U.G.: [1] Stability of linear systems with control dependent noise. SIAM J. on Control **11**, 382–394 (1973).

Haynes, G.W.: [1] On the optimality of a totally singular vector control: An extension of the Green's theorem approach to higher dimensions **4**, 662–685 (1966).

Hermes, H., La Salle, J.P.: [1] Functional analysis and time optimal control. Math., in Sci. and Engr. vol. 56, Academic Press, 1969.

Hestenes, M.R.: [1] Calculus of variations and optimal control theory. New York: Wiley, 1966.

Ho, Y.C., Chu, K.C.: [1] Information structure in dynamic multi-person control problems, Automatica, July 1974.

Holland, C.: [1] A numerical technique for small noise stochastic control problems. J. Optimization Theory Appl. **13**, 74–93 (1974).
[2] Small noise open loop control problems. SIAM J. Control **12** (1974).
[3] Gaussian open loop control problems. SIAM J. Control **13** (1975).

Hormander, L.: [1] Hypoelliptic second order differential equations. Acta Math. **119**, 147–171 (1968).

Hsu, J.C., Meyer, A.U.: [1] Modern control principles and applications. McGraw-Hill, 1968.

Hurewicz, W., Wallman, H.: [1] Dimension theory. Princeton: Princeton University Press, 1948.

Iglehart, D.L.: [1] Diffusion approximations in applied probability. Lectures in Applied Math. vol. 12, Math. in the Decision Sciences Part 2, 235–254, American Math. Soc., Providence, 1967.
[2] Diffusion approximations of models for certain congestion problems. J. Appl. Prob. **5**, 607–623 (1968).

Ito, K.: [1] Stochastic differentials. Internat. J. Appl. Math. Optimiz **1** (1974).

Jacobson, D.H.: [1] A new necessary condition of optimality for singular control problems. SIAM Journal on Control **7**, 578–595 (1969).

Jacobson, D.H., Speyer, J.L.: [1] Necessary and sufficient conditions for optimality for singular control problems: A limit approach. SIAM Journal on Control **8**, 403–423 (1970).

John, F.: [1] Partial differential equations. Springer-Verlag, 1971.

Johnson, C.D.: [1] Singular solutions in optimal control problems, in Advances in Control Systems, vol. 2, CT Leondes ed. New York: Academic Press, 1965.

Johnson, C.D., Gibson, J.E.: [1] Singular solutions in problems of optimal control. IEEE Transactions on Automatic Control, AC-8, 4–15 (1963).

Joseph, P.D., Tou, J.T.: [1] On linear control theory. AIEE Trans. Appl. and Ind., Part II, **80**, 193–196 (1961).

Kailath, T.: [1] An innovations approach to least-squares estimation. Part I: Linear filtering in additive white noise. IEEE Trans. Automatic Control, **AC-13**, 646–655 (1968).

Kailath, T., Frost, P.: [1] An innovations approach to least-squares estimation. Part II: Linear smoothing in additive white noise. IEEE Trans. Automatic Control, **AC-13**, 655–660 (1968).

Kalman, R.E.: [1] A new approach to linear filtering and prediction problems. ASME Transactions Part D (J. of Basic Engr.) **82**, 35–45 (1960).

Kalman, R.E., Bucy, R.S.: [1] New results in linear filtering and prediction theory. ASME Transactions **83**, Part D (J. of Basic Engr.) 95–108 (1961).

Kelley, H.J.: [1] Method of gradients, in Optimization Techniques, G. Leitman ed. Academic Press, 1962.
[2] Singular extremals in Lawden's problem of optimal rocket flight. J. AIAA **1**, 1578–1580 (1963).
[3] A transformation approach to singular subarcs in optimal trajectory and control problems. SIAM Journal on Control **2**, 234–246 (1964).

Kelley, H.J., Kopp, R.E., Moyer, H.G.: [1] "Singular extremals", in Topics in optimization, G. Leitman ed., pp. 63–101. New York: Academic Press, 1967.

Khasminskii, R.Z., Nevelson: [1] Stochastic approximation and recurrent estimation. (English translation. American Math. Soc. to appear.)

Kohn, J.J., Nirenberg, L.: [1] Degenerate elliptic-parabolic equations of second order. Comm. Pure Appl. Math. **20**, 797–872 (1967).

Kopp, R.E., Moyer, H.G.: [1] Necessary conditions for singular extremals. AIAA J. **3**, 1439–1444 (1965).

Krylov, N.V.: [1] On quasi diffusion processes. Theory of Prob. and its Appl. **11**, 373–389 (1966).

Krylov, N.V.: [2] On the uniqueness of the solution of Bellman's equation. Math. USSR Izvestija, **35**, 1387–1398 (1971).

[3] The control of the solution of a stochastic integral equation. Theory Probability Appl. **17** (1972).

[4] On the control of diffusion type processes, 2nd Japan-USSR Symp. Probab. Th. 1972.

[5] Lectures on the theory of elliptic differential equations. Izdat. Moskovskovo Univ. 1972 No. 34 (In Russian).

Kuo, H.H.: [1] On the stochastic maximum principle in Banach spaces, J. Funct. Anal. **14**, 146–161 (1973).

Kushner, H.J.: [1] Stochastic stability and control. New York: Academic Press, 1967.

[2] Introduction to stochastic control theory. New York: Holt, Rinehart, Winston, 1971.

[3] On the stochastic maximum principle: fixed time of control. J. Math. Anal. Appl. **11**, 78–92 (1965).

[4] Optimal discounted stochastic control for diffusion processes. SIAM J. Control **5**, 520–531 (1967).

[5] The Cauchy problem for a class of degenerate parabolic equations and asymptotic properties of the related diffusion process. J. Diff. Eq. **6**, 209–231 (1969).

[6] Necessary conditions for continuous parameter stochastic optimization problems. SIAM J. on Control **10**, 550–565 (1972).

[7] On the weak convergence of interpolated Markov chains to a diffusion. Annals of Probability **2**, 40–50 (1974).

[8] Existence results for optimal stochastic controls. J. Optimiz. Theory Appl. **15**, 347–360 (1975).

Kuznetzov, N.N., Siskin, A.A.: [1] On a many dimensional problem in the theory of quasilinear equations. Z. Vycisl. Mat. i. Mat. Fiz. **4** (1964), No. 4, Suppl. 192–205 (in Russian).

Ladyshenskaya, O.A., Solonnikov, V.A., Ural'seva, N.N.: [1] Linear and quasilinear equations of parabolic type. American Math. Soc., 1968.

Ladyzhenskaya, O.A., Uralseva, N.N.: [1] Linear and quasilinear elliptic equations. New York: Academic Press, 1968.

Lawden, D.F.: [1] Optimal intermediate-thrust arcs in a gravitional field. Astronautica Acta, **8**, 106–123 (1962).

[2] Optimal trajectories for space navigation. Washington: Butterworths, 1963.

Lee, E.B., Markus, L.: [1] Foundations of optimal control theory. New York: Wiley, 1967.

Lee, F.: [1] Dynamic programming and optimal stochastic control. Brown University Ph.D. Thesis, 1972.

Leondes, C.T., ed.: [1] Advances in control systems, v. 1–10. New York: Academic Press.

Levinson, N.: [1] Minimax, Liapunov, and "bang-bang". J. Differential Equs. **2**, 218–241 (1966).

Leitman, G.: [1] Optimization techniques, with applications to aerospace systems. New York: Academic Press, 1965.

[2] Topics in optimization. Academic Press, 1967.

Lions, J.L.: [1] Optimal control of systems governed by partial differential equations. Translated by S.K. Mitter, Springer-Verlag, 1971.

Lindquist, A.: [1] Optimal control of linear stochastic systems with applications to time lag systems. Infor. Sci. **5**, 81–126 (1973).

[2] On feedback control of linear stochastic systems. SIAM J. On Control **11**, 323–343 (1973).

Liptser, R.Sh., Shiryaev, A.N.: [1] Statistics of random processes. Izdat, "Nauka", Moscow, 1974, (in Russian) English translation, Springer-Verlag, to appear.

Loeve, M.: [1] Probability theory, 3rd ed. Princeton, N.J.: D. Van Nostrand Co., 1963.

McDanell, J., Powers, W.: [1] Necessary conditions for joining optimal singular and nonsingular subarcs. SIAM Journal on Control **9**, 161–173 (1971).

[2] New Jacobi-type necessary and sufficient conditions for singular optimization problems. AIAA J. **8**, 1416–1420 (1970).

McGill, R.: [1] Optimal control, inequality state constraints and the generalized Newton-Raphson Algon. SIAM Journal on Control **3**, 291–298 (1965).

McShane, E.J.: [1] Relaxed controls and variational problems. SIAM J. Control **5**, 438–485 (1967).

[2] Towards stochastic calculus, I, II. Proc. Nat. Acad. Sci. **63**, 275–280, 1084–1085 (1969).

[3] Stochastic integrals and stochastic functional equations. SIAM J. Appl. Math. **17**, 287–306 (1969).

[4] Stochastic differential equations and models of random processes. Proc. 6th Berkeley Symp. on Math. Stat. and Probability **3**, 263–294 (1970).

[5] Integration. Princeton, N.J.: Princeton Univ. Press, 1944.

[6] On multipliers for Lagrange problems. American Journal of Mathematics, **61**, 809–819 (1939).

[7] Stochastic calculus and stochastic models. New York, Academic Press, 1974.

Malinvaud, E.: [1] First order certainty equivalence. Econometrica **37**, 706–718 (1969).

Mancill, J.D.: [1] Identically non regular problems in the calculus of variations. Matematica y Fisica Teorica ser A **7**, 131–139 (1950).

Mandl, P.: [1] Analytical treatment of one-dimensional Markov processes. Die Grundlehren der math. Wissenschaften, Band 151, Springer-Verlag, 1968.

[2] Analytical methods in the theory of controlled Markov processes. Trans. Fourth Prague Conf. on Info. theory, Stat. Decision Ins., Random Processes (Prague, 1965, pp. 45–53). Academia, Prague, 1967.

[3] On the control of non-terminating diffusion processes. Theory Probability Appl. **9**, 591–603 (1964). [English transl.]

[4] On the control of the Wiener process with restricted number of switchings. Theory Probability Appl. **12**, 68–76 (1967) [English translation].

Meditch, J.S.: [1] On the problem of optimal thrust programming for a lunar soft landing. IEEE Transactions on Automatic Control, **AC-9**, 477–484 (1964).

[2] Stochastic optimal linear estimation and control. New York: McGraw Hill, 1969.

Merton, R.C.: [1] Optimal consumption and portfolio rules in a continuous-time model. J. Economic Theory **3**, 373–413 (1971).

Meyer, P.A.: [1] Probability and potentials. Waltham, Mass: Blaisdell, 1966.

Miele, A.: [1] The calculus of variations in applied aerodynamics and flight mechanics. In Optimization techniques, with applications to aerospace systems ed. G. Leitman. New York: Academic Press, 1960.

[2] Extremization of linear integrals by Green's theorem, 69–98 in Optimization techniques, with applications to aerospace systems (G. Leitmann ed.). Math. in Sci. and Engr. **5**, Academic Press, 1962.

Mieri, A.Z.: [1] A new approach to the general problem of optimal filtering and control of stochastic systems. Univ. of Calif., Berkeley Thesis, 1967.

Milnor, J.: [1] Morse theory. Princeton, N.J.: Princeton Univ. Press, 1963.

Morrey, C.B.: [1] Multiple integrals in the calculus of variations. Grundlehren Math. Wissenschaften. **130**, Springer-Verlag, 1966.

Morse, M.: [1] Introduction to analysis in the large, 1947 Lectures, Univ. Microfilms Inc, Ann Arbor, Mich., 1947.

Mortensen, R.E.: [1] Stochastic optimal control with noisy observations. Int. J. Control **4**, 455–464 (1966).

Morton, R.: [1] On the optimal control of stationary diffusion processes with inaccessible boundaries and no discounting. J. Appl. Probability **8**, 551–560 (1971).

Nelson, E.: [1] Dynamical theories of Brownian motion, Mathematical Notes. Princeton University Press, 1967.

Neustadt, L.W.: [1] An abstract variational theory with applications to a broad class of optimization problems I: General theory. SIAM Journal on Control **4** (1966).

[2] An abstract variational theory with applications to a broad class of optimization problems II: Applications. SIAM Journal on Control **5** (1967).

[3] Optimization: A theory of necessary conditions. Princeton University Press, Princeton, New Jersey, 1975.

Newell, G.F.: [1] Applications of queueing theory. London: Chapman and Hall, 1971.

[2] Approximate stochastic behavior of n-server service systems with large n. New York-Berlin: Springer-Verlag, 1973.

Nisio, M.: [1] On stochastic optimal control laws. Nagoya Math. J. **52**, 1–30 (1973).
Olech, C.: [1] Extremal solutions of a control system. J. Differential Equs. **2**, 74–101 (1966).
Oleinik, O. A.: [1] On linear equations of the second order with a non-negative characteristic form. Math. Sbornik **69**, 111–140 (1966).
Ornstein, L. S., Uhlenbeck, G. E.: [1] On the theory of Brownian motion. Physical Review **36**, 823–841 (1930). Reprinted in: Selected papers on noise and stochastic processes, ed. N. Wax. Dover, New York, 1954.
Phillips, R. S., Sarason, L.: [1] Elliptic-parabolic equations on the second order. J. Math. Mech. **17**, 891–918 (1968).
Pindyck, R. S.: [1] Optimal planning for economic stabilization. Amsterdam: North-Holland Publ. Co., 1973.
Pliska, S. R.: [1] Multiperson controlled diffusions. SIAM Journal on Control **11**, 563–586 (1973).
Polak, E.: [1] An historical survey of computational methods in optimal control. SIAM Review **15**, 553–584 (1973).
[2] Computational methods in optimization. New York, Academic Press, 1971.
Pontryagin, L. S.: [1] Optimal regulation processes. American Mathematical Society Translations, Series 2, **18**, 321–339 (1961).
[2] Boltyanskii, V. G., Gamkreledze, R. V., Mischenko, E. F.: The mathematical theory of optimal processes. Interscience, 1962.
Puterman, M. L.: [1] On the optimal control of diffusion processes. Stanford Univ. Operations Research TR 24, 1972.
Rado, T.: [1] On the problem of Plateau. New York: Chelsea, 1951.
Rishel, R. W.: [1] An extended Pontryagin principle for control systems whose control laws contain measures. SIAM J. on Control **2**, 191–205 (1965).
[2] Necessary and sufficient dynamic programming conditions for continuous-time stochastic optimal control. SIAM J. Control **8**, 559–571 (1970).
[3] Weak solutions of the partial differential equation of dynamic programming. SIAM J. on Control **9**, 519–528 (1971).
[4] Control of systems with jump Markov disturbances, IEEE Trans. Automatic Control, 241–244, AC-20 (1975).
[5] Dynamic programming and minimum principles for systems with jump Markov disturbances. SIAM J. on Control **13**, 338–371 (1975).
[6] A minimum principle for controlled jump processes, Proc. Internat. Symp. on Control, Numerical Methods and Computer Systems Modelling, IRIA, June 1974. Springer Lecture Notes in Econ. and Math. Systems **107** (1975).
Robbins, H. M.: [1] Optimality of intermediate thrust arcs of rocket trajectories. AIAA Journal **3**, 1094–1098 (1965).
Schach, S.: [1] Weak convergence results for a class of multivariate Markov processes. Annals of Math. Stat. **42**, 451–465 (1971).
Shell, K. ed.: [1] Essays on the theory of optimal economic growth. Cambridge, Mass: MIT Press, 1967.
Stone, L. D.: [1] Necessary and sufficient conditions for optimal control of semi-Markov jump processes. SIAM J. on Control **11**, 187–201 (1973).
Stratonovich, R. L.: [1] Topics in the theory of random noise, **1**, **2**. New York: Gordon and Breach, 1963.
[2] A new representation for stochastic integrals and equations. SIAM J. on Control **4**, 362–371 (1966).
[3] Conditional Markov processes and their application to the theory of optimal control. New York: American Elsevier, 1968.
Striebel, C.: [1] Sufficient statistics in the optimum control of stochastic systems. J. Math. Anal. Appl. **12**, 576–592 (1965).
Stroock, D. W., Varadhan, S. R. S.: [1] Diffusion processes with continuous coefficients. Comm. Pure Appl. Math. **20**, 345–400, 479–530 (1969).
Sworder, D. D.: [1] Feedback control of a class of linear systems with jump parameters. IEEE Trans. Autom. Control. **AC-14**, 9–14 (1969).
Tsai, C. P.: [1] Perturbed stochastic dynamical systems. Ph.D. Thesis, Brown University, 1974.

Valentine, F. A.: [1] Convex sets. New York: McGraw-Hill, 1964.

[2] The problem of Lagrange with differential inequalities as side conditions. In Contributions to Calculus of Variations, 1933–1937, pp. 407–448. Chicago: University of Chicago Press, 1937.

Varaiya, P. P.: [1] Optimal control of a partially observed stochastic system, in Stochastic Differential Equations. SIAM-AMS Proc. **6**, 173–188 (1973).

Warfield, V.: [1] A stochastic maximum principle, Ph. D. Thesis, Brown University. Providence, RI, 1971.

Warga, J.: [1] Optimal control of differential and functional equations. New York: Academic Press, 1972.

Witsenhausen, H. S.: [1] A counterexample in stochastic optimum control. SIAM J. Control, **6** 131–132 (1968).

[2] Separation of estimation and control for discrete time systems. Proc. IEEE **59**, 1557–1566 (1971).

[3] A standard form for sequential stochastic control. Math. Systems Th. **7**, 5–11 (1973).

Wong, E.: [1] Stochastic processes in information and dynamical systems. New York: McGraw-Hill, 1971.

[2] Representation of martingales, quadratic variation and applications. SIAM J. on Control **9**, 621–633 (1971).

[3] Martingale theory and applications to stochastic problems in dynamical systems. Publication 72/19, Imperial College Dept. of Computing and Control, 1972.

Wong, E., Zakai, M.: [1] On the convergence of ordinary integrals to stochastic integrals. Ann. Math. Stat. **36**, 1560–1564 (1965).

[2] On the relation between ordinary and stochastic differential equations. Internat. J. Engr. Sci. **3**, 213–229 (1965).

Wonham, W. M.: [1] A Liapunov method for the estimation of statistical averages. J. Diff. Equs. **2**, 365–377 (1966).

[2] Optimal stationary control of a linear system with state-dependent noise. SIAM J. Control **5**, 486–500 (1967).

[3] On the separation theorem of stochastic control. SIAM J. Control **6**, 312–326 (1968).

[4] Random differential equations in control theory, in A. T. Bharucha-Reid, ed., Probabilistic methods in applied math., vol. II. New York: Academic Press, 1969.

Yamada, T., Watanabe, S.: [1] On the uniqueness of solutions to stochastic differential equations. J. Math. Kyoto Univ. **11**, 155–167 (1971).

Yasuda, M.: [1] On the existence of optimal control in continuous time Markov decision processes. Bull. Math. Stat., Res. Assoc. of Stat. Sci., **15**, 7–17 (1972).

Young, L. C.: [1] Lectures on calculus of variations and optimal control theory. Philadelphia: Saunders, 1969.

Zadeh, L. A., Neustadt, L. W., Balakrishnan, A. V. (eds.): [1] Computing methods in optimization Problems. New York: Academic Press, 1969.

Index

Abnormal problems 28
Adjoint equations 27, 37
Admissible
– feedback control 90, 156
– nonanticipative control 162
– set of discontinuities 90
– variation 6
Autonomous
– Markov process 122
– stochastic control problem 163

Backward
– equation 128
– operator 128
Bolza problem 25
Boundary problems 129
Brownian motion
–, *n* dimensional 110
–, standard 109

Cauchy problem 128
Cell 90
Coarea formula 211
Conditional expectation 204
– distribution, regular 204
Cone of variations 46
Conical differential 46
Conjugate point 13
Control
–, feasible 81
–, feedback 36, 87, 153
–, impulsive 69
–, nonanticipative 162
–, zero memory 188
Controllable, completely 134
Convex
– functions 201
– sets 200

Determining function, of a cell 91
Differentiable function 83
Diffusion process 122
–, controlled 155
Dynamic programming equation 84, 154
–, generalized solutions of 178

Elliptic, uniformly 130
Euler's equation 8

Exit time 109
Extremal 7, 27

Feasible pair 24
– control 81
Forward operator 132
Free terminal point problem 39
Fundamental solution 133

Gateau differentiable 3
Generalized solutions 64, 177
Geodesic 17
Girsanov transformation formula 142
Gronwall-Bellman inequality 198
Gronwall's inequality 198

Hamiltonian 27

Inclusion probability, maximum 158
Independent increments, process with 109
Innovation process 137
Internal point 3
Ito conditions 118
Ito stochastic differential rule 114, 117

Jacobi necessary condition 13

Kalman-Bucy filter model 135

Lagrange problem 25
Least action, principle of 16
Linear regulator 23, 88
–, stochastic 165
Lipschitz condition 85
Local
– covariance matrix 122
– drift coefficient 122
– Lipschitz condition 85
– minimum 14

Markov process
–, transition density of 133
–, transition function of 122
Martingale 114
Mayer problem 25
Moon landing problem 21, 28
Multiplier rule, abstract 46

Optimal controls
–, continuity properties 74
–, existence of 63, 68, 166, 181
–, feedback, for stochastic problem 153
–, necessary conditions for 27, 42
–, piecewise constant 78
Optimal portfolio selection 160
Ornstein-Uhlenbeck model 126

Parabolic
–, degenerate 128
–, uniformly 128
Performance index (or criterion) 24
– function 85
Plateau's problem 10
Polynomial growth condition 124
Pontryagin's principle 27
–, for free terminal point problem 42

Reachable set 83
Radial point 3
Random vector
–, distribution of 202
–, independence 203
–, gaussian 203
Ramsey model 22
Regular point 182
Riccati equation, matrix 89

Selection lemma 199
Separation principle 188, 194
Simplest problem in calculus of variations 6
Singular optimization problem 37
Stochastic control problem
–, open loop 151, 187
–, with complete observations 151
–, with partial observations 151, 187
Stochastic differential equation 117
–, with random coefficients 120
Stochastic integral
–, Ito sense 111
–, Stratonovich sense 114
–, vector valued 116
Stochastic process
–, continuous 108
–, measurable 108
–, nonanticipative 108
–, sample function of 108
–, state space of 108
Stopping time 108
Strong regularity, region of 184
Submartingale 114
Switching surface 165, 185

Terminal set 81
Trajectory 24
Transition surface 185
Transversality conditions 27

Uniqueness in probability law 145

Value function 81
Variation
–, strong 39
–, weak 40
Variational integrand 5
–, regular 8
Verification theorem 88, 159

Weierstrass-Erdmann corner condition 8
White noise 110
Weiner measure 111